Lecture Notes in Physics

Volume 860

For further volumes:
http://www.springer.com/series/5304

Lecture Notes in Physics

The series Lecture Notes in Physics (LNP), founded in 1969, reports new developments in physics research and teaching—quickly and informally, but with a high quality and the explicit aim to summarize and communicate current knowledge in an accessible way. Books published in this series are conceived as bridging material between advanced graduate textbooks and the forefront of research and to serve three purposes:

- to be a compact and modern up-to-date source of reference on a well-defined topic
- to serve as an accessible introduction to the field to postgraduate students and nonspecialist researchers from related areas
- to be a source of advanced teaching material for specialized seminars, courses and schools

Both monographs and multi-author volumes will be considered for publication. Edited volumes should, however, consist of a very limited number of contributions only. Proceedings will not be considered for LNP.

Volumes published in LNP are disseminated both in print and in electronic formats, the electronic archive being available at springerlink.com. The series content is indexed, abstracted and referenced by many abstracting and information services, bibliographic networks, subscription agencies, library networks, and consortia.

Proposals should be sent to a member of the Editorial Board, or directly to the managing editor at Springer:

Christian Caron
Springer Heidelberg
Physics Editorial Department I
Tiergartenstrasse 17
69121 Heidelberg/Germany
christian.caron@springer.com

V. I. Kalikmanov

Nucleation Theory

 Springer

Dr. V. I. Kalikmanov
Twister Supersonic Gas Solutions BV
Rijswijk
The Netherlands

and

Faculty of Geosciences
Delft University of Technology
Delft
The Netherlands

ISSN 0075-8450   ISSN 1616-6361 (electronic)
ISBN 978-90-481-3642-1   ISBN 978-90-481-3643-8 (eBook)
DOI 10.1007/978-90-481-3643-8
Springer Dordrecht Heidelberg New York London

Library of Congress Control Number: 2012947396

Printed on acid-free paper

Springer is part of Springer Science+Business Media (www.springer.com)

Preface

One of the most striking phenomena in condensed matter physics is the occurrence of abrupt transitions in the structure of a substance at certain temperatures or pressures. These are first-order phase transitions, and examples such as the freezing of water and the condensation of vapors to form mist in the atmosphere are familiar in everyday life. A fascinating aspect of these phenomena is that the conditions at which the transformation takes place can sometimes vary. The freezing point of water is not always 0° C: the liquid can be supercooled considerably if it is pure enough and treated carefully. Similarly, it is possible to raise the pressure of a vapor above the so-called saturation vapor pressure, at which condensation ought to take place according to the thermodynamic properties of the separate phases. Both these phenomena occur because of the requirement for nucleation. In practice, the transformation takes place through the creation of small aggregates, or *clusters*, of the daughter phase out of the parent phase. In spite of the familiarity of the phenomena involved, accurate calculation of the rate of cluster formation for given conditions of the parent phase meets serious difficulties. This is because the properties of the small clusters are insufficiently well known.

The development from the 1980s onwards of increasingly accurate experimental measurements of the formation rate of droplets from metastable vapors has driven renewed interest in the problems of nucleation theory. Existing models, largely based upon versions of the *classical nucleation theory* developed in the 1920s–1940s, have on the whole explained the trends in nucleation behavior correctly, but have often failed spectacularly to account for this fresh data. The situation is more dramatic in the case of binary- or, more generally, multi-component nucleation where the trends predicted by the classical theory can be *qualitatively* in error leading to unphysical results.

This book, starting with the classical phenomenological description of nucleation, gives an overview of recent developments in nucleation theory. It also illustrates application of these various approaches to experimentally relevant problems focusing on the nonequilibrium gas–liquid transition, i.e., formation of liquid droplets from a metastable vapor. A monograph on nucleation theory would be incomplete without presenting the recent advances in computer simulations of nucleation on a molecular level, which is a powerful research tool complementing both theory and experiment. I was glad that my colleague and friend Dr. Thomas Kraska from the University of Cologne accepted my invitation to write the chapter on Monte Carlo and Molecular Dynamics simulation of nucleation (Chap. 8)—the field to which he made a number of significant contributions.

Obviously, in view of the modest size of the book it was not possible to cover all new approaches formulated in recent years. The choice of the topics, therefore, reflects the background and prejudices of the author.

This monograph is an introduction as well as a compendium to researchers in soft condensed matter physics and chemical physics, graduate and postgraduate students in physics and chemistry starting on research in the area of nucleation, and to experimentalists wishing to gain a better understanding of the efforts being made to account for their data.

I am grateful to a number of colleagues who collaborated with me at various stages of the work. I benefitted greatly from discussions of fundamental problems of nucleation with Howard Reiss, Joe Katz, and Gerry Wilemski, which advanced my understanding of the subject. Several years spent in the group of Rini van Dongen in Eindhoven University will remain an unforgettable experience of a remarkable scientific atmosphere and friendly environment; special thanks are due to the former Ph.D. students Carlo Luijten, Geert Hofmans, and Dima Labetski for numerous discussions at the seminars and help in understanding the subtleties of nucleation experiments. It is a pleasure to thank Ian Ford, Barbara Wyslouzil, Judith Wölk, Jan Wedekind, Dennis van Putten, and Anshel Gleyzer for constructive criticisms. I am indebted to my colleagues and friends Jos Thijssen, Lev Goldenberg, Bob Prokofiev, Leonid Neishtadt, Andrey Morozov, Lyudmila Tsareva, Dmitry Bulahov, Kees Tjeenk Willink, and Marco Betting for encouragement and help without which this book would not have been written. But above all, I am grateful to my family—Esta and Maria—for the constant support during the almost endless process of thinking, writing, and editing of the manuscript.

Delft, May 2012   V. I. Kalikmanov

Contents

Symbols

$\mathscr{A}_i^{\mathrm{v}}$	Vapor phase activity of component i
$\mathscr{A}_i^{\mathrm{l}}$	Liquid phase activity of component i
B_2	Second virial coefficient
c_p	Specific heat at constant pressure
c_v	Specific heat at constant volume
$\mathscr{F}$	Helmholtz free energy of the system
$\mathscr{F}_{\mathrm{int}}$	Intrinsic Helmholtz free energy
$\mathscr{F}_d$	Helmholtz free energy of hard spheres with diameter d
$\mathscr{F}_n$	Helmholtz free energy of the n-cluster
$\mathscr{F}_n^{\mathrm{conf}}$	Configurational Helmholtz free energy of the n-cluster
$\mathscr{F}^{(n)}$	Helmholtz free energy of the gas of n-clusters
G	Gibbs free energy of the system
$\Delta G(n)$	Gibbs free energy of n-cluster formation
ΔG^*	Nucleation barrier
$\hbar$	Planck constant
J_n	Net rate of cluster formation ($n \rightarrow n+1$)
J	Steady-state nucleation rate
J_0	Pre-exponential factor for the steady-state nucleation rate
k_{B}	Boltzmann constant
m_1	Mass of a molecule
n	Number of particles in a cluster
$n_{\mathrm{c}}; n^*$	Number of particles in a critical cluster
N_1	Coordination number in the liquid phase
p	Pressure
p^{l}	Liquid pressure
p^{v}	Vapor pressure
p_{c}	Critical pressure
p_d	Pressure of a hard sphere system
p_{sat}	Saturation pressure
q_n	Configuration integral of the n-cluster

$q_{n_a n_b}$	Configuration integral of the binary (n_a, n_b)-cluster
S	Supersaturation
$\mathscr{S}$	Entropy
$\mathscr{S}_n$	Entropy of the n-cluster
$\mathscr{S}_n^{\text{conf}}$	Configurational entropy of the n-cluster
T	Absolute temperature
T_c	Critical temperature
$u_{\text{LJ}}(r)$	Lennard–Jones interaction potential
$U_N(\mathbf{r}_1, \ldots, \mathbf{r}_N)$	Microscopic potential energy of a configuration of N particles
Z^{l}	Compressibility factor in the liquid
Z^{v}	Compressibility factor in the vapor
Z_n	Partition function of the n-cluster
$Z^{(n)}$	Partition function of the gas of n-clusters
$Z_{n_a n_b}$	Partition function of the binary (n_a, n_b)-cluster
$\mathscr{Z}$	Zeldovich factor
$\beta = 1/(k_B T)$	Inverse temperature
γ_∞	Surface tension of a flat interface
γ_{micro}	Helmholtz free energy per surface particle in the cluster (*microscopic surface tension*)
δ_{T}	Tolman length
$\kappa = c_p/c_v$	Ratio of specific heats
$\varepsilon_{\text{LJ}}, \sigma_{\text{LJ}}$	Parameters of a Lennard–Jones potential
Λ	de Broglie wavelength of a particle
μ	Chemical potential
μ_n	Chemical potential of the n-cluster
μ_d	Chemical potential of a hard sphere with a diameter d
μ_{sat}	Chemical potential of a substance at vapor–liquid equilibrium (saturation chemical potential)
ν	Impingement rate per unit surface
ν_i	Impingement rate per unit surface of component i in binary nucleation
ν_{av}	Average impingement rate per unit surface in binary nucleation
ρ^{l}	Number density in the bulk liquid
ρ^{v}	Number density in the bulk vapor
ρ_c	Critical number density
τ_t	Line tension
$\rho(n)$	Number density of n-clusters
$\rho_{\text{sat}}(n)$	Number density of n-clusters at saturation
Ω	Grand potential of the system
θ_∞	Reduced surface tension of a flat interface
θ_{micro}	Reduced Helmholtz free energy per surface particle (reduced microscopic surface tension)
CKE	Classical Kelvin equation
CAMS	Constant angle Mie scattering

CGNT	Coarse-grained nucleation theory
CNT	Classical nucleation theory
BCNT	Binary classical nucleation theory
MKNT	Mean-field kinetic nucleation theory
EoS	Equation of state
EMLD	Extended modified liquid drop model
DNT	Dynamic nucleation theory
DFT	Density functional theory
FPE	Fokker-Planck equation
GKE	Generalized Kelvin equation
HPS	High-pressure section of the shock tube
LPS	Low-pressure section of the shock tube
MC	Monte Carlo
MD	Molecular dynamics
NPC	Nucleation pulse chamber
NVT	Canonical (NVT) ensemble
NVE	Microcanonical (NVE) ensemble
RESS method	Rapid expansion of supercritical solution
tWF	ten Wolde–Frenkel cluster definition
SAFT	Statistical associating fluid theory
SANS	Small-angle neutron scattering
SAXS	Small-angle X-ray scattering
SSN	Laval supersonic nozzle
MFPT	Mean first passage time cluster definition
WCA	Weeks–Chandler–Anderson theory

Chapter 1
Introduction

Condensation of a vapor, evaporation of a liquid, melting of a crystal, crystallization of a liquid are examples of the processes called *phase transitions*. Generally speaking, they reflect the ability of physical systems to explore a huge range of microscopic configurations in accordance with the second law of thermodynamics. A characteristic feature of a phase transition is an abrupt change of certain properties. When ice is heated its state first changes continuously up to the moment when the temperature 0° C (at normal pressure) is achieved, at which ice begins transforming into liquid water with absolutely different properties. Another example is gas cooled at constant pressure: its state first changes continuously up to a certain temperature at which condensation begins transforming gas into a liquid. The states of a substance between which a phase transition takes place are called phases. If the difference between phases is of quantitative nature, it can in principle be detected through a microscope. In this case one speaks of a *first-order* transition. Condensation of gas is an example of such a transition in which coexisting vapor and liquid phases have essentially different densities. On the opposite, if the difference between phases is of qualitative nature it can not be detected by examination of a microscopic sample of the substance. Phase transition is associated then with a change in symmetry ("symmetry breaking"): the two phases are characterized by different internal symmetries (e.g. structural transitions in crystals result in formation of crystal lattices with different symmetries). This change is also abrupt, although the state of the system changes continuously; a transition in this case is called continuous or *second-order*.

Close to the two-phase coexistence lines of a first-order phase transition one can find domains of metastable states. In particular, it is possible to raise the vapor pressure above the saturation pressure, so that in the domain, where the liquid phase is thermodynamically stable, a metastable *supersaturated vapor* can exist. Similarly, at certain conditions in the domain, where the vapor phase is stable, a metastable *superheated liquid* can exist, and in the domain, where the crystalline phase is stable, a metastable *supercooled liquid* can exist. The occurrence of these metastable states originates from the requirement for *nucleation*. The phenomenon of nucleation is

V. I. Kalikmanov, *Nucleation Theory*, Lecture Notes in Physics 860,
DOI: 10.1007/978-90-481-3643-8_1, © Springer Science+Business Media Dordrecht 2013

associated with the *nonequilibrium first-order transitions* transforming a metastable parent phase to a thermodynamically stable daughter phase. The transformation takes place through the creation of small clusters of molecules of the daughter phase out of the parent phase by thermal fluctuations. Although the parent phase is metastable with respect to the *bulk* daughter phase, it can remain stable with respect to *small clusters* of the daughter phase. The reason for this is that the thermodynamic properties of the clusters differ from those of the bulk because of the presence of an interface between the phases. Whenever there is an abrupt change in the parent and daughter phases, there is a thermodynamic cost in creating such an interface. Due to this reason the parent phase can remain present in conditions where the bulk daughter phase should be more stable.

Nucleation has many practical consequences in science and technology. For example, a transition from dry to wet steam in turbines, which proceeds by nucleation of small droplets due to the presence of pressure and temperature gradients, can lead to undesirable effects on the performance of the machine and causes erosion of the turbine blades [1]. Control of nucleation in rocket and jet engines, wind tunnels, and combustion processes is important for achieving efficient, ecologically sound operation. Nucleation is the key process in the supersonic gas-liquid separator Twister™ [2] aimed at removal of water and heavy hydrocarbons from the natural gas without use of chemicals. In atmospheric science formation of water droplets and ice crystals in the atmosphere, proceeding by the same mechanism, affects the weather. In the long-term, these processes play an important role in understanding global warming (or cooling) [3]. In biology, there is much interest in bypassing nucleation of ice in the cryopreservation of human tissues [4].

In an attempt to classify the existing theoretical approaches to nucleation one can conventionally distinguish four groups of models:

- phenomenological models: *Classical Nucleation Theory* and its modifications

  The tool most often used in nucleation studies is the phenomenological Classical Nucleation Theory (CNT) formulated in the first half of the twentieth century by Volmer, Weber, Becker, Döring and Zeldovich [5–7] (see also [8, 9]). Its cornerstone is the capillarity approximation considering a cluster, however small, as a macroscopic droplet of the condensed phase. Chapter 3 outlines foundations of the CNT and its most important implications. One of the successful modifications of the classical approach—the *Extended modified liquid drop model* developed by Reiss and co-workers [10–13]—is discussed in Chap. 6

- density functional theory

  Density Functional Theory (DFT) of nonuniform fluids [14] was applied to nucleation by Oxtoby and Evans [15] and later developed in a number of publications by Oxtoby and coworkers. Chapter 5 presents the fundamentals of the DFT and its application to nucleation studies.

- semi-phenomenological models

  The semi-phenomenological approach [16], discussed in Chap. 7, bridges the microscopic and macroscopic description of nucleation combining statistical mechanical treatment of clusters and empirical data.

- direct computer simulations

  Simulations of nucleation on molecular level by means of Monte Carlo and Molecular Dynamics methods is a technique that complements theoretical and experimental studies and as such may be regarded as a virtual (computer) experiment. Chapter 8 gives an introduction to molecular simulation methods that are relevant for modelling of the nucleation process and presents their application to various nucleation problems.

An important link between theory and nucleation experiment is provided by the so called *nucleation theorems* discussed in Chap. 4. Chapter 9 outlines the peculiar features of nucleation behavior at deep quenches near the upper limit of metastability. Chapter 10 is devoted to argon nucleation because of the exceptional role argon plays in various areas of soft condensed matter physics; here a comparison is presented of the predictions of theoretical models (outlined in the previous chapters), computer simulation and available experimental data on argon nucleation. Extensions of theoretical models to the case of *binary* nucleation are discussed in Chaps. 11–13; a general approach to the *multi-component* nucleation is outlined in Chap. 14.

Chapters 3–14 refer to *homogeneous* nucleation (unary, binary, multi-component). If the nucleation process involves the presence of pre-existing surfaces (foreign bodies, dust particles, etc.) on which clusters of the new phase are formed, the process is termed *heterogeneous* nucleation; it is discussed in Chap. 15.

Though the main aim of the book is to present various theoretical approaches to nucleation, the general picture would be incomplete without a reference to experimental methods. Chapter 16 gives a short insight into the experimental techniques used to measure nucleation rates.

References

1. F. Bakhtar, M. Ebrahami, R. Webb, Proc. Instn. Mech. Engrs. **209**(C2), 115 (1995)
2. V. Kalikmanov, J. Bruining, M. Betting, D. Smeulders, in *SPE Annual Technical Conference and Exhibition* (Anaheim, California, USA, 2007), pp. 11–14. Paper No: SPE 110736
3. P.E. Wagner, G. Vali (eds) *Atmospheric Aerosols and Nucleation* (Springer, Berlin, 1988)
4. M. Toner, E.G. Cravalho, M. Karel, J. Appl. Phys. **67**, 1582 (1990)
5. R. Becker, W. Döring. Ann. Phys. **24**, 719 (1935)
6. M. Volmer, *Kinetik der Phasenbildung* (Steinkopf, Dresden, 1939)
7. Ya. B. Zeldovich, Acta physicochim. URSS **18**, 1 (1943)
8. J.E. McDonald, Am. J. Phys. **30**, 870 (1962)
9. J.E. McDonald, Am. J. Phys. **31**, 31 (1963)
10. H. Reiss, A. Tabazadeh, J. Talbot, J. Chem. Phys. **92**, 1266 (1990)

11. H.M. Ellerby, C.L. Weakliem, H. Reiss, J. Chem. Phys. **95**, 9209 (1991)
12. H.M. Ellerby, H. Reiss, J. Chem. Phys. **97**, 5766 (1992)
13. D. Reguera et al., J. Chem. Phys. **118**, 340 (2003)
14. R. Evans, Adv. Phys. **28**, 143 (1979)
15. D.W. Oxtoby, R. Evans, J. Chem. Phys. **89**, 7521 (1988)
16. V.I. Kalikmanov, J. Chem. Phys. **124**, 124505 (2006)

Chapter 2
Some Thermodynamic Aspects of Two-Phase Systems

In this chapter we briefly recall the basic features of equilibrium thermodynamics of a two-phase system, i.e. a system consisting of two coexisting bulk phases, which will serve as ingredients for the nucleation models discussed in this book.

2.1 Bulk Equilibrium Properties

Two phases (1 and 2) can coexist of they are in thermal and mechanical equilibrium. The former implies that there is no heat flux and therefore $T_1 = T_2$, and the latter implies that there is no mass flux, which yields equal pressures $p_1 = p_2$. However, this is not sufficient. Let N be the total number of particles in the two-phase system $N = N_1 + N_2$. The number of particles in either phase can vary while N is kept fixed. If the whole system is at equilibrium, its total entropy $\mathscr{S} = \mathscr{S}_1 + \mathscr{S}_2$ is maximized, which means in particular that

$$\frac{\partial \mathscr{S}}{\partial N_1} = 0$$

Using the additivity of $\mathscr{S}$, this condition can be expressed as

$$\frac{\partial \mathscr{S}_1}{\partial N_1} = \frac{\partial \mathscr{S}_2}{\partial N_2} \tag{2.1}$$

The basic thermodynamic relationship reads:

$$dU = T d\mathscr{S} - p dV + \mu dN \tag{2.2}$$

Rewriting it in the form

$$d\mathscr{S} = \frac{dU}{T} + \frac{p}{T} dV - \frac{\mu}{T} dN$$

V. I. Kalikmanov, *Nucleation Theory*, Lecture Notes in Physics 860,
DOI: 10.1007/978-90-481-3643-8_2, © Springer Science+Business Media Dordrecht 2013

we find

$$\frac{\partial \mathscr{S}}{\partial N} = -\frac{\mu}{T}$$

From (2.1): $\mu_1/T_1 = \mu_2/T_2$ and since $T_1 = T_2$, the chemical potentials of the coexisting phases must be equal. Hence, two phases in equilibrium at a temperature T and pressure p must satisfy the equation

$$\mu_1(p, T) = \mu_2(p, T) \tag{2.3}$$

which implicitly determines the $p(T)$-phase equilibrium curve. Thus, T and p cannot be fixed independently, but have to provide for equality of the chemical potentials of the two phases. Differentiating this equation with respect to the temperature and bearing in mind that $p = p(T)$, we obtain:

$$\frac{\partial \mu_1}{\partial T} + \frac{\partial \mu_1}{\partial p}\frac{dp}{dT} = \frac{\partial \mu_2}{\partial T} + \frac{\partial \mu_2}{\partial p}\frac{dp}{dT} \tag{2.4}$$

From the Gibbs–Duhem equation (see e.g. [1])

$$\mathscr{S} \, dT - V \, dp + N \, d\mu = 0 \tag{2.5}$$

we find

$$d\mu = -s \, dT + v \, dp \tag{2.6}$$

where $s = \frac{\mathscr{S}}{N}$ and $v = \frac{V}{N}$ are entropy and volume per particle, implying that

$$\left(\frac{\partial \mu}{\partial T}\right)_p = -s, \qquad \left(\frac{\partial \mu}{\partial p}\right)_T = v$$

Using (2.4), we obtain the Clapeyron equation describing the shape of the (p, T)-equilibrium curve:

$$\frac{dp}{dT} = \frac{s_2 - s_1}{v_2 - v_1} \tag{2.7}$$

In the case of vapor-liquid equilibrium this curve is called a *saturation line* and the pressure of the vapor in equilibrium with its liquid is called a *saturation pressure*, p_{sat}. Liquid and vapor coexist along the saturation line connecting the triple point corresponding to three-phase coexistence (solid, liquid and vapor) and the critical point. Below the critical temperature T_c one can discriminate between liquid and vapor by measuring their density. At T_c the difference between them disappears. The line of liquid-solid coexistence has no critical point and goes to infinity since the difference between the symmetric solid phase and the asymmetric liquid phase can not disappear.

Most of the first-order phase transitions are characterized by absorption or release of the *latent heat*. According to the first law of thermodynamics (expressing the principle of conservation of energy), the amount of heat supplied to the system, δQ, is equal to the change in its internal energy δU plus the amount of work performed by the system on its surroundings $p\delta V$

$$\delta Q = \delta U + p\delta V \tag{2.8}$$

In the theory of phase transitions the quantity δQ is called the latent heat L. For processes *at constant pressure* the latent heat is given by the change in the enthalpy:

$$L \equiv \delta Q = \delta(\mathcal{U} + pV) = \delta H(N, p, \mathcal{S}) = T(\delta\mathcal{S})_{p,N}$$

The latent heat per molecule $l = L/N$ is then

$$l = T(s_2 - s_1) \tag{2.9}$$

Using (2.9) the Clapeyron equation at $T < T_c$ can be written as

$$\frac{dp}{dT} = \frac{l}{T(v_2 - v_1)} \tag{2.10}$$

For the gas–liquid transition at temperatures far from T_c the molecular volume in the liquid phase $v_1 \equiv v^l$ is much smaller than in the vapor $v_2 \equiv v^v$. Neglecting v_1 and applying the ideal gas equation for the vapor

$$p v_2 = k_B T \tag{2.11}$$

we present Eq. (2.10) as

$$\frac{dp}{dT} = \frac{lp}{k_B T^2} \tag{2.12}$$

where

$$k_B = 1.38 \times 10^{-16} \text{erg/K} \tag{2.13}$$

is the Boltzmann constant. Considering the specific latent heat to be constant, which is usually true for a wide range of temperatures and various substances,[1] and integrating (2.12) over the temperature, we obtain:

$$p_{\text{sat}}(T) = p_\infty e^{-\beta l}, \quad \beta = \frac{1}{k_B T} \tag{2.14}$$

where p_∞ is a constant.

---

[1] For example, for water in the temperature interval between 0 and 100° C, l changes by only 10 %.

2.2 Thermodynamics of the Interface

Let us discuss the *interface* between the two bulk phases in equilibrium. For concreteness we refer to the coexistence of a liquid with its saturated vapor at the temperature T. Equilibrium conditions are characterized by equality of temperature, pressure, and chemical potentials in both bulk phases. The density, however, is not constant but varies continuously along the interface between two bulk equilibrium values $\rho^v(T)$ and $\rho^l(T)$. Note, that local fluctuations of density take place even in homogeneous fluid, where, however, they are small and short-range. In the two-phase system these fluctuations are macroscopic: for vapor–liquid systems at low temperatures the bulk densities ρ^v and ρ^l can differ by 3–4 orders of magnitude.

2.2.1 Planar Interface

Consider a two-phase system contained in a volume V with a planar interface between the vapor and the liquid. Inhomogeneity is along the z direction; $z \to +\infty$ corresponds to bulk vapor, and $z \to -\infty$ to bulk liquid (see Fig. 2.1). Variations in the density give rise to an extra contribution to the thermodynamic functions: they are modified to include the work $\gamma\, dA$ which has to be imposed by external forces in order to change the interface area A by dA:

$$d\mathscr{F} = -p\,dV - \mathscr{S}\,dT + \gamma\,dA + \mu\,dN \tag{2.15}$$

$$dG = V\,dp - \mathscr{S}\,dT + \gamma\,dA + \mu\,dN \tag{2.16}$$

$$d\Omega = -p\,dV - \mathscr{S}\,dT + \gamma\,dA - N d\mu \tag{2.17}$$

(in (2.17) N is the average number of particles in the system). The coefficient γ is the surface tension; its thermodynamic definition follows from the above expressions:

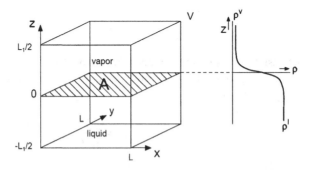

Fig. 2.1 Schematic representation of the vapor–liquid system contained in a volume $V = L^2 L_1$. Inhomogeneity is along the z axis (Reprinted with permission from Ref. [1], copyright (2001), Springer-Verlag.)

Fig. 2.2 Gibbs dividing surface

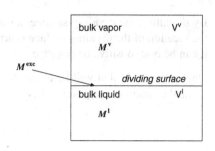

Following Gibbs [2] we introduce a *dividing surface*, being a mathematical surface of zero width which establishes a boundary between the bulk phases as shown in Fig. 2.2. Although its position is arbitrary, it is convenient to locate it somewhere in the transition zone. Once the position of a dividing surface is chosen, the volumes of the two phases are fixed, and satisfy

$$\gamma = \left(\frac{\partial \mathscr{F}}{\partial A}\right)_{N,V,T} \tag{2.18}$$

$$\gamma = \left(\frac{\partial G}{\partial A}\right)_{N,p,T} \tag{2.19}$$

$$\gamma = \left(\frac{\partial \Omega}{\partial A}\right)_{\mu,V,T} \tag{2.20}$$

Following Gibbs [2] we introduce a *dividing surface*, being a mathematical surface of zero width which establishes a boundary between the bulk phases as shown in Fig. 2.2. Although its position is arbitrary, it is convenient to locate it somewhere in the transition zone. Once the position of a dividing surface is chosen, the volumes of the two phases are fixed, and satisfy

$$V^{\mathrm{v}} + V^{\mathrm{l}} = V$$

The idea of Gibbs was that any extensive thermodynamic quantity $\mathscr{M}$ (the number of particles, energy, entropy, etc.) can be written as a sum of bulk contributions $\mathscr{M}^{\mathrm{v}}$ and $\mathscr{M}^{\mathrm{l}}$ and an excess contribution $\mathscr{M}^{\mathrm{exc}}$ that is assigned to the chosen dividing surface:

$$\mathscr{M} = \mathscr{M}^{\mathrm{v}} + \mathscr{M}^{\mathrm{l}} + \mathscr{M}^{\mathrm{exc}} \tag{2.21}$$

Equation (2.21) is in fact a definition of $\mathscr{M}^{\mathrm{exc}}$; its value depends on the location of the dividing surface, and so do the values of $\mathscr{M}^{\mathrm{v}}$ and $\mathscr{M}^{\mathrm{l}}$ (as opposed to $\mathscr{M}$, which is an actual physical property and as such can not depend on the location of the Gibbs surface). Several important examples are

$$N = N^{\mathrm{v}} + N^{\mathrm{l}} + N^{\mathrm{exc}} \tag{2.22}$$
$$\mathscr{S} = \mathscr{S}^{\mathrm{v}} + \mathscr{S}^{\mathrm{l}} + \mathscr{S}^{\mathrm{exc}}$$
$$\Omega = \Omega^{\mathrm{v}} + \Omega^{\mathrm{l}} + \Omega^{\mathrm{exc}}$$
$$\mathscr{F} = \mathscr{F}^{\mathrm{v}} + \mathscr{F}^{\mathrm{l}} + \mathscr{F}^{\mathrm{exc}}$$
$$V = V^{\mathrm{v}} + V^{\mathrm{l}}$$

By definition the dividing surface has a zero width implying that $V^{\text{exc}} = 0$. Since the location of the dividing surface is arbitrary, the excess quantities accumulated on it can be both positive or negative.

One special case that will be useful for future discussions is the *equimolar surface* defined through the requirement $N^{\text{exc}} = 0$. The surface density of this quantity

$$\Gamma = \frac{N^{\text{exc}}}{A} \tag{2.23}$$

is called *adsorption*. Thus, the equimolar surface corresponds to zero adsorption. The thermodynamic potentials, such as $\mathscr{F}$, Ω, G, are homogeneous functions of the first order with respect to their extensive variables. We can derive their expressions for the two-phase system by integrating Eqs. (2.15)–(2.17) using Euler's theorem for homogeneous functions (see e.g. [1], Sect. 1.4). In particular, integration of (2.17) results in

$$\Omega = -pV + \gamma A \tag{2.24}$$

whereas in each of the bulk phases $\Omega^{\text{v}} = -p V^{\text{v}}$, $\Omega^{\text{l}} = -p V^{\text{l}}$, where we used the equality of pressures in the coexisting phases. Thus,

$$\Omega^{\text{exc}} = \gamma A \tag{2.25}$$

irrespective of the choice of the dividing surface. Independence of Ω^{exc} on the location of a dividing surface gives rise to the most convenient thermodynamic route for determination of the surface tension. Equation (2.25) is used in density functional theories of fluids (discussed in Chap. 5) to determine γ from the form of the intermolecular potential.

By definition

$$\Omega^{\text{exc}} = \Omega - \Omega^{\text{v}} - \Omega^{\text{l}} \tag{2.26}$$

For each of the bulk phases

$$d\Omega^{\text{v}} = -p \, dV^{\text{v}} - \mathscr{S}^{\text{v}} \, dT - N^{\text{v}} d\mu \tag{2.27}$$
$$d\Omega^{\text{l}} = -p \, dV^{\text{l}} - \mathscr{S}^{\text{l}} \, dT - N^{\text{l}} d\mu \tag{2.28}$$

Differentiating (2.26) using (2.17) and (2.27)–(2.28) yields

$$d\Omega^{\text{exc}} = -\mathscr{S}^{\text{exc}} \, dT + \gamma \, dA - N^{\text{exc}} \, d\mu$$

On the other hand, from (2.25)

$$d\Omega^{\text{exc}} = \gamma \, dA + A \, d\gamma$$

Comparison of these two equalities leads to the *Gibbs adsorption equation*

$$A \, d\gamma + \mathscr{S}^{\text{exc}} \, dT + N^{\text{exc}} \, d\mu = 0 \qquad (2.29)$$

describing the change of the surface tension resulting from the changes in T and μ. An important consequence of (2.29) is the expression for adsorption:

$$\Gamma = -\left(\frac{\partial \gamma}{\partial \mu}\right)_T \qquad (2.30)$$

where the surface tension refers to a particular dividing surface.

2.2.2 Curved Interface

Gibbs' notion of a dividing surface is a useful concept for thermodynamic description of an interface. At the same time, as we saw in Sect. 2.2.1, the planar surface tension is not affected by a particular location of a dividing surface since the surface area A remains constant at any position of the latter. The situation drastically changes when we discuss a *curved* interface. Here *the position of the dividing surface determines* not only the volumes of the two bulk phases but also *the interfacial area*. An arbitrary curved surface is characterized by two radii of curvature. Consider a liquid droplet inside a fixed total volume V of the two-phase system containing in total N molecules at the temperature T. The "radius" of the droplet is smeared out on the microscopic level since it can be defined to within the width of the interfacial zone, which is of the order of the correlation length. Let us choose a spherical dividing surface with a radius R. The sizes of the two phases and the surface area are fully determined by a set of four variables for which it is convenient to use R, A, V^{l} and V^{v} [3], where V^{l} and V^{v} are the bulk liquid and vapor volumes and A is the surface area:

$$V^{\text{l}} = \frac{4\pi}{3}R^3, \quad V^{\text{v}} = V - \frac{4\pi}{3}R^3, \quad A = 4\pi R^2$$

A sketch of a spherical interface is shown in Fig. 2.3. The change of the Helmholtz free energy $\mathscr{F}$ of the two-phase system "droplet + vapor" when its variables change at isothermal conditions is given by [3]:

$$(\text{d}\mathscr{F})_T = -p^{\text{l}}(\text{d}V^{\text{l}})_T - p^{\text{v}}(\text{d}V^{\text{v}})_T + \mu(\text{d}N)_T + \gamma(\text{d}A)_T + A\left[\frac{\text{d}\gamma}{\text{d}R}\right](\text{d}R)_T \quad (2.31)$$

Here by a differential in square brackets we denote a *virtual change* of a thermodynamic parameter, corresponding to a change in R. The pressure p^{l} inside the liquid phase refers to the bulk liquid held at the same chemical potential as the surrounding vapor with the pressure p^{v}: $\mu^{\text{v}}(p^{\text{v}}) = \mu^{\text{l}}(p^{\text{l}})$. The surface tension $\gamma(R)$ refers to the

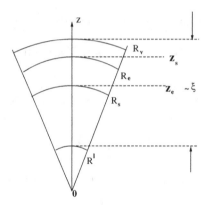

Fig. 2.3 Sketch of a spherical interface. The z axis is perpendicular to the interface pointing away from the center of curvature. R_e and $R_s \equiv R_t$ denote, respectively, the location of the equimolar surface and the surface of tension (see the text). The width of the transition zone between bulk vapor and bulk liquid is of the order of the correlation length ξ (Reprinted with permission from Ref. [1], copyright (2001), Springer-Verlag.)

dividing surface of the radius R; the term in the square brackets gives the change of γ with respect to a mathematical displacement of the dividing surface. It is important to stress that the physical quantities $\mathscr{F}$, p^v, p^l, μ, N, V, do not depend on the location of a dividing surface. So they remain unchanged when only R is changed and from (2.31)

$$0 = [\mathrm{d}\mathscr{F}] = -\Delta p\, 4\pi R^2\, [\mathrm{d}R] + 8\pi R\, \gamma\, [\mathrm{d}R] + 4\pi R^2 \left[\frac{\mathrm{d}\gamma}{\mathrm{d}R}\right][\mathrm{d}R]$$

where $\Delta p = p^l - p^v$. Dividing by $4\pi R^2\, [\mathrm{d}R]$ we obtain the *generalized Laplace equation*:

$$\Delta p = \frac{2\gamma[R]}{R} + \left[\frac{\mathrm{d}\gamma}{\mathrm{d}R}\right] \tag{2.32}$$

It is clear that since Δp, as a physical property of the system, is independent of R, the surface tension *must* depend on the choice of dividing surface. A particular choice $R = R_t$, such that

$$\left[\frac{\mathrm{d}\gamma}{\mathrm{d}R}\right]_{R=R_t} = 0, \tag{2.33}$$

corresponds to the so-called surface of tension; it converts (2.32) into the standard *Laplace equation*

$$\Delta p = \frac{2\gamma_t}{R_t} \tag{2.34}$$

where $\gamma_t = \gamma[R_t]$. One can relate the surface tension taken at an arbitrary dividing surface of a radius R to γ_t. To this end let us write (2.32) in the form

$$\Delta p \, R^2 = \left[\frac{d}{dR}\right] R^2 \gamma[R]$$

and integrate it from R_t to R. Using (2.34) for Δp we obtain the Ono-Kondo equation [4]

$$\gamma[R] = \gamma_t \, f_{OK}\left(\frac{R}{R_t}\right), \quad \text{with} \quad f_{OK} = \frac{1}{3}\frac{1}{x^2} + \frac{2}{3}x \qquad (2.35)$$

Elementary analysis shows that f_{OK} has a minimum at $x = 1$ corresponding to $R = R_t$. Thus, γ_t is the minimum surface tension among all possible choices of the dividing surface:

$$\gamma[R] = \gamma_t \left[1 + O\left(\frac{R - R_t}{R_t}\right)^2\right] \qquad (2.36)$$

When R differs from R_t by a small value, $\gamma[R]$ remains constant to within terms of order $1/R_t^2$.

Among various dividing surfaces we distinguished two special cases—the equimolar surface R_e and the surface of tension R_t—which are related to the certain physical properties of the system. Let us introduce a quantity describing the separation between them

$$\delta = R_e - R_t$$

The limiting value of δ at the planar limit

$$\delta_T = \lim_{R_t \to \infty} \delta = z_e - z_t \qquad (2.37)$$

is called the *Tolman length*. Its sign can be both positive and negative depending on the relative location of the two dividing surfaces. By definition δ_T does not depend on either radius R_t, or R_e (whereas δ does) but can depend on the temperature. Both dividing surfaces lie in the interfacial zone implying that δ_T is of the order of the correlation length. Let Γ_t be the adsorption at the surface of tension. From the Gibbs adsorption equation

$$\Gamma_t = -\left(\frac{\partial \gamma_t}{\partial \mu}\right)_T$$

Using the thermodynamic relationship (2.6) in both phases we rewrite this result as

$$d\gamma_t = -\Gamma_t \, d\mu = -\Gamma_t \frac{dp^v}{\rho^v} = -\Gamma_t \frac{dp^l}{\rho^l}$$

From the second and the third equations of this chain

$$dp^l = dp^v \frac{\rho^l}{\rho^v}$$

resulting in

$$d(\Delta p) = \Delta \rho \, d\mu$$

Substituting Δp from the Laplace equation (2.34) we obtain

$$d\gamma_t = -\frac{\Gamma_t}{\Delta \rho} d \left(\frac{2\gamma_t}{R_t} \right) \tag{2.38}$$

For a curved surface Tolman [5] showed (see also [3]) that

$$\frac{\Gamma_t}{\Delta \rho} = \delta_T \left(1 + \frac{\delta_T}{R_t} + \frac{1}{3} \frac{\delta_T^2}{R_t^2} \right) \tag{2.39}$$

but the terms in (δ_T/R_t) and $(\delta_T/R_t)^2$ can be omitted to the order of accuracy we need. This means that in all derivations below we need to keep only the linear terms in δ_T. With this in mind Eqs. (2.38) and (2.39) give

$$d\gamma_t = -\delta_T \, d \left(\frac{2\gamma_t}{R_t} \right)$$

which after simple algebra yields

$$d \ln \gamma_t = \frac{2\delta_T}{R_t(R_t + 2\delta_T)} dR_t = \left[\frac{1}{R_t} - \frac{1}{R_t + 2\delta_T} \right] dR_t$$

Integrating from the planar limit $(R_t \rightarrow \infty)$ to R_t we obtain:

$$\frac{\gamma_t}{\gamma_\infty} = \frac{R_t}{R_t + 2\delta_T} \tag{2.40}$$

where γ_∞ is the planar surface tension discussed in Sect. 2.2.1. Keeping the linear term in δ_T we finally obtain the *Tolman equation*

$$\gamma_t = \gamma_\infty \left(1 - \frac{2\delta_T}{R_t} + \ldots \right) \tag{2.41}$$

It is important to emphasize that the second order term in the δ_T can not be obtained from (2.40) since this equation is derived to within the linear accuracy in the Tolman length.

Equation (2.41) represents the expansion of the surface tension of a curved interface (droplet) in powers of the curvature. Its looses its validity when the radius of the droplet becomes of the order of molecular sizes. The concept of a curvature dependent surface tension frequently emerges in nucleation studies. It is therefore important to estimate the minimal size of the droplet for which the Tolman equation holds. For simple fluids (characterized by the Lennard-Jones and Yukawa intermolecular potentials) near their triple points the density functional calculations [6] reveal that the Tolman equation is valid for droplets containing more than 10^6 molecules.

References

1. V.I. Kalikmanov, in *Statistical Physics of Fluids. Basic Concepts and Applications* (Springer, Berlin, 2001)
2. J.W. Gibbs, in *The Scientific Papers* (Ox Bow, Woodbridge, 1993)
3. J.S. Rowlinson, B. Widom, *Molecular Theory of Capillarity* (Clarendon Press, Oxford, 1982)
4. S. Ono, S. Kondo, in *Encyclopedia of Physics*, vol. 10, ed. by S. Flugge (Springer, Berlin, 1960), p. 134
5. R.C. Tolman, J. Chem. Phys. **17**(118), 333 (1949)
6. K. Koga, X.C. Zeng, A.K. Schekin, J. Chem. Phys. **109**, 4063 (1998)

Chapter 3
Classical Nucleation Theory

3.1 Metastable States

Nucleation refers to the situation when a system (parent phase) is put into a nonequilibrium *metastable state*. Experimentally it can be achieved by a number of ways (for definiteness we refer to the vapor-liquid transition): e.g. by isothermally compressing vapor up to a pressure p^v exceeding the saturation vapor pressure at the given temperature $p_{sat}(T)$. At this state, characterized by p^v and T, the chemical potential in the bulk liquid $\mu^l(p^v, T)$ is lower than in the bulk vapor at the same conditions $\mu^v(p^v, T)$, which makes it thermodynamically favorable to perform a transformation from the parent phase (vapor) to the daughter phase (liquid). The driving thermodynamic force for this transformation is the chemical potential difference

$$\Delta\mu = \mu^v(p^v, T) - \mu^l(p^v, T) > 0 \tag{3.1}$$

Physically a metastable state (supersaturated vapor) corresponds to a *local minimum* of the free energy (see Fig. 3.1 for a schematic illustration) as a function of the appropriate order parameter. Metastability means that the system in this state (state A in Fig. 3.1) is stable to small fluctuations in the thermodynamic variables but after a certain time will evolve to state B corresponding to the *global minimum* of the free energy (bulk liquid). In order to perform this transformation, this system has to overcome a barrier, representing a *local maximum* of the free energy. The latter corresponds to an *unstable equilibrium* state in which the system is unstable with respect to small fluctuations of the thermodynamic variables. To overcome the energy barrier a fluctuation is required which causes a formation of a small quantity (cluster) of the new phase called *nucleus*. Such process of *homogeneous nucleation* is thermally achieved (the case of *heterogeneous* nucleation by impurities is discussed in Sect. 15.1). Let us rewrite (3.1)

$$\Delta\mu = [\mu^v(p^v, T) - \mu_{sat}(T)] + \{\mu_{sat}(T) - \mu^l(p^v, T)\}$$

V. I. Kalikmanov, *Nucleation Theory*, Lecture Notes in Physics 860,
DOI: 10.1007/978-90-481-3643-8_3, © Springer Science+Business Media Dordrecht 2013

Fig. 3.1 Sketch of the free
energy as a function of the
order parameter. Local minimum corresponds to the
metastable state A (*supersaturated vapor*). The global
minimum corresponds to the
stable state B (*bulk liquid*).
Transition between the states
A and B involves overcoming
the free energy barrier

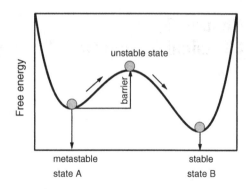

where $\mu_{sat}(T)$ is the chemical potential at saturation at the temperature T. Expression
in the curl brackets is the difference in chemical potential in the *liquid* phase, taken
at saturation (i.e. at the pressure p_{sat}) and at the actual pressure p^v. Assuming the
incompressibility of the liquid and applying the thermodynamic relationship (2.6) to
the liquid phase, we rewrite (3.1) as

$$\Delta\mu \approx [\mu^v(p^v, T) - \mu_{sat}(T)] - \frac{1}{\rho^l_{sat}}(p^v - p_{sat}) \qquad (3.2)$$

where ρ^l_{sat} is the liquid number density at saturation. Using the compressibility factor
of the liquid phase at saturation

$$Z^l_{sat} = \frac{p_{sat}}{\rho^l_{sat} k_B T} \qquad (3.3)$$

we present $\Delta\mu$ as

$$\Delta\mu = k_B T \ln S - k_B T Z^l_{sat}\left(\frac{p^v}{p_{sat}} - 1\right)$$

where the dimensionless quantity

$$S = \exp\left[\frac{\mu^v(p^v, T) - \mu_{sat}(T)}{k_B T}\right] \qquad (3.4)$$

is called the *supersaturation ratio*, or simply the *supersaturation*. The chemical
potentials are not directly accessible in experiments. Assuming the ideal gas law for
the vapor

$$\rho^v(p^v) = \frac{p^v}{k_B T}$$

and using Eq. (2.6) for the vapor phase, the supersaturation can be approximated as

$$S = \frac{p^v}{p_{sat}} \qquad (3.5)$$

so that

$$\Delta\mu = k_B T \ln S - k_B T \, Z^l_{sat} (S - 1) \qquad (3.6)$$

At temperatures not too close to T_c, $Z^l_{sat} \sim 10^{-6} - 10^{-3}$, and the second term in (3.6) can be safely neglected resulting in

$$\Delta\mu = k_B T \ln S \qquad (3.7)$$

We expressed the driving force to nucleation in terms of the supersaturation S containing the experimentally controllable quantities: pressure and temperature. The quantity S characterizes the degree of metastability of the system; $S \geq 1$, where the equality sign corresponds to equilibrium.

3.2 Thermodynamics

If the lifetime of the metastable state is much larger than the relaxation time necessary for the system to settle in this state, we can apply the concept of quasi-equilibrium treating the metastable state as if it were an equilibrium. Instead of a thermodynamic probability of occurrence of a nucleus we shall discuss the "equilibrium" (in the above sense) distribution function of n-clusters, $\rho_{eq}(n)$, which is proportional to it. Considerations based on the thermodynamic fluctuation theory [1] yield:

$$\rho_{eq}(n) = \rho_1 \exp\left[-\frac{W_{min}(n)}{k_B T}\right] \qquad (3.8)$$

where ρ_1 is a temperature dependent constant and $W_{min}(n)$ is a minimum (reversible) work required to form an n-cluster in the surrounding vapor at the pressure p^v and the temperature T.

To calculate W_{min} we use general thermodynamic considerations following Debenedetti [2]. Consider the system put in contact with a reservoir (heat bath). The initial state of the system is pure vapor with the internal energy U^v_0 (the subscripts "0" and "f" denote the initial and final conditions, respectively). The final state (after the droplet was formed) is the "droplet + vapor". Within the framework of Gibbs thermodynamics we introduce a dividing surface of a radius R and write the internal energy at the final state as (cf. Eq. (2.21))

$$U_f = U^v_f + U^l + U^{exc}$$

where the first term refers to the bulk vapor, the second term refers to the bulk liquid and the third one gives the surface (excess) contribution. The total change in the internal energy $\Delta U = U_f - U_0$, caused by the change in the physical state, includes:

- the work W exerted on the system by an external source (creating pressure);
- the work performed by the heat bath to create a droplet, and
- the heat received by the system from the heat bath.

The heat bath is considered to be large enough so that its pressure p_r and temperature T_r remain constant (quantities referring to the heat bath are denoted with a subscript "r"). The work performed by the heat bath is $p_r \Delta V_r$, and the heat given by it is $-T_r \Delta \mathscr{S}_r$. Thus,

$$\Delta U = W + p_r \Delta V_r - T_r \Delta \mathscr{S}_r \tag{3.9}$$

The total volume of heat bath and the system remains unchanged, $\Delta V_r = -\Delta V$. According to the second law of thermodynamics the total change of entropy (heat bath + system) should be nonnegative:

$$\Delta \mathscr{S}_r + \Delta \mathscr{S} \geq 0 \tag{3.10}$$

From (3.9) and (3.10) we obtain

$$W \geq \Delta U - T_r \Delta \mathscr{S} + p_r \Delta V$$

Equality sign corresponds to the reversible process (in which the total entropy remains unchanged) yielding the minimum work

$$W_{\min} = \Delta U - T_r \Delta \mathscr{S} + p_r \Delta V \tag{3.11}$$

If furthermore we assume that the process of transformation to the final state (i.e. the droplet formation) takes place at a constant temperature $T = T_r$ and pressure $p^v = p_r$ then

$$W_{\min} = \Delta U - T \Delta \mathscr{S} + p^v \Delta V = \Delta G \tag{3.12}$$

where $G = U - T\mathscr{S} + p^v V$ is the Gibbs free energy of the two-phase system "droplet + vapor"; V is its volume. Thus, the minimum work to form an n-cluster is equal to the change of the Gibbs free energy. That is why in nucleation theories it is common to speak about the "*Gibbs free energy of droplet formation*".

Let us calculate ΔG from Eq. (3.12). The change in the internal energy is

$$\Delta U = (U_f^v - U_0^v) + U^l + U^{exc} \tag{3.13}$$

Integrating (2.2) using Euler's theorem for homogeneous functions we obtain:

$$U_0^v = T \mathscr{S}_0^v - p^v V_0^v + \mu^v N_0^v \tag{3.14}$$

$$U_f^v = T \mathscr{S}_f^v - p^v V_f^v + \mu^v N_f^v \tag{3.15}$$

$$U^l = T \mathscr{S}^l - p^l V^l + \mu^l N^l \tag{3.16}$$

$$U^{exc} = T \mathscr{S}^{exc} + \gamma(R) A(R) + \mu^{exc} N^{exc} \tag{3.17}$$

Since the vapor pressure and temperature are constant the vapor chemical potential $\mu^v(p^v, T)$ does not change during the transformation of vapor from the initial to the final state. In the last expression $\gamma(R)$ is a surface tension at the dividing surface R with a surface area $A(R)$. The change of the system volume is

$$\Delta V = \left(V_f^v + V^l \right) - V_0^v \tag{3.18}$$

Substituting (3.13)–(3.18) into Eq. (3.12) and taking into account conservation of the number of molecules $N_0^v = N_f^v + N^l + N^{exc}$, we obtain

$$\Delta G = (p^v - p^l)V^l + \gamma A + N^l \left[\mu^l(p^l) - \mu^v(p^v) \right] + N^{exc} \left[\mu^{exc} - \mu^v(p^v) \right] \tag{3.19}$$

This is an exact general expression for the Gibbs free energy of cluster formation. The location of a dividing surface is not specified. Choosing the equimolar surface R_e (with $\gamma = \gamma_e$, $A = A_e$), the last term in (3.19) vanishes. Considering the liquid phase to be incompressible we write

$$\mu^l(p^l) = \mu^l(p^v) + v^l(p^l - p^v) \tag{3.20}$$

where v^l is a molecular volume of the liquid phase (usually taken at coexistence $v^l = v_{sat}^l(T)$). Substitution of (3.20) into (3.19) yields

$$\Delta G = -n \, \Delta \mu + \gamma_e \, A_e \tag{3.21}$$

where $n \equiv N^l$ and $\Delta \mu$ is defined in (3.1). Within the classical approach small droplets causing nucleation of the bulk liquid from the vapor are treated as macroscopic objects. This is the essence of the so called *capillarity approximation* which is the fundamental assumption of the *Classical Nucleation Theory* (CNT) developed in 1926–1943 by Volmer, Weber, Becker, Döring and Zeldovich [3–5]. Within the capillarity approximation

- a cluster is viewed as a *large* homogeneous spherical droplet of a well defined radius with the bulk liquid properties inside it and the bulk vapor density outside it;
- the liquid is considered incompressible and
- the surface energy of the cluster containing n molecules (*n-cluster*) is presented as the product of the planar interfacial tension at the temperature T, $\gamma_\infty(T)$, and the surface area of the cluster $A(n)$.

With these assumptions Eq. (3.21) becomes

$$\Delta G = -n \, \Delta \mu + \gamma_\infty A(n) \tag{3.22}$$

where $A(n) = 4\pi \, r_n^2$, r_n is the radius of the cluster

$$r_n = r^1 n^{1/3}$$

and

$$r^1 = \left(\frac{3 v^1}{4\pi} \right)^{1/3} \tag{3.23}$$

is the average intermolecular distance in the bulk liquid. Thus,

$$A(n) = s_1 n^{2/3}$$

where

$$s_1 = (36\pi)^{1/3} \left(v^1 \right)^{2/3} \tag{3.24}$$

is the "surface area of a monomer". Using (3.7) the reduced free energy of cluster formation reads:

$$\beta \Delta G(n) = -n \, \ln S + \theta_\infty n^{2/3} \tag{3.25}$$

with the reduced surface tension

$$\theta_\infty = \frac{\gamma_\infty s_1}{k_B T} \tag{3.26}$$

The function $\Delta G(n)$ is schematically shown in Fig. 3.2. At small n the positive surface term dominates making it energetically unfavorable to create a very small droplet in view of the large uncompensated surface energy. For large n the negative bulk contribution prevails. The function has a maximum at

$$n_c = \left[\frac{2}{3} \frac{\theta_\infty}{\ln S} \right]^3 \tag{3.27}$$

The cluster containing n_c molecules is called a *critical cluster*; $\Delta G(n_c) = \Delta G^*$ represents an *energy barrier* which a system has to overcome to form a new stable (liquid) phase. Droplets with $n < n_c$ molecules *on average* dissociate whereas those with $n > n_c$ *on average* grow into the new phase. Substituting (3.27) into (3.22) we obtain the CNT free energy barrier, called also the *nucleation barrier*:

$$\Delta G^* = \frac{1}{3} \gamma_\infty A(n_c) = \frac{16\pi}{3} \frac{(v^1)^2 \gamma_\infty^3}{(k_B T \, \ln S)^2} \tag{3.28}$$

Fig. 3.2 Gibbs free energy of cluster formation $\Delta G(n)$. n_c is the critical cluster; clusters below n_c are *on average* dissociating; clusters beyond n_c are *on average* growing to the new bulk phase. Within the critical region, characterized by $\Delta G(n_c) - \Delta G(n) \leq k_B T$, the fluctuation development of clusters occurs

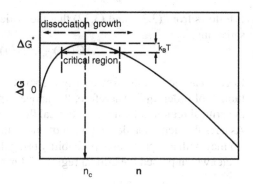

or in the dimensionless quantities:

$$\beta \Delta G^* = \frac{4}{27} \frac{\theta_\infty^3}{(\ln S)^2} \qquad (3.29)$$

The critical cluster is in metastable (quasi-) equilibrium with the surrounding vapor yielding

$$\mu^v(p^v) = \mu^l(p^l)$$

This equality together with Eq. (3.20) leads to an alternative form of the nucleation barrier:

$$\Delta G^* = \frac{16\pi}{3} \frac{\gamma_\infty^3}{(\Delta p)^2} \qquad (3.30)$$

where $\Delta p = p^l - p^v$; note that p^l refers to the bulk liquid held at the same temperature T and the same chemical potential μ as the supersaturated vapor.

Maximum of $\Delta G(n)$ corresponds to the exponentially sharp minimum of the distribution function (3.8). Therefore instead of speaking about the critical point $n = n_c$ it would be more correct to discuss the critical *region* around n_c, where $\Delta G(n)$ to a good approximation has the parabolic form:

$$\Delta G(n) \approx \Delta G^* + \frac{1}{2} \Delta G''(n_c)(n - n_c)^2, \quad ' = \frac{d}{dn} \qquad (3.31)$$

(the term linear in $(n - n_c)$ vanishes). This quadratic expansion yields the Gaussian form for $\rho_{eq}(n)$, centered at n_c and having the width

$$\Delta = \left[-\frac{1}{2} \beta \Delta G''(n_c) \right]^{-1/2} \qquad (3.32)$$

It follows from (3.31) and (3.32) that the critical region corresponds to the clusters satisfying

$$\Delta G(n_c) - \Delta G(n) \leq k_B T$$

Recall that the average free energy associated with an independent fluctuation in a fluid is of order $k_B T$. Therefore, the above expression indicates that the clusters in the critical region fall within the typical fluctuation range around the critical cluster. As a result fluctuation development of nuclei in the domain of cluster sizes $(n_c, n_c + \Delta)$ may with an appreciable probability bring them back to the subcritical region but nuclei which passed the critical region will irreversibly develop into the new phase.

3.3 Kinetics and Steady-State Nucleation Rate

The equilibrium cluster distribution

$$\rho_{eq}(n) = \rho_1 \exp\left[-\frac{\Delta G(n)}{k_B T}\right], \tag{3.33}$$

is limited by the stage *before* the actual phase transition and therefore it can not predict the development of this process. At large n, the Gibbs formation energy is dominated by the negative bulk contribution $-n\Delta\mu$ implying that for large clusters $\rho_{eq}(n)$ diverges

$$\rho_{eq} \to \infty \quad \text{as} \quad n \to \infty$$

The true number density $\rho(n, t)$ (nonequilibrium cluster distribution), as any other physical quantity, should remain finite for any n and at any moment of time t. To determine $\rho(n, t)$ it is necessary to discuss kinetics of nucleation. Within the CNT the following assumptions are made:

- the elementary process which changes the size of a nucleus is the attachment to it or loss by it of one molecule
- if a monomer collides a cluster it sticks to it with probability unity
- there is no correlation between successive events that change the number of particles in a cluster.

The last assumptions means that nucleation is a *Markov process*. Its schematic illustration is presented in Fig. 3.3. Let $f(n)$ be a forward rate of attachment of a molecule to an n-cluster (condensation) as a result of which it becomes an $(n + 1)$-cluster, and $b(n)$ be a backward rate corresponding to loss of a molecule by an n-cluster (evaporation) as a result of which it becomes and $(n - 1)$-cluster. Then the kinetics of the nucleation process is described by the set of coupled rate equations

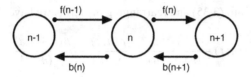

Fig. 3.3 Schematic representation of kinetics of homogeneous nucleation; $f(n)$—forward rate (*condensation*), $b(n)$—backward rate (*evaporation*)

$$\frac{\partial \rho(n,t)}{\partial t} = f(n-1)\,\rho(n-1,t) - b(n)\,\rho(n,t) - f(n)\,\rho(n,t) + b(n+1)\,\rho(n+1,t)$$

(3.34)

A net rate at which n-clusters become $(n+1)$-clusters is defined as

$$J(n,t) = f(n)\,\rho(n,t) - b(n+1)\,\rho(n+1,t)$$

(3.35)

implying that

$$\frac{\partial \rho(n,t)}{\partial t} = J(n-1,t) - J(n,t)$$

(3.36)

The set of equations (3.36) for various n with $J(n,t)$ given by (3.35) was proposed by Becker and Döring [3] and is called the *Becker-Döring equations*. The expression for the forward rate $f(n)$ depends on the nature of the phase transition. For gas-to-liquid transition $f(n)$ is determined by the rate of collisions of gas monomers with the surface of the cluster

$$f(n) = \nu\,A(n)$$

(3.37)

Here the monomer flux to the unit surface ν, called an *impingement rate*, is found from the gas kinetics [1]:

$$\nu = \frac{p^{\mathrm{v}}}{\sqrt{2\pi\,m_1 k_{\mathrm{B}} T}},$$

(3.38)

m_1 is the mass of a molecule. Thus, $f(n)$ is proportional to the pressure of the supersaturated vapor.

The backward (evaporation) rate $b(n)$, at which a cluster *looses* molecules, *a priori* is not known. It is feasible to assume that this quantity is to a large extent determined by the surface area of the cluster rather than by the properties of the surrounding vapor. Therefore $b(n)$ can be assumed to be independent on the actual vapor pressure. In order to find it CNT uses the detailed balance condition at a so called *constrained equilibrium* state [6, 7], which would exist for a vapor at the same temperature T and the supersaturation $S > 1$ as the vapor in question. In the constrained equilibrium the net flux is absent $J(n,t) = 0$ since it corresponds to the stage *before* the nucleation process starts, and the cluster distribution is given by $\rho_{\mathrm{eq}}(n)$. From (3.35) this implies

$$b(n+1) = f(n)\,\frac{\rho_{\mathrm{eq}}(n)}{\rho_{\mathrm{eq}}(n+1)}$$

(3.39)

Substituting (3.39) into (3.35) and rearranging the terms we have:

$$J(n, t) \frac{1}{f(n)\, \rho_{eq}(n)} = \frac{\rho(n, t)}{\rho_{eq}(n)} - \frac{\rho(n + 1, t)}{\rho_{eq}(n + 1)} \tag{3.40}$$

The kinetic process described by Eq. (3.34) rapidly reaches a steady state: a characteristic relaxation time, τ_{tr}, is usually $\sim 1\ \mu s$ (we briefly discuss the transient nucleation behavior in Sect. 3.8) which is much smaller than a typical experimental time scale. In the steady *nonequilibrium* state the number densities of the clusters no longer depend on time. This implies that all fluxes are equal

$$J(n, t) = J \qquad \text{for all } n,\ (t \to \infty)$$

The flux J, called the *steady-state nucleation rate*, is a number of nuclei (of any size) formed per unit volume per unit time. Summation of both sides of Eq. (3.40) from $n = 1$ to a sufficiently large N yields (due to mutual cancelation of successive terms):

$$J \sum_{n=1}^{N} \left[\frac{1}{f(n)\, \rho_{eq}(n)} \right] = \frac{\rho(1)}{\rho_{eq}(1)} - \frac{\rho(N + 1)}{\rho_{eq}(N + 1)} \tag{3.41}$$

For small clusters the free energy barrier is dominated by the positive surface contribution $\theta\, n^{2/3}$, implying that the number of small clusters continues to have its equilibrium value in spite of the constant depletion by the flux J:

$$\frac{\rho(n)}{\rho_{eq}(n)} \to 1 \qquad \text{as } n \to 1+ \tag{3.42}$$

For large n the forward rate exceeds the reverse: the system evolves into the new, thermodynamically stable, phase. As n grows $\rho_{eq}(n)$ increases without limit whereas the true distribution $\rho(n)$ remains finite. Thus,

$$\frac{\rho(n)}{\rho_{eq}(n)} \to 0 \qquad \text{as } n \to \infty \tag{3.43}$$

By choosing large enough N we can neglect the second term in (3.41). Extending summation to infinity we rewrite it as

$$J = \left[\sum_{n=1}^{\infty} \frac{1}{f(n)\, \rho_{eq}(n)} \right]^{-1} \tag{3.44}$$

Examine the terms of this series. The cluster distribution at constrained equilibrium is given by Eqs. (3.33) and (3.25)

$$\rho_{eq}(n) = \rho_1\, S^n \, \exp[-\theta_\infty\, n^{2/3}] \tag{3.45}$$

In this form it was first discussed by Frenkel [8, 9] and is called the *Frenkel distribution*. Initially (for small n) $\rho_{eq}(n)$ decreases due to the surface contribution, then reaches a minimum at $n = n_c$ beyond which it exponentially grows due to the bulk contribution S^n.

The major contribution to the series (3.44) comes from the terms in the vicinity of n_c. If n_c is large enough (and the validity of CNT requires large n_c) the number of these terms is large while the difference between the successive terms for n and $n+1$ is small. This makes it possible to replace summation by an integral:

$$J = \left[\int_1^\infty dn \, \frac{1}{f(n)\rho_{eq}(n)} \right]^{-1} \tag{3.46}$$

The integral can be calculated using the steepest descent method if one takes into account the sharp exponential minimum of ρ_{eq} at n_c. Expanding ρ_{eq} about n_c we write:

$$\rho_{eq}(n) \approx \rho_{eq}(n_c) \exp\left[-\frac{1}{2}\frac{1}{k_B T} \Delta G''(n_c)(n - n_c)^2 \right], \quad \Delta G''(n_c) < 0$$

The Gaussian integration in (3.46) results in

$$J = \mathscr{Z} f(n_c)\rho_{eq}(n_c) \tag{3.47}$$

where

$$\mathscr{Z} = \sqrt{-\frac{1}{2\pi}\frac{1}{k_B T} \Delta G''(n_c)} \tag{3.48}$$

is called the *Zeldovich factor*. Comparison of (3.48) with (3.32) shows that $\mathscr{Z}$ is inversely proportional to the width of the critical region:

$$\Delta = \frac{1}{\sqrt{\pi}\,\mathscr{Z}} \tag{3.49}$$

Using (3.48) and (3.25) the Zeldovich factor takes the form:

$$\mathscr{Z} = \frac{1}{3}\sqrt{\frac{\theta_\infty}{\pi}}\, n_c^{-2/3} \tag{3.50}$$

or equivalently

$$\mathscr{Z} = \sqrt{\frac{\gamma_\infty}{k_B T}}\, \frac{1}{2\pi \rho^l r_c^2} \tag{3.51}$$

where $r_c = r^l n_c^{1/3}$ is the radius of the critical cluster.

Summarizing, the main result of the CNT states that the steady state nucleation rate is an exponential function of the energy barrier

$$J = J_0 \exp\left(-\frac{\Delta G^*}{k_B T}\right) \qquad (3.52)$$

$$\Delta G^* = \frac{16\pi}{3} \frac{(v^l)^2 \gamma_\infty^3}{|\Delta\mu|^2} \qquad (3.53)$$

where the pre-exponential factor is

$$J_0 = \mathscr{Z} \, v \, A(n_c) \, \rho^v \qquad (3.54)$$

with $\rho^v \approx \rho_1$ being the density of the supersaturated vapor. Here $v \, A(n_c) = f(n_c)$ is the rate at which molecules attach to the critical cluster causing it to grow. However, for a cluster in the critical region there exist a chance that it will not cross the barrier but dissociate back the mother phase. The Zeldovich factor stands for the probability of a critical cluster to cross the energy barrier; therefore, the rate at which the cluster actually crosses the barrier and grows into the new phase is not $f(n_c)$, but $\mathscr{Z} f(n_c)$. From (3.50): $\mathscr{Z} \cong n_c^{-2/3}$, thus for the critical clusters with $n_c = 10 - 100$ molecules $\mathscr{Z}$ is of order $0.1 - 0.01$. The pre-exponential factor can be presented in a simple form containing measurable quantities if we apply the ideal gas model for the vapor. Then, from (3.37), (3.38), (3.51) and (3.54):

$$J_0 \cong \frac{(\rho^v)^2}{\rho^l} \sqrt{\frac{2\gamma_\infty}{\pi m_1}} \qquad (3.55)$$

It is important to realize that the steady state flux J does not depend on size, but *is observed* at the critical cluster size.

Nucleation of water vapor plays an exceptionally important role in a number of environmental processes and industrial applications. That is why water can be chosen as a first example to analyze the predictions of the CNT. Two experimental groups, Wölk et al. [10] and Labetski et al. [11] reported the results of nucleation experiments for water in helium (as a carrier gas) in the wide temperature range: 220–260 K in [10] and 200–235 K in [11]. Though, the two groups used different experimental setups—an expansion chamber in [10] and a shock wave tube in [11]—the results obtained are consistent. Figure 3.4 shows the relative—experiment to theory (CNT)—nucleation rate of water for the temperature range $200 < T < 260$ K. Circles correspond to the experiment of [10], squares—to the experiment of [11]. The thermodynamic data for water are given in Appendix A. Figure 3.4 demonstrates a clear trend: the CNT underestimates the experiment (up to 4 orders of magnitude) for lower temperatures, and slightly overestimates it for higher temperatures. The dashed line ("ideal line") corresponds to $J_{exp} = J_{cnt}$; the predictions of CNT coincide with experiment for water at temperatures around 240 K. These results indicate a general feature of the CNT: while its predictions of the nucleation rate dependence on S are quantitatively correct, the dependence of J on the temperature are in many cases in error—as illustrated by the long-dashed line in Fig. 3.4. The

Fig. 3.4 Relative nucleation rate $J_{rel} = J_{exp}/J_{cnt}$ for water. *Circles*: experiment of Wölk et al. [10]; *squares*: experiment of Labetski et al. [11]. The *long-dashed line*, shown to guide the eye, indicates the temperature dependence of the relative nucleation rate. Also shown is the "ideal line" (*dashed line*) $J_{exp} = J_{cnt}$

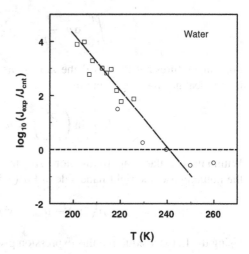

discrepancy between the CNT and experiment grows as the temperature decreases. This is not surprising since the critical cluster at $200 < T < 220$ K at experimental conditions (supersaturation) contains only ≈ 15–20 molecules as follows from Eq. (3.27), implying that the purely phenomenological approach, based on the capillarity approximation becomes fundamentally in error. In the next chapters we discuss alternative models of nucleation which do not invoke the capillarity approximation.

3.4 Kelvin Equation

Consider in a more detail the metastable equilibrium between the liquid droplet of a radius R and the surrounding supersaturated vapor at the pressure p^v and the temperature T. We characterize the droplet radius by its value at the surface of tension $R = R_t$. Equilibrium implies that the chemical potentials of a molecule outside and inside the droplet are equal:

$$\mu_R^v(p^v) = \mu_R^l(p^l) \tag{3.56}$$

Here p^l is the pressure inside the cluster. The same expression written for the *bulk* equilibrium at the temperature T yields

$$\mu_{bulk}^v = \mu_{bulk}^l = \mu_{sat}(T) \tag{3.57}$$

In this case the pressure in both phases is equal to the saturation pressure: $p_{sat}(T)$. Note that Eq. (3.57) can be viewed as an asymptotic form of (3.56) when the droplet radius $R \to \infty$. Subtracting (3.56) from (3.57) and using the thermodynamic relationship (2.6) we obtain

$$\int_{p_{\text{sat}}}^{p^{\text{v}}} dp \, \frac{1}{\rho(p)} = \mu_R^l(p^l) - \mu_{\text{sat}} \tag{3.58}$$

For temperatures not close to T_c the vapor density on the left-hand side can be written in the ideal gas form resulting in

$$k_B T \, \ln \left(\frac{p^{\text{v}}}{p_{\text{sat}}} \right) = \mu_R^l(p^l) - \mu_{\text{sat}} \tag{3.59}$$

Within the capillarity approximation using the thermodynamic relationship (2.6) for the liquid phase, the right-hand side of Eq. (3.59) becomes

$$\mu_R^l(p^l) - \mu_{\text{sat}} = v^l \, (p^l - p_{\text{sat}})$$

Using the Laplace equation this expression gives

$$\ln \left(\frac{p^{\text{v}}}{p_{\text{sat}}} \right) = \frac{2\gamma_t \, v^l}{k_B T \, R_t} + Z_{\text{sat}}^l \left[\left(\frac{p^{\text{v}}}{p_{\text{sat}}} \right) - 1 \right] \tag{3.60}$$

For low temperatures the liquid compressibility factor at saturation $Z_{\text{sat}}^l \ll 1$ and the last term in (3.60) can be safely neglected yielding

$$p^{\text{v}} = p_{\text{sat}} \, \exp \left[\frac{2\gamma_\infty \, v^l}{k_B T \, R_t} \right] \tag{3.61}$$

This result, called the *Kelvin equation*, was formulated by Sir William Thomson (Lord Kelvin) in 1871 [12]. It relates the vapor pressure, p^{v}, over a spherical liquid drop to its radius.

In the original Kelvin's work nucleation was not discussed. Later on the same equation naturally appeared in the formulation of the CNT since the *critical cluster* is in the metastable equilibrium with the surrounding vapor which can be expressed in the form of Eq. (3.56). At the same time in nucleation theory this equilibrium corresponds to the maximum of the Gibbs energy of cluster formation

$$\mu_R^{\text{v}}(p^{\text{v}}) = \mu_R^l(p^l) \Leftrightarrow \max_n \Delta G(n) \tag{3.62}$$

Thus, the alternative way to derive the Kelvin equation is to maximize the free energy of cluster formation. Using this equivalent formulation and applying the capillarity approximation to $\Delta G(n)$, we derive the classical Kelvin equation (3.61). The latter has long played a very important role in nucleation theory (for a detailed discussion see [13]). Since the radius of a spherical n-cluster is $R = r^l n^{1/3}$, we can rewrite it in the form containing only dimensionless quantities which is more suitable for nucleation studies:

$$n_c = \left[\frac{2}{3}\frac{\theta_\infty}{\ln S}\right]^3 \tag{3.63}$$

which coincides with Eq. (3.27).

3.5 Katz Kinetic Approach

To find the evaporation rates CNT uses the concept of constrained equilibrium, which would exist for a vapor at the same temperature T and supersaturation $S > 1$ as the vapor in question. Such a fictitious state can be achieved by introducing "Maxwell demons" which ensure that monomers are continuously replenished by artificial dissociation of clusters which grow beyond a certain (critical) size. The necessity of such an artificial construction clearly follows the fact that a supersaturated vapor is not a true equilibrium state. In an alternative procedure formulated by Katz and coworkers [14, 15] and called a "kinetic theory of nucleation", the evaporation rate is obtained from the detailed balance condition at the (true) stable equilibrium of the *saturated* vapor at the same temperature T. Within this procedure no artificial construction is needed.

Assuming, as in the CNT, that $b(n)$ is independent of the gas pressure, we apply the detailed balance condition at the saturation point, where $J = 0$ and the cluster distribution $\rho_{sat}(n)$ is independent of time. From (3.35) this results in

$$b(n+1) = f_{sat}(n)\frac{\rho_{sat}(n)}{\rho_{sat}(n+1)} \tag{3.64}$$

Since the chemical potentials of liquid and vapor at saturation are equal, the Gibbs formation energy of a cluster at saturation contains only the positive surface contribution

$$\Delta G_{sat}^{CNT}(n) = \gamma_\infty s_1 n^{2/3} \tag{3.65}$$

implying that

$$\rho_{sat}(n) = \rho_{sat}^v \exp[-\theta_\infty n^{2/3}] \tag{3.66}$$

Let us divide both sides of Eq. (3.35) by $f(n)\rho_{sat}(n)S^n$. Since the forward rate is proportional to the pressure, we have:

$$S = f(n)/f_{sat}(n)$$

Then from (3.64):

$$\frac{J(n,t)}{f(n)\rho_{sat}(n)\,S^n} = \frac{1}{S^n}\frac{\rho(n,t)}{\rho_{sat}(n)} - \frac{1}{S^{n+1}}\frac{\rho(n+1,t)}{\rho_{sat}(n+1)}$$

Summation from $n = 1$ to an arbitrary large $N - 1$ yields (due to mutual cancelation of successive terms):

$$\sum_{n=1}^{N-1} \frac{J(n, t)}{f(n)\, \rho_{\text{sat}}(n)\, S^n} = 1 - \frac{\rho(N, t)}{S^N\, \rho_{\text{sat}}(N)} \tag{3.67}$$

Examine the second term on the right-hand side for large N. In the nominator $\rho(N, t)$ is limited for any N (as any other physical quantity). In the denominator the first term exponentially diverges as $e^{N \ln S}$ while the second term vanishes exponentially, but slower than the first one—see (3.66). As a whole the last term on the right-hand side becomes asymptotically small as $N \to \infty$. Extending summation to infinity we obtain for the steady state nucleation rate:

$$J = \left[\sum_{n=1}^{\infty} \frac{1}{f(n)\, S^n\, \rho_{\text{sat}}(n)} \right]^{-1} \tag{3.68}$$

This result looks almost similar to the CNT expression (3.46). The difference between the two expressions is in the prefactor of the cluster distribution function. Nucleation rates given by the kinetic approach, $J_{\text{kin,phen}}$ and by the CNT, J_{CNT}, differ by a factor $1/S$ known as the *Courtney correction* [16]:

$$J_{\text{kin,phen}} = \left(\frac{1}{S} \right) J_{\text{CNT}} \tag{3.69}$$

Both theories yield the same critical cluster size. Thus, if in the kinetic approach one uses the same as in the CNT (phenomenological) model for ΔG, two approaches become identical in all respects except for the $1/S$ correction in the prefactor J_0.

This conclusion, however, looses its validity if within the kinetic approach a different expression for ΔG is chosen. It gives rise to the different form of the cluster distribution function. The importance of the kinetic approach is, thus, in setting the methodological framework for nucleation models with other than classical forms of the Gibbs formation energy.

3.6 Consistency of Equilibrium Distributions

The equilibrium Frenkel distribution employed in the CNT, based on the capillarity approximation, has the form (3.45)

$$\rho_{\text{eq}}(n) = \rho_1 \exp[n \ln S - \theta_\infty n^{2/3}] \tag{3.70}$$

Here ρ_1 is the monomer concentration of the supersaturated vapor; in terms of the vapor concentrations the supersaturation can be written as

$$S = \frac{p^v}{p_{sat}(T)} = \frac{\rho_1}{\rho_{sat}^v(T)} \tag{3.71}$$

where $\rho_{sat}^v(T)$ is the monomer concentration of the vapor at saturation. The *law of mass action* (see e.g. [17], Chap. 6) written for the "chemical reaction" of formation of the n-cluster E_n from n monomers E_1

$$n E_1 \rightleftharpoons E_n \tag{3.72}$$

states that the equilibrium cluster distribution function should have the form:

$$\rho_{eq}(n) = (\rho_1)^n K_n(T) \tag{3.73}$$

where $K_n(T)$ is the *equilibrium constant* for the reaction (3.72) which can depend on n and T but can not depend on ρ_1, or equivalently on the actual pressure p^v. The Frenkel distribution does not satisfy this requirement:

$$\frac{\rho_{eq}(n)}{(\rho_1)^n} = (\rho_1)^{1-n} S^n e^{-\theta_\infty n^{2/3}} = \rho_1 \left(\frac{1}{\rho_{sat}^v(T)}\right)^n e^{-\theta_\infty(T) n^{2/3}}$$

Katz's kinetic approach replaces the Frenkel distribution (3.70) by the *Courtney distribution*

$$\rho_{eq}(n) = \rho_{sat}^v \exp[n \ln S - \theta_\infty n^{2/3}] \tag{3.74}$$

for which the equilibrium constant

$$K_n(T) = (\rho_{sat}^v(T))^{1-n} e^{-\theta_\infty(T) n^{2/3}}$$

satisfies the law of mass action. Weakliem and Reiss [18] showed that the Courtney distribution is not unique but represents one of the possible corrections to the Frenkel distribution which converts it to a form compatible with the law of mass action.

Although the Courtney distribution satisfies the law of mass action, it does not satisfy the *limiting consistency* requirement [13]: in the limit $n \to 1$ it does not return the identity

$$\rho_1 = \rho_1, \text{ for } n = 1$$

The same refers to the Frenkel distribution. At the same time, the limiting consistency is not a fundamental property, to which a cluster distribution should obey, but rather a mathematical convenience to have a single formula which could be valid for all cluster sizes [13]. However, the fact that the CNT does not satisfy the limiting consistency can not be considered as its "weakness", since CNT is valid for relatively large clusters which can be treated as macroscopic objects. However, for nucleation models which are constructed to be valid for small clusters the requirement of limiting consistency deserves special attention.

3.7 Zeldovich Theory

Nucleation and growth of clusters can be viewed as the *flow in the space of cluster sizes*. This space is one-dimensional if the cluster size is determined by the number of molecules, n, in the cluster. The flow in this space is characterized by the "density" $\rho(n, t)$ and the "flow rate"

$$v = dn/dt \equiv \dot{n}(n)$$

In the "cluster language", $\rho(n, t)$ is the cluster distribution function and $\dot{n}(n)$ is the cluster growth law. By definition $\rho(n, t)dn$ is the number of clusters (in the unit physical volume) having the sizes between n and $n + dn$; thus, the "mass" of the cluster fluid inside the (one-dimensional) cluster space volume V_n is

$$\int_{V_n} \rho(n, t)\, dn$$

Using the analogy with hydrodynamics we can apply general hydrodynamic considerations [19] to calculate the density of the "cluster fluid" $\rho(n, t)$. In the absence of nucleation $\rho(n, t)$ satisfies the continuity equation

$$\frac{\partial}{\partial t} \int_{V_n} \rho(n, t)\, dn = -\oint \rho v\, dA_n \tag{3.75}$$

where the zero-dimensional surface A_n bounds the one-dimensional cluster space volume V_n. This equation means that the change of the mass of the cluster fluid inside an arbitrary volume V_n is equal to flow of the cluster fluid through its boundary A_n. In the differential form Eq. (3.75) reads

$$\frac{\partial \rho}{\partial t} + \frac{\partial}{\partial n} (\rho\, \dot{n}) = 0 \tag{3.76}$$

Equations (3.75)–(3.76) describe the evolution of the cluster distribution *in the absence of nucleation*.

Nucleation introduces an extra, source term in (3.75), describing an *additional flux in the cluster space with the density i*. Its role is analogous to diffusion in the physical space for a real fluid. Using the standard hydrodynamic considerations (see [19], Chap. 6), Eq. (3.75) is modified to:

$$\frac{\partial}{\partial t} \int_{V_n} \rho(n, t)\, dn = -\oint \rho v\, dA_n - \oint i\, dA_n \tag{3.77}$$

or in the differential form

$$\frac{\partial \rho}{\partial t} = -\frac{\partial}{\partial n} (\rho\, \dot{n}) - \frac{\partial}{\partial n} i \tag{3.78}$$

Similar to hydrodynamics we write the flux i using Fick's law (now in the cluster size space):

$$i = -B \frac{\partial \rho}{\partial n}$$

where B is the diffusion coefficient. Equation (3.78) becomes

$$\frac{\partial \rho(n, t)}{\partial t} = - \frac{\partial J(n, t)}{\partial n} \tag{3.79}$$

where the flux $J(n, t)$ is:

$$J(n, t) = -B \frac{\partial \rho(n, t)}{\partial n} + \dot{n}\, \rho(n, t) \tag{3.80}$$

This is the *Fokker-Planck equation* (FPE) [20] for diffusion in the cluster size space. The first term describes diffusion in the n-space, with B being the corresponding diffusion coefficient, while the second term describes a drift with the velocity $\dot{n}$ under the action of an external force. The necessary input parameters to solve FPE are:

- the growth law $\dot{n}(n)$, which takes into account both condensation (growth) of supercritical clusters (positive drift) and evaporation of subcritical ones (negative drift) and
- the model for the Gibbs free energy of the cluster formation $\Delta G(n)$

The idea to apply the Fokker-Planck equation to describe kinetics of cluster formation belongs to Zeldovich [5]. This formalism can be viewed as a continuous analogue of the set of Becker-Döring equations (3.36) and leads to the alternative formulation of the CNT, called the *Zeldovich theory*, which we discuss below. In (constrained) equilibrium: $J(n, t) = 0$ for all clusters and $\rho(n, t) = \rho_{eq}(n)$, with $\rho_{eq}(n)$ given by the general expression (3.33). This implies that the drift coefficient in (3.80) is related to the diffusion coefficient by

$$\dot{n}(n) = -B(n)\, g_1(n), \quad g_1 \equiv \frac{\partial \beta \Delta G(n)}{\partial n} \tag{3.81}$$

This implies that Eq. (3.80) describes diffusion in the field of force $-\frac{\partial}{\partial n} \Delta G(n)$. As in the previous sections, we will be interested in the steady state solution $J(n, t) = J = $ const, $\rho(n, t) = \rho_s(n)$ of the kinetic equation (3.79). It is convenient to introduce a new unknown function

$$y = \rho_s(n)/\rho_{eq}(n)$$

Using (3.81) and (3.33), Eq. (3.80) takes the form

$$-B \rho_{eq} \frac{\partial y}{\partial n} = J \tag{3.82}$$

which after integration results in:

$$y = -J \int_0^n dn' \frac{1}{B(n')\rho_{eq}(n')} + C$$

This equation contains two unknown constants—C and J—which can be found from the standard boundary conditions in the limit of small and large clusters (3.42)–(3.43):

$$y \to 1, \text{ for } n \to 0, \quad y \to 0, \text{ for } n \to \infty$$

The solution of Eq. (3.82) satisfying these boundary conditions is:

$$\frac{\rho_s(n)}{\rho_{eq}(n)} = J \int_n^\infty dn' \frac{1}{B(n')\rho_{eq}(n')} \tag{3.83}$$

and the steady-state nucleation rate is given by

$$J = \left[\int_0^\infty dn' \frac{1}{B(n')\rho_{eq}(n')} \right]^{-1} \tag{3.84}$$

Exploring the exponential dependence of the integrand on n we use the second order expansion of the Gibbs free energy $g(n) \equiv \beta \Delta G(n)$ around the critical cluster:

$$g(n) \approx g(n_c) + \frac{1}{2} g''(n_c)(n - n_c)^2, \quad \text{with } g''(n_c) = \frac{d^2 g}{dn^2}\bigg|_{n_c} \tag{3.85}$$

Following the same steps as in Sect. 3.3 we obtain

$$J = \mathscr{Z} B(n_c) \rho_{eq}(n_c) \tag{3.86}$$

where $\mathscr{Z}$ is the Zeldovich factor.

This result looks similar to (3.47); however, we have not yet specified the "diffusion coefficient" for the critical cluster $B(n_c)$. If we were able to determine it from independent *macroscopic* considerations, it would give a possibility to estimate the nucleation rate without using microscopic information—the possibility which could be very advantageous from experimental point of view. Following Zeldovich [5], notice that in the supercritical region $n \gg n_c$ the distribution function is practically constant: a nucleus, after finding itself here, starts monotonically increasing in size, practically never coming back to the subcritical domain. In view of the considerations presented in Sect. 3.3, one can state that with a high degree of accuracy the supercritical region corresponds to $n > n_c + \Delta$, where Δ is given by Eq. (3.49). In this domain we can neglect the term with $\frac{\partial \rho}{\partial n}$ in the flux (3.80) and set:

$$J = \dot{n}\rho, \quad n > n_c + \Delta \tag{3.87}$$

From the physical meaning of $J(n, t)$ as a *flux* in the n-space, we identify the coefficient $\dot{n}$ as a *velocity* in this space:

$$\dot{n} = \left(\frac{dn}{dt}\right)_{macro} \tag{3.88}$$

where the subscript "macro" indicated that the growth of the supercritical nucleus follows a certain *macroscopic* equation (e.g. diffusion *in the real space*). Then from (3.88) and (3.81) we find

$$B(n) = -\frac{1}{g_1(n)} \left(\frac{dn}{dt}\right)_{macro} \tag{3.89}$$

This fundamental *Zeldovich relation* makes it possible to calculate the nucleation rate without referring to the microscopic description but using the deterministic (macroscopic) growth law—ballistic, diffusion, or combination of both. Rigorously speaking, this result is valid for $n > n_c$, whereas we are interested in $B(n_c)$. However, since the function $B(n)$ does not have a singularity at n_c we can use it there. Indeed, at $n = n_c$ the growth rate becomes zero indicating that the critical cluster is in a metastable equilibrium (and therefore does not grow). In the vicinity of n_c we have from (3.85)

$$g_1(n) = g''(n_c)(n - n_c) + O(n - n_c)^2$$

Similarly for the drift velocity

$$\dot{n} = \frac{1}{\tau}(n - n_c) + O(n - n_c)^2$$

where the parameter τ, introduced by Zeldovich (*Zeldovich time*), has a dimensionality of time and is defined as

$$\tau^{-1} = \left.\frac{d\dot{n}}{dn}\right|_{n_c} \tag{3.90}$$

Using l'Hopital's rule, we find:

$$B(n_c) = -\lim_{n \to n_c+} \left(\frac{\dot{n}(n)}{g_1(n)}\right) = -\frac{1}{\tau\, g''(n_c)}$$

At the critical cluster both $\dot{n}$ and g_1 vanish, while $B(n_c)$ remains finite. Using (3.48) and (3.49) we rewrite this result in terms of the Zeldovich factor

$$\tau = \frac{1}{B(n_c)\, 2\pi \mathscr{Z}^2} = \frac{1}{2B(n_c)}\Delta^2 \tag{3.91}$$

The macroscopic growth rate is determined by the mechanism of mass exchange between the cluster and its surroundings. The widely used growth models are: the ballistic (surface limited) and the diffusion limited model. Growth rates $\dot{n}(n)$, referring to both of these mechanisms can be written in a unified form in terms of the reduced radius

$$\bar{r} = \left(\frac{n}{n_c}\right)^{1/3}$$

With neglect of discreteness effects [21]:

$$\dot{\bar{r}} = \frac{1}{\tau\,\bar{r}^{\vartheta}}\left(1 - \frac{1}{\bar{r}}\right) \tag{3.92}$$

Here $\vartheta = 0$ and $\vartheta = 1$ for the ballistic and diffusion limited cases, respectively; note that $\vartheta = -1$ corresponds to cavitation [5].

3.8 Transient Nucleation

In the previous section we discussed the steady state regime. The latter is preceded by the transient non-stationary regime characterized by a characteristic relaxation time τ_{tr}. Strictly speaking the steady regime can be reached only at infinite time when all transient effects have disappeared. However, one can pose a question: how much time is required for the flux to reach *an appreciable fraction* of the steady state value J? To answer this question we start with the expression for the time-dependent flux (3.80) rewritten in the form:

$$\frac{\partial}{\partial n}\left[\frac{\rho(n,t)}{\rho_{\mathrm{eq}}(n)}\right] = -\frac{J(n,t)}{B(n)\rho_{\mathrm{eq}}(n)} \tag{3.93}$$

Integrating it from some small n_1 to a large n_2 $(n_2 \gg n_c)$ and taking into account the boundary conditions (3.42)–(3.43) we obtain:

$$\int_{n_1}^{n_2}\frac{J(n,t)}{B(n)\rho_{\mathrm{eq}}(n)}\,\mathrm{d}n = 1 \tag{3.94}$$

We discuss times at which the flux at the point $n = n_c$ is a noticeable fraction of the steady state value J. Due to the sharp maximum of the integrand at n_c the following expansions are plausible:

$$J(n,t) = J(n_c,t) + \frac{1}{2}J''(n_c,t)(n - n_c)^2$$

$$\rho_{\mathrm{eq}}(n) = \rho_{\mathrm{eq}}(n_c)\exp\left[\pi\,\mathscr{Z}^2(n - n_c)^2\right]$$

Then Eq. (3.94) reads:

$$J(n_c, t) \int_{n_1}^{n_2} \frac{1}{B(n)\rho_{eq}(n)} \, dn + \frac{1}{2} \frac{J''(n_c, t)}{B(n_c)\rho_{eq}(n_c)} \int_{n_1}^{n_2} (n - n_c)^2 \exp\left[-\pi \mathscr{L}^2 (n - n_c)^2\right] dn = 1$$

In the first integral one recognizes J^{-1} whereas the second integral can be calculated using (3.47) and the Gaussian identity [22]

$$\int_{-\infty}^{+\infty} x^2 e^{-a^2 x^2} \, dx = \frac{\sqrt{\pi}}{2a^3}, \qquad a > 0$$

resulting in

$$J(n_c, t) + \frac{1}{4\pi \mathscr{L}^2} J''(n_c, t) = J \tag{3.95}$$

Differentiation of the continuity equation (3.79) with respect to n gives:

$$\frac{\partial^2 J(n, t)}{\partial n^2} = -\frac{\partial}{\partial t} \left[\frac{\partial \rho(n, t)}{\partial n}\right] \tag{3.96}$$

At $n = n_c$ the flux contains only the diffusion term:

$$J(n_c, t) = -B(n_c) \left(\frac{\partial \rho}{\partial n}\right)_{n_c} \tag{3.97}$$

From (3.95)–(3.97) we derive the linear differential equation:

$$J(n_c, t) + \tau_{tr} \frac{dJ(n_c, t)}{dt} = J$$

with

$$\tau_{tr} = \frac{1}{4\pi B(n_c) \mathscr{L}^2} \tag{3.98}$$

whose solution is:

$$J(n_c, t) = J \left(1 - e^{-t/\tau_{tr}}\right) \tag{3.99}$$

The parameter τ_{tr} characterizes the relaxation period to a steady state, or a *time-lag*. For times $t > \tau_{tr}$ one can speak about the steady regime. Comparing (3.91) and (3.98) one can see that the time-lag is related to the Zeldovich time τ as:

$$\tau_{tr} = \frac{1}{4B(n_c)} \Delta^2 = \frac{\tau}{2} \tag{3.100}$$

Typical values of τ_{tr} are of the order of $1 \div 10 \, \mu$s.

3.9 Phenomenological Modifications of Classical Theory

In a number of physically relevant applications critical clusters predicted by CNT turn out to be small in contradiction with the assumptions of the classical theory. Various modifications of CNT were discussed in the literature aiming to propose an expression for free energy applicable to *all* cluster sizes. Dufour and Defay [23] suggested to replace $\theta_\infty(T)$ by a size-dependent surface tension $\theta(n; T)$ such that $\theta(1; T) = 0$. Girshick and Chiu [24] proposed a model in which the surface energy of the cluster is reduced by the "*surface energy of a monomer*"

$$\beta \Delta G_{GC}^{surf}(n) = \beta \Delta G_{CNT}^{surf}(n) - \beta \Delta G_{CNT}^{surf}(n = 1) = \theta_\infty (n^{2/3} - 1) \qquad (3.101)$$

One can notice that by writing

$$\beta \Delta G_{GC}^{surf}(n) = \theta(n, T) n^{2/3}$$

with

$$\theta(n, T) = \theta_\infty \left(1 - n^{-2/3}\right) \qquad (3.102)$$

the Girshick-Chiu expression reduces to a particular realization of the Dufour-Defay conjecture. Since the radius of the (spherical) cluster scales as $n^{1/3}$, Eq. (3.102) shows that the first non-vanishing term in the curvature correction to the plain layer surface tension scales as $\sim 1/R^2$ implying that the Tolman length δ_T for all substances and all temperatures is identically zero (cf. Eq. (2.41)).

Indeed it is known that δ_T vanishes *for symmetric systems*, such as lattice-gas models [25, 26], however for real fluids with asymmetry in vapor-liquid coexistence δ_T is nonzero, though of molecular sizes [27–29]. The Girshick-Chiu cluster distribution reads:

$$\rho_{eq}(n) = \rho_{sat}^v \exp\left[n \ln S - \theta_\infty (n^{2/3} - 1)\right] \qquad (3.103)$$

It is straightforward to see that the resulting nucleation model, termed the Internally Consistent Classical Theory (ICCT) [24] results in

$$J_{ICCT} = \left(\frac{e^{\theta_\infty}}{S}\right) J_{CNT} \qquad (3.104)$$

Compared to the modest Courtney ($1/S$) correction the ICCT correction to the classical theory, e^{θ_∞}/S can be very large. Expression (3.101) implicitly suggests that the CNT form of the free energy barrier is valid down to the cluster containing just one molecule. That is why ICCT can be viewed as a rather arbitrary choice which may, however, empirically improve the fit to experiment [30].

References

1. L.D. Landau, E.M. Lifshitz, *Statistical Physics* (Pergamon, Oxford, 1969)
2. P.G. Debenedetti, *Metastable Liquids* (Princeton University Press, Princeton, Concepts and Principles, 1996)
3. R. Becker, W. Döring. Ann. Phys. **24**, 719 (1935)
4. M. Volmer, *Kinetik der Phasenbildung* (Steinkopf, Dresden, 1939)
5. Ya. B. Zeldovich, Acta Physicochim. URSS **18**, 1 (1943)
6. J.E. McDonald, Am. J. Phys. **30**, 870 (1962)
7. J.E. McDonald, Am. J. Phys. **31**, 31 (1963)
8. J. Frenkel, J. Chem. Phys. **7**, 538 (1939)
9. J. Frenkel, *Kinetic Theory of Liquids* (Clarendon, Oxford, 1946)
10. J. Wölk, R. Strey, J. Phys. Chem. B **105**, 11683 (2001)
11. D.G. Labetski, V. Holten, M.E.H. van Dongen, J. Chem. Phys. **120**, 6314 (2004)
12. W.T. Thomson, Phil. Mag. **42**, 448 (1871)
13. G. Wilemski, J. Chem. Phys. **103**, 1119 (1995)
14. J.L. Katz, H. Wiedersich, J. Colloid Interface Sci. **61**, 351 (1977)
15. J.L. Katz, M.D. Donohue, Adv. Chem. Phys. **40**, 137 (1979)
16. J. Courtney, J. Chem. Phys. **35**, 2249 (1961)
17. C. Garrod, *Statistical Mechanics and Thermodynamics* (Oxfor University Press, New York, 1995)
18. C.L. Weakliem, H. Reiss, J. Phys. Chem. **98**, 6408 (1994)
19. L.D. Landau, E.M. Lifshitz, *Fluid Dynamics* (Pergamon, Oxford, 1986)
20. E.M. Lifshitz, L.P. Pitaevski, *Physical Kinetics* (Pergamon, Oxford, 1981)
21. V.A. Shneidman, J. Chem. Phys. **115**, 8141 (2001)
22. I.S. Gradshtein, I.M. Ryzhik, *Tables of Integrals, Series, and Produscts* (Academic Press, New York, 1980)
23. L. Dufour, R. Defay, *Thermodynamics of Clouds* (Academic, New York, 1963)
24. S. Girshick, C.-P. Chiu, J. Chem. Phys. **93**, 1273 (1990)
25. J.S. Rowlinson, J. Phys. Condens. Matter **6**, A1 (1994)
26. M.P.A. Fisher, M. Wortis, Phys. Rev. B **29**, 6252 (1984)
27. E. Blokhuis, J. Kuipers, J. Chem. Phys. **124**, 074701 (2006)
28. J. Barrett, J. Chem. Phys. **124**, 144705 (2006)
29. M.A. Anisimov, Phys. Rev. Lett. **98**, 035702 (2007)
30. D.W. Oxtoby, J. Phys. Cond. Matt. **4**, 7627 (1992)

Chapter 4
Nucleation Theorems

4.1 Introduction

Various nucleation models use their own set of approximations, have their own range
of validity and certain fundamental and technical limitations. Therefore it is desirable
to formulate some general, *model-independent* statements which would establish the
relationships between the physical quantities characterizing the nucleation behavior.
One of such statements was proposed by Kashchiev [1] later on termed the *Nucleation
Theorem* (NT). In 1996 Ford [2] derived another general statement which was termed
the *Second Nucleation Theorem*. Since then Kashchiev's result and its generalization
is sometimes also referred to as the *First Nucleation Theorem*.

Following Kashchiev, consider a general form of the Gibbs free energy of n-cluster
formation

$$\Delta G(n, \Delta\mu) = -n\,\Delta\mu + F_s(n, \Delta\mu) \tag{4.1}$$

Here $F_s(n, \Delta\mu)$ is the excess beyond the first (bulk) term free energy of cluster
formation. For our present purposes we do not need to specify it. The critical cluster
satisfies:

$$\left.\frac{\partial \Delta G}{\partial n}\right|_{T,\Delta\mu} = 0$$

resulting in

$$-\Delta\mu + \left.\frac{\partial F_s}{\partial n}\right|_{n_c} = 0 \tag{4.2}$$

The work of the critical cluster formation is

$$W^* \equiv \Delta G(n_c(\Delta\mu), \Delta\mu) = -n_c(\Delta\mu)\,\Delta\mu + F_s(n_c(\Delta\mu), \Delta\mu)$$

V. I. Kalikmanov, *Nucleation Theory*, Lecture Notes in Physics 860,
DOI: 10.1007/978-90-481-3643-8_4, © Springer Science+Business Media Dordrecht 2013

Taking the full derivative with respect to $\Delta\mu$

$$\frac{dW^*}{d\Delta\mu} = -n_c + \frac{\partial F_s}{\partial\Delta\mu} + \frac{\partial n_c}{\partial\Delta\mu}\left(-\Delta\mu + \frac{\partial F_s}{\partial n}\bigg|_{n_c}\right)$$

and noticing that the expression in the round brackets vanishes in view of (4.2), we obtain the Nucleation Theorem in the form given in Ref. [1]:

$$\frac{dW^*}{d\Delta\mu} = -n_c + \frac{\partial F_s}{\partial\Delta\mu}\bigg|_{n_c} \qquad (4.3)$$

This result is particularly useful when F_s is only weakly dependent on the supersatura-·
tion; for example in the CNT within the capillarity approximation $F_s = \gamma_\infty(T)A(n)$ is totally independent of $\Delta\mu$. In this case one can determine the size of the critical cluster from the nucleation experiments measuring the nucleation rates and finding the slope of $\ln J - \ln S$ curves (cf. Eq. (3.52)).

4.2 First Nucleation Theorem for Multi-Component Systems

In 1994 Oxtoby and Kashchiev [3] extended original Kashchiev's treatment to a general form which is valid for multi-component systems and applicable to various types of nucleation phenomena. Consider an arbitrary first order phase transition from the mother phase "v" to the new phase "l" (the phases, as mentioned earlier, should not necessarily be vapor and liquid). The mother phase is at constant pressure p^v and temperature T and contains inside itself a cluster of the new phase. Using Gibbs thermodynamics, we introduce an arbitrary dividing surface and write the total volume V of the system "cluster inside a metastable phase" as

$$V = V^v + V^l$$

(see Fig. 4.1). Here V^l encloses the cluster of the new phase "l" and V^v contains the original phase "v". In a mixture of q components the total number of molecules of component i is an extensive quantity and according to (2.22) is given by

$$N_i = N_i^v + N_i^l + N_i^{exc}, \quad i = 1,\ldots,q$$

where N_i^v is the number of molecules of type i in the homogeneous phase "v" occupying the volume V^v, N_i^l is the number of molecules of type i in the homogeneous phase "l" occupying the volume V^l, N_i^{exc} is the excess number of molecules of type i accumulated on the dividing surface. The Gibbs free energy of cluster formation

Fig. 4.1 Sketch of the system "cluster inside a metastable phase"

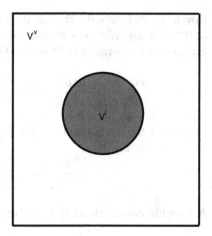

in a multi-component case is given by the straightforward extension of Eq. (3.19) to the multi-component case:

$$\Delta G = (p^{\text{v}} - p^{\text{l}})V^{\text{l}} + \sum_{i=1}^{q} N_i^{\text{l}} \left[\mu_i^{\text{l}}(p^{\text{l}}) - \mu_i^{\text{v}}(p^{\text{v}}) \right]$$

$$+ \sum_{i=1}^{q} N_i^{\text{exc}} \left[\mu_i^{\text{exc}} - \mu_i^{\text{v}}(p^{\text{v}}) \right] + \phi(V^{\text{l}}, \{\mu_i^{\text{v}}\}, T) \qquad (4.4)$$

The last term, $\phi(V^{\text{l}}, \{\mu_i^{\text{v}}\}, T)$, is the total surface energy of the cluster. We do not specify here its functional form (e.g. by introducing the surface tension and the surface area of the cluster) which implies that the general form (4.4) can be applied to small clusters (for which the physical meaning of the surface tension looses its validity). The critical cluster (denoted by the subscript "c") is in the mechanical and chemical equilibrium (though metastable) with the mother phase resulting in the extremum of the Gibbs energy with respect to $\{N_i^{\text{l}}\}$, $\{N_i^{\text{surf}}\}$ and V^{l}:

$$\mu_i^{\text{l}}(p_c^{\text{l}}) = \mu_i^{\text{v}}(p^{\text{v}}) = \mu_i^{\text{exc}} \quad \text{for all } i \qquad (4.5)$$

$$p_c^{\text{l}} = p^{\text{v}} + \left. \frac{\partial \phi}{\partial V^{\text{l}}} \right|_{\{\mu_i^{\text{v}}\}, T} \qquad (4.6)$$

Substituting these expressions into (4.4) we find the work of formation of the *critical cluster* at the given external conditions—the temperature T and the set of the vapor phase chemical potentials $\{\mu_i^{\text{v}}\}$:

$$W^* \equiv \Delta G^* = (p^{\text{v}} - p_c^{\text{l}}) V_c^{\text{l}} + \phi_c \qquad (4.7)$$

where $\phi_c \equiv \phi(V_c^{\text{l}}, \{\mu_i^{\text{v}}\}, T)$.

Now let us study how W^* changes if we change the external conditions. The variation of W^* with respect to the variation of the chemical potential of the component i while keeping the temperature and the rest chemical potentials fixed, reads:

$$\frac{\partial W^*}{\partial \mu_i^v} = V_c^l \frac{\partial (p^v - p_c^l)}{\partial \mu_i^v} + \frac{\partial V_c^l}{\partial \mu_i^v}\left[(p^v - p_c^l) + \frac{\partial \phi_c}{\partial V_c^l}\right] + \frac{\partial \phi_c}{\partial \mu_i^v}$$

In view of the equilibrium condition (4.6) the expression in the square brackets vanishes resulting in

$$\frac{\partial W^*}{\partial \mu_i^v} = V_c^l \frac{\partial (p^v - p_c^l)}{\partial \mu_i^v} + \frac{\partial \phi_c}{\partial \mu_i^v} \tag{4.8}$$

A straightforward extension of the Gibbs–Duhem Eq. (2.5) for a mixture reads

$$\mathcal{S}dT - Vdp + \sum_{k=1}^{q} N_k \, d\mu_k = 0 \tag{4.9}$$

We write (4.9) for both of the bulk phases at isothermal conditions:

$$- V_c^l \, dp_c^l + \sum_k N_{k,c}^l \, d\mu_k^l(p_c^l) = 0, \quad -V^v \, dp^v + \sum_k N_k^v \, d\mu_k^v(p^v) = 0 \tag{4.10}$$

Similarly, the Gibbs adsorption equation (2.29) for a mixture at isothermal conditions is:

$$d\phi_c + \sum_{k=1}^{q} N_k^{\text{exc}} \, d\mu_k^{\text{exc}} = 0 \tag{4.11}$$

In view of the equality of the chemical potentials (4.5) these relations can be written as:

$$V_c^l \, dp_c^l = N_{i,c}^l \, d\mu_i^v(p^v) + \left[\sum_{k\neq i} N_{k,c}^l \, d\mu_k^v(p^v)\right]$$

$$V^v \, dp^v = N_i^v \, d\mu_i^v(p^v) + \left[\sum_{k\neq i} N_k^v \, d\mu_k^v(p^v)\right] \tag{4.12}$$

$$d\phi_c = -N_i^{\text{exc}} \, d\mu_i^v(p^v) - \left[\sum_{k\neq i} N_k^{\text{exc}} \, d\mu_k^v(p^v)\right]$$

In all of these relations the sums in the square brackets has to be set to zero since in Eq. (4.8) all $d\mu_k^v = 0$ except for $k = i$. Substituting (4.12) into (4.8), we find:

$$\frac{\partial W^*}{\partial \mu_i^v} = -N_{i,c}^l - N_i^{exc} + \left(\frac{V_c^l}{V^v}\right) N_i^v \qquad (4.13)$$

This result can be simplified if we introduce the number densities of the component i in the both phases "v" and "l":

$$\rho_i^l = \frac{N_{i,c}^l}{V_c^l}, \qquad \rho_i^v = \frac{N_i^v}{V^v}$$

Then

$$\left(\frac{V_c^l}{V^v}\right) N_i^v = \rho_i^v V_c^l$$

is the number of molecules of component i that existed in the volume V_c^l before the critical cluster was formed. Eq. (4.13) becomes

$$\frac{\partial W^*}{\partial \mu_i^v}\bigg|_{\{\mu_j^v\},\, j \neq i} = -\left[\left(1 - \frac{\rho^v}{\rho^l}\right) N_{i,c}^l + N_i^{exc}\right] \equiv -\Delta n_{i,c} \qquad (4.14)$$

The quantity $\Delta n_{i,c}$ is the *excess number of molecules* of component i in the cluster beyond that present in the same volume (V^l) of the mother phase before the cluster was formed. While the quantities $N_{i,c}^l$, N_i^v, N_i^{exc} depend on the location of the dividing surface the *excess* number $\Delta n_{i,c}$ is independent of this choice and of the cluster shape (which for small clusters can have a fractal structure). Thus, $\frac{\partial W^*}{\partial \mu_i^v}$ is also independent of the location of the dividing surface. This is consistent with the fact that the nucleation barrier W^* itself is invariant with respect to the dividing surface (which is a mathematical abstraction rather than a measurable physical quantity).

Equation (4.14) represents the *Nucleation Theorem for multi-component systems*. The most important feature of this result is that it is derived without any assumptions concerning the size, shape and composition of the critical cluster and thus is of general validity.[1] Although the presented proof is based on thermodynamics, it makes no assumptions about the size of the critical cluster and thus is valid down to atomic size critical nuclei.

For a *unary* system Eq. (4.14) reads:

$$\frac{dW^*}{d\mu^v} = -\left[\left(1 - \frac{\rho^v}{\rho^l}\right) N_c^l + N^{exc}\right] \equiv -\Delta n_c \qquad (4.15)$$

---

[1] Bowles et al. [4] showed that the Nucleation Theorem is a powerful result which is not restricted to nucleation, as its name suggests, but refers to all equilibrium systems containing local nonuniform density distributions stabilized by external field (not only a nucleus in nucleation theory).

We can rewrite this result in terms of the supersaturation ratio S, recalling that

$$k_B T \ln S = \mu^v(p^v, T) - \mu_{sat}(T)$$

The Nucleation Theorem then becomes

$$\frac{d(\beta W^*)}{d \ln S} = -\Delta n_c \tag{4.16}$$

If the mother phase is dilute—$\rho^v/\rho^l \ll 1$—and the number of molecules in the critical cluster is defined as $n_c = N_c^l + N^{exc}$, Eq. (4.16) reduces to the original Kashchiev's expression (4.3) with $\frac{\partial F_s}{\partial \mu} = 0$.

Nucleation Theorem provides a general, model-independent tool for the analysis of nucleation phenomena. However, in experiments the directly measurable quantity is not the work of the critical cluster formation but the nucleation rate J. For the steady state

$$J = J_0 e^{-\beta W^*} \tag{4.17}$$

where the prefactor J_0 depends on a particular nucleation model and on the dimensionality of the problem. Taking in both sides of (4.17) the derivative with respect to $\ln S$, we obtain

$$\frac{d(\beta W^*)}{d \ln S} = -\left(\frac{\partial \ln J}{\partial \ln S}\right)_T + \left(\frac{\partial \ln J_0}{\partial \ln S}\right)_T \tag{4.18}$$

where first term on the right-hand side can be directly measured in nucleation experiments. In the CNT the prefactor J_0 is given by Eq. (3.55)

$$J_0 = \psi(T) S^2$$

where $\psi(T)$ does not depend on S. This implies that

$$\left(\frac{\partial \ln J_0}{\partial \ln S}\right)_T = 2 \tag{4.19}$$

and the Nucleation Theorem for the single-component systems takes a particularly simple form:

$$\Delta n_c = \left(\frac{\partial \ln J}{\partial \ln S}\right)_T - 2 \tag{4.20}$$

Within the kinetic approach to nucleation, discussed in Sect. 3.5, the prefactor J_0 contains the Courtney $(1/S)$ correction

$$J_0 = \psi(T) S$$

so that the Nucleation Theorem becomes

$$\Delta n_c = \left(\frac{\partial \ln J}{\partial \ln S}\right)_T - 1 \qquad (4.21)$$

For *binary* nucleation of components a and b the prefactor J_0 has a more complex form than in the single-component case. An important feature of J_0 is that it depends on the *composition of the critical cluster*; a simple expression describing this dependence is not available. Applying (4.14) to the binary case we find:

$$\Delta n_{i,c} = \left[\frac{\partial \ln J}{\partial(\beta\mu_i^v)}\right]_T - \left[\frac{\partial \ln J_0}{\partial(\beta\mu_i^v)}\right]_T, \qquad i = a, b$$

A contribution of the second term to $\Delta n_{i,c}$ is small and typically ranges from 0 to 1 [3] resulting in

$$\Delta n_{i,c} = \left[\frac{\partial \ln J}{\partial(\beta\mu_i^v)}\right]_T - (0 \text{ to } 1), \qquad i = a, b \qquad (4.22)$$

4.3 Second Nucleation Theorem

Nucleation theorem studied in the previous section describes the variation of W^* with respect to the variation of the chemical potential of one of the species at a constant temperature. In [2, 5] Ford derived what is called now the *Second Nucleation Theorem* which describes the variation of W^* with respect to the temperature *at the fixed chemical potentials*. From Eq. (4.7) we obtain:

$$\frac{\partial W^*}{\partial T} = V_c^l \frac{\partial(p^v - p_c^l)}{\partial T} + \left[(p^v - p_c^l)\frac{\partial V_c^l}{\partial T} + \frac{\partial\phi_c}{\partial V_c^l}\frac{\partial V_c^l}{\partial T}\right] + \frac{\partial\phi_c}{\partial T}$$

In view of the equilibrium condition (4.6) the expression in the square brackets vanishes resulting in

$$\frac{\partial W^*}{\partial T} = V_c^l \frac{\partial(p^v - p_c^l)}{\partial T} + \frac{\partial\phi_c}{\partial T}\bigg|_{V_c^l, \{\mu_i^v\}} \qquad (4.23)$$

The Gibbs–Duhem equations for both of the bulk phases at the *fixed chemical potentials* read

$$-V_c^l \, dp_c^l + S^l \, dT = 0, \qquad -V^v \, dp^v + S^v \, dT = 0 \qquad (4.24)$$

The Gibbs adsorption equation at the *fixed chemical potentials* becomes:

$$d\phi_c + S^{\text{exc}} \, dT = 0 \qquad (4.25)$$

From (4.23)–(4.25) we find:

$$\frac{\partial W^*}{\partial T}\bigg|_{\{\mu_i^v\}} = -\left[\mathcal{S}^l + \mathcal{S}^{exc} - \left(\frac{V_c^l}{V^v}\right)\mathcal{S}^v \right] \equiv -\Delta\mathcal{S}_c \qquad (4.26)$$

where $\Delta\mathcal{S}_c$ is the excess entropy due to the critical cluster formation. (Note that $\mathcal{S}^v/V^v$ is the entropy per unit volume in the vapor phase). Equation (4.26) represents the *Second Nucleation Theorem*.

As in the case of the First Nucleation Theorem, the application of the Second Theorem to the analysis of experiments requires its representation in the form containing measurable quantities. Since the nucleation barrier enters the nucleation rate in the form of the Boltzmann factor, we combine Eq. (4.26) with the identity

$$\frac{\partial(\beta W^*)}{\partial T} = \beta\left(\frac{\partial W^*}{\partial T}\right) - \frac{\beta W^*}{T}$$

which results in

$$\frac{\partial(\beta W^*)}{\partial T} = -\beta\,\Delta\mathcal{S}_c - \frac{\beta W^*}{T} = -\frac{(W^* + T\Delta\mathcal{S}_c)}{k_B T^2} \qquad (4.27)$$

According to the thermodynamic consideration of Sect. 3.2 (see Eq. (3.12)) the expression in the round brackets is the excess internal energy of the critical cluster, i.e. the change in the internal energy beyond that present in the same volume (V^l) of the mother phase before the cluster was formed:

$$\frac{\partial(\beta W^*)}{\partial T}\bigg|_{\{\mu_i\}} = -\frac{\Delta U^*}{k_B T^2} \qquad (4.28)$$

Then, Eq. (4.28) becomes

$$\frac{\partial \ln(J/J_0)}{\partial T}\bigg|_{\{\mu_i\}} = \frac{\Delta U^*}{k_B T^2} \qquad (4.29)$$

The contribution of the pre-exponential term can be found easily for the one component case. Approximating J_0 by the classical expression (3.55) we find for the leading temperature dependence:

$$\frac{\partial \ln J_0}{\partial T}\bigg|_S = 2\frac{d \ln p_{sat}}{dT} - \frac{2}{T}$$

The first term can be worked out using the Clausius–Clapeyron equation (2.14)

$$\frac{d \ln p_{sat}}{dT} = \frac{l}{k_B T^2}$$

where l is the latent heat of evaporation per molecule. Thus, for a unary system the Second Nucleation Theorem reads:

$$\frac{\partial \ln J}{\partial T}\bigg|_S = \frac{2(l - k_B T) + \Delta U^*}{k_B T^2} \tag{4.30}$$

With its help one can find the excess internal energy of the critical cluster from the nucleation rate measurements and the known specific latent heat.

4.4 Nucleation Theorems from Hill's Thermodynamics of Small Systems

An alternative derivation of both Nucleation Theorems (as well as several other useful forms of NT) can be obtained within the framework of *thermodynamics of small systems* developed by Hill [6, 7]. Hill's fundamental result states that for a q-component system the change of the work of critical cluster formation in terms of $q + 1$ independent variables $T; \mu_1^v, \ldots, \mu_q^v$ can be presented in the following exact form:

$$dW^*(T; \mu_1^v, \ldots, \mu_q^v) = -\Delta S_c \, dT - \sum_{i=1}^{q} \Delta n_{i,c} \, d\mu_i^v \tag{4.31}$$

where ΔS_c is the excess entropy of the critical nucleus and $\Delta n_{i,c}$ is the excess number of molecules of component i in it. This result does not depend on the choice of a dividing surface.

It is straightforward to see that both the First and the Second Nucleation theorems follow from Eq. (4.31). In particular, fixing the temperature and all but one chemical potential in the mother phase, we obtain

$$\frac{\partial W^*}{\partial \mu_i^v}\bigg|_{T; \{\mu_j^v\}, j \neq i} = -\Delta n_{i,c} \tag{4.32}$$

which is the First Nucleation Theorem (cf. (4.14)). Varying T and keeping all chemical potentials in (4.31) fixed results in the Second Nucleation Theorem:

$$\frac{\partial W^*}{\partial T}\bigg|_{\{\mu_j^v\}} = -\Delta S_c \tag{4.33}$$

(cf. (4.26)). There exist various other forms of NT resulting from Hill's theory (see [8] for the detailed discussion). Below we consider one such form which can be particularly useful for the analysis of the effect of total pressure on multi-component nucleation. Let us choose the following set of independent variables: the temperature

T, the total pressure p^v and the molar fractions of components in the mother phase $\{y_j\}$, $j = 1, \ldots, q$. All these parameters are measurable in experiments. In view of normalization

$$\sum_{j=1}^{q} y_j = 1$$

the number of independent variables is still $q + 1$. Presenting

$$\mu_i^v = \mu_i^v(p^v, T; \{y_j\}_{j \neq i})$$

we write its full differential as

$$d\mu_i^v = -s_i^v \, dT + v_i^v \, dp^v + \sum_{j \neq i} \left(\frac{\partial \mu_i^v}{\partial y_j} \right) dy_j \tag{4.34}$$

where the second term results from the Maxwell relation

$$\frac{\partial \mu_i^v}{\partial p^v} \bigg|_{\{y_j\}, T} = v_i^v \tag{4.35}$$

and v_i^v and s_i^v are, respectively, the partial molecular volume and entropy in the mother phase

$$v_i^v \equiv \frac{\partial V^v}{\partial N_i^v} \bigg|_{p^v, T, N_{j, j \neq i}^v} \quad , \quad i = 1, 2, \ldots \tag{4.36}$$

$$s_i^v \equiv \frac{\partial S^v}{\partial N_i^v} \bigg|_{p^v, T, N_{j, j \neq i}^v} \quad , \quad i = 1, 2, \ldots \tag{4.37}$$

Substituting (4.34) into Hill's fundamental equation (4.31) we obtain its alternative form in the $p^v, T, \{y_j\}$ variables:

$$dW^* = -\left(\Delta S_c - \sum_{i=1}^{q} s_i^v \, \Delta n_{i,c} \right) dT - \left(\sum_{i=1}^{q} v_i^v \, \Delta n_{i,c} \right) dp^v$$

$$- \sum_{i=1}^{q} \left(\sum_{j \neq i} \frac{\partial \mu_i^v}{\partial y_j} dy_j \right) \Delta n_{i,c} \tag{4.38}$$

(for a single-component case summation over j in the last term should be omitted). Let us differentiate the general expression for the steady-state nucleation rate

$$\ln \left(\frac{J}{J_0} \right) = -\beta \, W^*$$

with respect to $\ln p^v$:

$$\left. \frac{\partial \ln(J/J_0)}{\partial \ln p^v} \right|_{\{y_k\},T} = -\beta \left. \frac{\partial W^*}{\ln p^v} \right|_{\{y_k\},T}$$

Applying (4.38), we find

$$\left. \frac{\partial \ln(J/J_0)}{\partial \ln p^v} \right|_{\{y_k\},T} = \frac{p^v}{k_B T} \sum_i v_i^v \, \Delta n_{i,c} \qquad (4.39)$$

This result, termed *Pressure Nucleation Theorem* for multi-component systems [9], makes it possible to study the effect of total pressure on the nucleation rate in mixtures.

References

1. D. Kashchiev, J. Chem. Phys. **76**, 5098 (1982)
2. I.J. Ford, J. Chem. Phys. **105**, 8324 (1996)
3. D.W. Oxtoby, D. Kashchiev, J. Chem. Phys. **100**, 7665 (1994)
4. R.K. Bowles, D. Reguera, Y. Djikaev, H. Reiss, J. Chem. Phys. **115**, 1853 (2001)
5. I.J. Ford, Phys. Rev. E **56**, 5615 (1997)
6. T.L. Hill, J. Chem. Phys. **36**, 3182 (1962)
7. T.L. Hill, *Thermodynamics of Small Systems* (Dover, New York, 1994)
8. D. Kashchiev, J. Chem. Phys. **125**, 014502 (2006)
9. V.I. Kalikmanov, D.G. Labetski, Phys. Rev. Lett. **98**, 085701 (2007)

Chapter 5
Density Functional Theory

5.1 Nonclassical View on Nucleation

Classical phenomenological description of nucleation is based on the capillarity approximation treating all droplets (clusters) as if they were macroscopic objects characterized by a well defined rigid boundary of radius R with a bulk liquid density inside R and bulk vapor density outside R. Moreover, the surface free energy of the cluster is the same as for the planar interface at the same temperature, and therefore is characterized by the planar surface tension. Making a step from the purely macroscopic to a microscopic view, it is plausible to consider the system "*a droplet in a supersaturated vapor*" as an inhomogeneous fluid with some (unknown) density profile $\rho(r)$ continuously changing from the liquid-like value in the center of the droplet to the bulk vapor density far from it as shown in Fig. 5.1. This density profile gives rise to the *free energy functional* $\mathscr{F}[\rho(r)]$ which can be expressed in terms of molecular interactions and therefore does not invoke information about the macroscopic properties. As a result the capillarity approximation is avoided. The critical cluster in this picture is given by the *density profile* $\rho_c(r)$ being an extremum (saddle point) of the corresponding free energy (grand potential) functional over all admissible functions $\rho(r)$, while the nucleation barrier is the value of the functional at $\rho_c(r)$.

In the *theory of nonuniform fluids* the approach based on the concept of functionals of arbitrary distribution functions is called the *density functional theory* (DFT). It was proposed by Ebner et al. [1] and developed by Evans [2] (for an excellent review see the paper of Evans [3]).[1] Application of DFT to nucleation was formulated by Oxtoby and Evans [6] and later developed in a number of publications by

---

[1] Historically the DFT in the theory of fluids originates from the quantum mechanical ideas formulated by Hohenberg and Kohn [4] and Kohn and Sham [5]; these authors showed that the intrinsic part of the ground state energy of an inhomogeneous electron liquid can be cast in the form of a unique functional of the electron density $\rho_e(\mathbf{r})$. By doing so the quantum *many-body* problem—the solution of the many-electron Shrödinger equation—is replaced by a variational *one-body* problem for an electron in an effective potential field.

V. I. Kalikmanov, *Nucleation Theory*, Lecture Notes in Physics 860, DOI: 10.1007/978-90-481-3643-8_5, © Springer Science+Business Media Dordrecht 2013

Fig. 5.1 Representation of
a cluster in the density functional theory: continuous
density profile $\rho(r)$, changing
between the liquid-like value
in the center of a cluster to
the bulk vapor density far
from it. R_e indicates the Gibbs
equimolar dividing surface

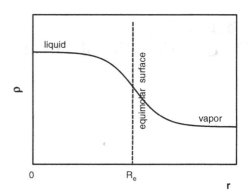

Oxtoby and coworkers [7–11]. The general feature of all density functional models
is the assumption that the thermodynamic potential of a *nonuniform* system can be
approximated using the knowledge of structural and thermodynamic properties of
the corresponding *uniform* system. Various DFT models differ from each other in the
way this approximation is formulated [12]. Although the results of DFT calculations
can not be presented in a closed form, these calculations are much faster than the
purely microscopic computer simulations (Monte Carlo, molecular dynamics). One
can thus classify the DFT as a semi-microscopic approach.

An important feature of DFT in nucleation is that calculation of the nucleation barrier does not invoke the a priori information about the surface tension. The curvature
effects of the surface free energy are incorporated into the DFT so that no *ad hoc*
assumptions are necessary. This is the consequence of the fact that DFT uses the
microscopic (interaction potential) rather the macroscopic input. The DFT approach
naturally recovers the CNT when the system is close to equilibrium (low supersaturations, large droplets). However, it considerably deviates from CNT at higher S.
In particular, the DFT predicts vanishing of the nucleation barrier at some finite S
(which signals the spinodal) while the CNT barrier remains finite even as the spinodal
is approached.

Before studying the application of DFT to nucleation we briefly formulate its fundamentals for the theory of nonhomogeneous fluids.

5.2   Fundamentals of the Density Functional Approach in the Theory of Liquids

5.2.1   General Principles

The cornerstone of DFT is the statement that the free energy of an inhomogeneous
fluid is a *functional of the density profile* $\rho(\mathbf{r})$. On the basis of the knowledge of

this functional one can calculate interfacial tensions, properties of confined systems, adsorption properties, determine depletion forces, study phase transitions, etc. From these introductory remarks it is clear that DFT represents an alternative—variational—formulation of statistical mechanics. Determination of the exact form of the free energy functional is equivalent to calculating the partition function, which, as it is well known, is not possible for realistic potentials. Therefore, one has to formulate approximations which could lead to computationally tractable results, and at the same time be applicable to a number of practical problems. However, even without knowing the exact form of the functional one can formulate several rigorous statement about it.

Let us consider the N-particle system in the volume V at the temperature T. Being concerned with classical systems, we can always consider the momentum part of the Hamiltonian to be described by the equilibrium Maxwellian distribution. This implies that the arbitrariness in the distribution function refers only to the configurational part of the Hamiltonian. We require that an arbitrary N-body distribution function, $\hat{\rho}^N(\mathbf{r}^N)$, should be positive and satisfy the normalization

$$\int \hat{\rho}^{(N)}(\mathbf{r}^N)\, d\mathbf{r}^N = N!, \quad \hat{\rho}^{(N)}(\mathbf{r}^N) > 0 \qquad (5.1)$$

The "hat" indicates that the distribution does not necessarily have its equilibrium form; the corresponding equilibrium function will be denoted without the "hat". The requirement (5.1) shows that functions $\hat{\rho}^{(N)}$ are subject to the same normalization as the *equilibrium* function $\rho^{(N)}(\mathbf{r}^N)$ (see e.g. [13] Chap. 2). To limit the class of functions $\hat{\rho}^{(N)}$, we employ the following considerations [14]. For a given interatomic interaction potential $u(r)$ every fixed external field $u_{\text{ext}}(\mathbf{r})$ gives rise to a certain equilibrium one-particle distribution function

$$\rho^{(1)}(\mathbf{r})[u_{\text{ext}}(\mathbf{r})] \equiv \rho(\mathbf{r})[u_{\text{ext}}(\mathbf{r})]$$

i.e. only one $u_{\text{ext}}(\mathbf{r})$ can determine a given $\rho(\mathbf{r})$. Here the square brackets denote that $\rho(\mathbf{r})$ is a *functional* of $u_{\text{ext}}(\mathbf{r})$. Keeping the form of this functional and varying $u_{\text{ext}}(\mathbf{r})$ we can generate a set of singlet density functions $\hat{\rho}(\mathbf{r})$ so that each of them will be an equilibrium density corresponding to some other external field, $\hat{u}_{\text{ext}}(\mathbf{r})$,

$$\hat{\rho}(\mathbf{r}) = \rho[\hat{u}_{\text{ext}}(\mathbf{r})]$$

but is nonequilibrium with respect to the original field $u_{\text{ext}}(\mathbf{r})$. For every $\hat{\rho}(\mathbf{r})$ there exists a unique N-particle distribution function $\hat{\rho}^{(N)}$ (for a proof see [2]). Thus, any functional of $\hat{\rho}^{(N)}(\mathbf{r}^N)$ can be equally considered a functional of the one-particle distribution function $\hat{\rho}(\mathbf{r})$.

Let us formally define the functional of *intrinsic free energy*

$$\mathscr{F}_{\text{int}}[\hat{\rho}] = \frac{1}{N!} \int d\mathbf{r}^N\, \hat{\rho}^{(N)}[U_N + k_B T \ln(\Lambda^{3N}\hat{\rho}^{(N)})] \qquad (5.2)$$

where

$$\Lambda = \sqrt{\frac{2\pi \hbar^2}{m_1 k_B T}}$$

is the thermal de Broglie wavelength of a particle, m_1 is its mass, $\hbar$ is the Planck constant, $U_N(\mathbf{r}^N)$ is the potential energy of a N-particle configuration $\mathbf{r}^N = (\mathbf{r}_1, \ldots, \mathbf{r}_N)$. As we have established, $\mathscr{F}_{int}[\hat{\rho}]$ is a unique functional of $\hat{\rho}(\mathbf{r})$ for a given interaction potential $u(\mathbf{r})$. This means that $\mathscr{F}_{int}[\hat{\rho}]$ has the *same dependence* on $\hat{\rho}(\mathbf{r})$ for all systems with the same $u(\mathbf{r})$ irrespective of the external field u_{ext} producing inhomogeneity. As an implication of this statement $\mathscr{F}_{int}$ will look the same for the vapor–liquid or liquid–solid interface as soon as the substances are characterized by the same interaction potential. The term *intrinsic free energy* becomes clear if we apply the functional (5.2) to the equilibrium function $\rho^{(N)}(\mathbf{r}^N)$ given by the Boltzmann distribution. In the absence of external fields it reads

$$\rho^{(N)} = \frac{e^{-\beta U_N}}{\Lambda^{3N} Z_N} \tag{5.3}$$

where Z_N is the canonical partition function. Substitution of (5.3) into (5.2) results in

$$\mathscr{F}_{int}[\rho] = -k_B T \ln Z_N = \mathscr{F} \tag{5.4}$$

which is the Helmholtz free energy of the system in the absence of external fields (intrinsic free energy). In the presence of an external field the free energy functional can be defined as a straightforward extension of Eq. (5.2):

$$\mathscr{F}[\hat{\rho}] = \frac{1}{N!} \int d\mathbf{r}^N \, \hat{\rho}^{(N)}[U_N + U_{N,ext} + k_B T \ln(\Lambda^{3N} \hat{\rho}^{(N)})] \tag{5.5}$$

where

$$U_{N,ext} = \sum_{i=1}^{N} u_{ext}(\mathbf{r}_i) \tag{5.6}$$

is the total external field energy of the configuration $\mathbf{r}^N$. Substituting (5.6) into (5.5) we obtain

$$\mathscr{F}[\hat{\rho}] = \mathscr{F}_{int}[\hat{\rho}] + \int d\mathbf{r}^N \, \frac{\hat{\rho}^{(N)}}{N!} \sum_{i=1}^{N} u_{ext}(\mathbf{r}_i)$$

It is easy to see that the integral contains N equal terms

$$\mathscr{F}[\hat{\rho}] = \mathscr{F}_{int}[\hat{\rho}] + N \int d\mathbf{r}_1 \, u_{ext}(\mathbf{r}_1) \int d\mathbf{r}_2 \ldots d\mathbf{r}_N \, \frac{\hat{\rho}^{(N)}(\mathbf{r}^N)}{N!}$$

Integration of the N-particle probability density function $\hat{\rho}^{(N)}/N!$ over all possible positions of $(N-1)$ particles results in the singlet distribution function:

$$\int d\mathbf{r}_2 \ldots d\mathbf{r}_N \frac{\hat{\rho}^{(N)}(\mathbf{r}^N)}{N!} = \frac{\hat{\rho}^{(1)}(\mathbf{r}_1)}{N}$$

implying that

$$\mathscr{F}[\hat{\rho}] = \mathscr{F}_{int}[\hat{\rho}] + \int d\mathbf{r}\, u_{ext}(\mathbf{r})\hat{\rho}(\mathbf{r}) \tag{5.7}$$

Thus, the free energy functional in the presence of an external field is a sum of the intrinsic free energy functional and the (average) energy of the system in an external field. Similarly to Eq. (5.4), for equilibrium conditions $\mathscr{F}[\rho]$ recovers the Helmholtz free energy of the system in an external field.

Finally, we define the *grand potential functional*:

$$\Omega[\hat{\rho}; u_{ext}] = \mathscr{F}[\hat{\rho}] - \mu \int d\mathbf{r}\, \hat{\rho}(\mathbf{r}) \tag{5.8}$$

where μ is the chemical potential. Obviously, for equilibrium conditions it reduces to the grand potential of the system:

$$\Omega[\rho; U_{ext}] = \mathscr{F} - \mu N = \Omega$$

The functionals of arbitrary distribution functions possess two important properties:

- they reach extrema when the distribution functions are those of the equilibrium state, and
- those extremal values are the equilibrium values of the corresponding thermodynamic potentials.

The are summarized in the following:
Theorem. Among all density profiles with the normalization

$$\int \hat{\rho}(\mathbf{r})\, d\mathbf{r} = N \tag{5.9}$$

the equilibrium profile $\rho(\mathbf{r})$ minimizes the functional of the free energy. The proof is presented elsewhere (see e.g. [13], Chap. 9). Using the Lagrange multipliers the minimizing property for $\mathscr{F}$ under the condition (5.9) can be cast in the form of the *unconditional* minimum of the grand potential functional:

$$\left. \frac{\delta\Omega}{\delta\hat{\rho}(\mathbf{r})} \right|_{\rho(\mathbf{r})} = 0 \tag{5.10}$$

The variational equation (5.10) using (5.8) and (5.9) yields:

$$\mu = \mu_{\text{int}}(\mathbf{r}) + u_{\text{ext}}(\mathbf{r}) \tag{5.11}$$

where

$$\mu_{\text{int}}(\mathbf{r}) \equiv \left. \frac{\delta \mathscr{F}_{\text{int}}[\hat{\rho}]}{\delta \hat{\rho}(\mathbf{r})} \right|_{\rho(\mathbf{r})} \tag{5.12}$$

is the *intrinsic chemical potential*. The spatial dependence of μ_{int} must be exactly canceled by the radial dependence of $u_{\text{ext}}(r)$, since the "full" chemical potential μ (which is the Lagrange parameter in this variational problem) is constant. Equations (5.11)–(5.12) represent the fundamental result of DFT. If we had means to determine $\mathscr{F}_{\text{int}}$, then (5.11) would be an exact equation for the equilibrium density.

5.2.2 Intrinsic Free Energy: Perturbation Approach

For realistic interactions the exact expression for the functional $\mathscr{F}_{\text{int}}$ is not available, and one has to invoke approximations. To this end let us study the response of the system to a small change in the pair potentials $\delta u(r_{ij})$ that alters the total interaction energy (assumed to be pairwise additive)

$$\delta U_N(\mathbf{r}^N) = \sum_{i<j} \delta u(r_{ij})$$

We describe this macroscopic reaction by the change in the grand potential

$$\Omega(\mu, V, T) = -k_B T \ln \varXi(\mu, V, T)$$

where

$$\varXi(\mu, V, T) = \sum_{N \geq 0} \lambda^N Z_N$$

is the grand partition function of the system, $\lambda = e^{\beta \mu}$ is the activity. We have

$$\delta \Omega = -k_B T \frac{\delta \varXi}{\varXi} = \frac{1}{\varXi} \sum_{N \geq 0} \lambda^N \frac{1}{\Lambda^{3N} N!} \int d\mathbf{r}^N \, e^{-\beta(U_N + U_{N,\text{ext}})} \delta U_N(\mathbf{r}^N)$$

Examination of the right-hand side reveals that it represents the thermal average (in the grand canonical ensemble) of δU_N

$$\delta \Omega = \overline{\delta U_N}$$

In view of the pairwise additivity of δU_N, this result can be transformed by means of the standard argument (*theorem of averaging* [13]) to give

$$\delta\Omega = \frac{1}{2}\int d\mathbf{r}_1\, d\mathbf{r}_2\, \rho^{(2)}(\mathbf{r}_1, \mathbf{r}_2)\, \delta u(r_{12})$$

where $\rho^{(2)}(\mathbf{r}_1, \mathbf{r}_2)$ is the pair distribution function. Using the definition of a variational derivative, this result can be expressed as

$$\frac{\delta\Omega}{\delta u(r_{12})} = \frac{1}{2}\rho^{(2)}(\mathbf{r}_1, \mathbf{r}_2) \tag{5.13}$$

From the definition of $\Omega[\rho]$ it follows that the same expression is valid for $\mathscr{F}_{int}$:

$$\frac{\delta\mathscr{F}_{int}[\rho]}{\delta u(r_{12})} = \frac{1}{2}\rho^{(2)}(\mathbf{r}_1, \mathbf{r}_2) \tag{5.14}$$

Let us decompose the interaction potential

$$u = u_0 + u_1 \tag{5.15}$$

where u_0 is some reference interaction and u_1 is a perturbation, and introduce a family of "test systems" characterized by potentials

$$u_\alpha(r_{12}) = u_0(r) + \alpha\, u_1(r), \quad 0 \le \alpha \le 1 \tag{5.16}$$

which gradually change from u_0 to u when the formal parameter α changes from zero to unity. The functional integration of (5.14) then gives:

$$\mathscr{F}_{int}[\rho] = \mathscr{F}_{int,0}[\rho] + \frac{1}{2}\int_0^1 d\alpha \int d\mathbf{r}_1\, d\mathbf{r}_2 \rho_\alpha^{(2)}(\mathbf{r}_1, \mathbf{r}_2)u_1(r_{12}) \tag{5.17}$$

where we have expressed δu_α as

$$\delta u_\alpha = \frac{\partial u_\alpha}{\partial \alpha}\, d\alpha = u_1\, d\alpha$$

The first term in (5.17) is the reference contribution—the intrinsic free energy of the system with the interaction potential $u_0(r)$. The second term refers to the perturbative part. The pair distribution function $\rho_\alpha^{(2)}$ is that of a system with the density ρ and the interaction potential u_α. Equation (5.17) is the second fundamental equation of the DFT which gives the *exact* (though intractable!) expression for the intrinsic free energy. To make the theory work we imply the *perturbation approach* in which u_1 is considered a small perturbation. Expanding (5.17) in u_1 to the first order we obtain

$$\mathscr{F}_{int}[\rho] = \mathscr{F}_{int,0}[\rho] + \frac{1}{2}\int d\mathbf{r}_1\, d\mathbf{r}_2\, \rho_0^{(2)}(\mathbf{r}_1, \mathbf{r}_2)\, u_1(r_{12}) + O(u_1^2)$$

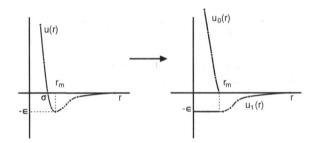

Fig. 5.2 Weeks–Chandler–Andersen decomposition of the interaction potential $u(r)$; $u_0(r)$ is the reference interaction, $u_1(r)$ is the perturbation

where the distribution function $\rho_0^{(2)}$ is now that of the reference system with the density $\rho(\mathbf{r})$ and interaction potential $u_0(r)$. One can go further and treat the reference part in the *local density approximation* (LDA):

$$\mathscr{F}_{int,0}[\rho] \approx \int d\mathbf{r}\, \psi_0(\rho(\mathbf{r})) \tag{5.18}$$

and

$$\rho_0^{(2)}(\mathbf{r}_1, \mathbf{r}_2) \approx \rho(\mathbf{r}_1)\, \rho(\mathbf{r}_2)\, g_0(\bar{\rho}; r_{12}) \tag{5.19}$$

where $\psi_0(\rho)$ is the free energy density of the uniform reference system with number density ρ; $g_0(\bar{\rho}; r_{12})$ is the pair correlation function of the uniform reference system evaluated at some mean density $\bar{\rho}$, e.g. $\bar{\rho} = [\rho(\mathbf{r}_1) + \rho(\mathbf{r}_2)]/2$. The LDA is valid for weakly inhomogeneous systems, such as a liquid–vapor interface. For strongly inhomogeneous systems, e.g. liquid at a wall, it becomes too crude and one has to use a nonlocal approximation, such as the weighted density [15] or the modified weighted-density approximation [12].

The most widely used decomposition of the interaction potential, (5.15) is given by the Weeks–Chandler–Anderson theory (WCA) [16] (see Fig. 5.2):

$$u_0(r) = \begin{cases} u(r) + \varepsilon & \text{for } r < r_m \\ 0 & \text{for } r \geq r_m \end{cases} \tag{5.20}$$

$$u_1(r) = \begin{cases} -\varepsilon & \text{for } r < r_m \\ u(r) & \text{for } r \geq r_m \end{cases} \tag{5.21}$$

where ε is the depth of the potential $u(r)$ and r_m is the corresponding value of r: $u(r_m) = -\varepsilon$. The advantage of the WCA scheme is that all strongly varying parts of the potential are subsumed by the reference model describing the harshly repulsive interaction, whereas $u_1(r)$ varies slowly and therefore the importance of fluctuations in the free energy expansion (represented by the second order term) is reduced.

The free energy of the reference model can be expressed as the free energy of a hard-sphere system with a suitably defined effective diameter d at the same temperature T and the number density ρ as the original system. The thermodynamic properties of a hard-sphere system are readily available from the Carnahan–Starling theory [17]. In particular the pressure and the chemical potential are (see e.g. [13]):

$$\frac{p_d}{\rho k_B T} = \frac{1 + \phi_d + \phi_d^2 - \phi_d^3}{(1 - \phi_d)^3} \tag{5.22}$$

$$\frac{\mu_d}{k_B T} = \ln(\rho \Lambda^3) + \frac{8\phi_d - 9\phi_d^2 + 3\phi_d^3}{(1 - \phi_d)^3} \tag{5.23}$$

where $\phi_d = (\pi/6)\rho d^3$ is the volume fraction of effective hard spheres. Then, from the standard thermodynamic relationship the Helmholtz free energy density of the hard-sphere system reads

$$\psi_d = \rho \mu_d - p_d \tag{5.24}$$

In the WCA theory it equals the free energy density of the reference model: $\psi_0(\rho) = \psi_d(\rho)$.

One can go further and *ignore all correlations* between particles in the perturbative term which results in setting $g_0 = 1$ in (5.19); this is the *random phase approximation* (RPA), equivalent to the mean field (van der Waals-like) theory. A number of studies showed that for systems with weak inhomogeneities, the RPA is sufficient when the system is not too close to the critical point. Combining LDA and RPA, we obtain the intrinsic free energy in its simplest form:

$$\mathscr{F}_{\text{int}}[\rho] = \int d\mathbf{r} \, \psi_d(\rho(\mathbf{r})) + \frac{1}{2} \int d\mathbf{r}_1 \, d\mathbf{r}_2 \, \rho(\mathbf{r}_1) \, \rho(\mathbf{r}_2) \, u_1(r_{12}) \tag{5.25}$$

In the same approximation the DFT equation (5.11) becomes

$$\mu_d(\rho(\mathbf{r})) = \mu - \int d\mathbf{r}' \, \rho(\mathbf{r}') \, u_1(|\mathbf{r} - \mathbf{r}'|) - u_{\text{ext}}(\mathbf{r}) \tag{5.26}$$

where $\mu_d(\rho(\mathbf{r}))$ is the local chemical potential of the hard-sphere fluid. The integral equation (5.26) can be solved iteratively for $\rho(\mathbf{r})$ starting with some initial profile satisfying the boundary conditions corresponding to bulk equilibrium:

$$\rho(z) \to \rho_{\text{sat}}^{\text{v}} \quad \text{in the bulk vapor}$$
$$\rho(z) \to \rho_{\text{sat}}^{\text{l}} \quad \text{in the bulk liquid}$$

In turn, the bulk equilibrium properties $\mu = \mu_{\text{sat}}(T)$, $\rho_{\text{sat}}^{\text{v}}(T)$, $\rho_{\text{sat}}^{\text{l}}(T)$, $p_{\text{sat}}(T)$ can be found from the DFT applied to a *uniform* system ($\rho = \text{const}$). Equations (5.25)–(5.26) (with $u_{\text{ext}}(\mathbf{r}) \equiv 0$) in this case become

$$\mathscr{F}[\rho] = \mathscr{F}_d[\rho] - \rho^2 a V \tag{5.27}$$

$$\mu = \mu_d(\rho) - 2\rho a, \tag{5.28}$$

where

$$a = -\frac{1}{2} \int d\mathbf{r}\, u_1(r) \tag{5.29}$$

is the background interaction parameter. Bulk equilibrium properties satisfy the equations

$$\mu^l(\rho_{\text{sat}}^l, T) = \mu^v(\rho_{\text{sat}}^v, T) \equiv \mu_{\text{sat}}(T) \tag{5.30}$$

$$p^l(\rho_{\text{sat}}^l, T) = p^v(\rho_{\text{sat}}^v, T) \equiv p_{\text{sat}}(T) \tag{5.31}$$

Given μ_{sat} one can perform an iteration process for the density profile $\rho(z)$ starting with an initial guess, say a step-function $\rho(z)$ with $\rho(z) = \rho_{\text{sat}}^l$ for $z < 0$ and $\rho(z) = \rho_{\text{sat}}^v$ for $z > 0$. This profile is put into the rhs of Eq. (5.26) and the latter is solved by inversion of the function $\mu_d(\rho)$ (given by the Carnahan-Starling approximation) for each point z of the interface. The density is then put back into the rhs of (5.26) and the process continues. One can be sure that iterations will eventually converge to the equilibrium profile describing the gas-liquid interface since it corresponds to the *minimum* of the grand potential functional:

$$\frac{\delta^2 \Omega}{\delta\rho(z)\delta\rho(z')} > 0$$

Differentiation of $\mathscr{F}$ with respect to the volume yields the virial equation of state:

$$p = p_d - \rho^2 a \tag{5.32}$$

where p_d is the pressure of the hard-sphere system.

For a Lennard–Jones fluid with the interaction potential

$$u_{\text{LJ}}(r) = 4\varepsilon \left[\left(\frac{\sigma}{r}\right)^{12} - \left(\frac{\sigma}{r}\right)^{6} \right] \tag{5.33}$$

the WCA decomposition of $u_{\text{LJ}}(r)$ yields for the parameter a

$$a_{\text{LJ}}^{\text{WCA}} = \frac{16\pi\sqrt{2}}{9} \varepsilon\sigma^3 \tag{5.34}$$

There is an obvious resemblance between (5.32) and the van der Waals equation:

$$p = \frac{\rho k_B T}{1 - b^{\text{vdW}}} - \rho^2 a^{\text{vdW}}$$

where for the Lennard–Jones fluid the van der Waals parameters are:

$$a_{LJ}^{vdW} = \frac{16\pi}{9}\,\varepsilon\sigma^3, \quad b_{LJ}^{vdW} = \frac{2\pi}{3}\,\sigma^3$$

At the same time the DFT uses a more sophisticated approach to describe the repulsive part of the potential than the free volume considerations of van der Waals. Furthermore, due to the different decomposition schemes, the background interaction parameters are different: $a_{LJ}^{WCA} = \sqrt{2}\,a_{LJ}^{vdW}$.

5.2.3 Planar Surface Tension

Consider an inhomogeneous (vapor-liquid) system characterized by a density profile $\rho(z)$ (inhomogeneity is in the z direction). The thermodynamic relationship (2.25) for the surface tension reads:

$$\gamma = \frac{\Omega^{exc}}{A} = \frac{\Omega[\rho] + pV}{A} \tag{5.35}$$

where A is the surface area. The grand potential functional (5.8) with $u_{ext} = 0$ then reads

$$\Omega[\rho] = -\int d\mathbf{r}\, p_d(\rho) + \int d\mathbf{r}\, \rho\, \mu_d(\rho)$$
$$+ \frac{1}{2}\int d\mathbf{r}\, \rho(\mathbf{r}) \int d\mathbf{r}'\, \rho(\mathbf{r}')\, u_1(|\mathbf{r} - \mathbf{r}'|) - \mu \int d\mathbf{r}\, \rho(\mathbf{r})$$

where we used the intrinsic free energy functional in the RPA-form (5.25). Using the DFT equation (5.26) and taking into account that $d\mathbf{r} = A\, dz$ we find

$$\gamma = -\int dz \left\{ p_d(\rho(z)) + \frac{1}{2}\rho(z) \int d\mathbf{r}'\, \rho(z')\, u_1(|\mathbf{r} - \mathbf{r}'|) - p_{sat} \right\} \tag{5.36}$$

where the equilibrium vapor–liquid density profile $\rho(z)$ satisfies Eq. (5.26).

Let us apply the DFT to a Lennard–Jones fluid [9, 18] characterized by the interaction potential $u_{LJ}(r)$. We begin by searching for the bulk equilibrium conditions at a given temperature $T < T_c$. Performing the WCA decomposition, we determine the effective hard-sphere diameter d for the reference model. Densities in the bulk phases, together with the equilibrium chemical potential and pressure, are found from the coupled nonlinear equations (5.28)–(5.32)

$$\mu = \mu_d(\rho^v) - 2\rho^v a = \mu_d(\rho^l) - 2\rho^l a$$
$$p = p_d(\rho^v) - (\rho^v)^2 a = p_d(\rho^l) - (\rho^l)^2 a$$

Fig. 5.3 Density profiles for
the two-phase Lennard–Jones
fluid at various dimensionless
temperatures $t = k_B T / \varepsilon$

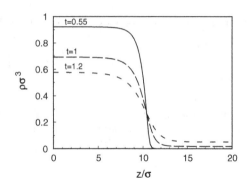

Fig. 5.4 Surface tension of
the Lennard–Jones fluid. *Solid
line*: DFT predictions; *stars*:
simulation results of Chapela
et al. [19]

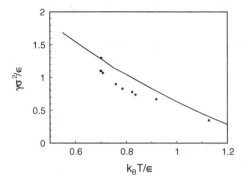

The density profiles for different dimensionless temperatures $t \equiv k_B T / \varepsilon$ are shown
in Fig. 5.3. The higher the temperature, the smaller the difference between the two
bulk densities and the broader the transition zone: for $t = 0.55$ it is $\approx 2\sigma$, while
for $t = 1.2$ it is about 7σ. Using the equilibrium $\rho(z)$ for each temperature, we find
the surface tension by integration in (5.36) (note that the integrand in this expression
vanishes outside the transition zone). The DFT predictions of the surface tension are
shown in Fig. 5.4 together with the simulation results of Chapela et al. [19]. The latter
are located somewhat lower then the DFT line in view of truncation of the interaction
potential in computer simulations, while in DFT the untruncated potential is used.

One important remark concerning these results must be made. The presented approximate theory is strictly mean-field in character, and therefore does not take into account
fluctuations, which become increasingly important at high temperatures close to the
critical point of the gas–liquid transition. This means that this treatment is not valid
in the critical region. Several models have been proposed to improve this approach
(for a review see [3]). In spite of this difficulty, the perturbation DFT turns out to be
very productive in giving an insight into various problems of the liquid state, such
as adsorption and wetting phenomena [20], phase transitions in confined fluids [21],
depletion interactions [22], etc.

5.3 Density Functional Theory of Nucleation

5.3.1 Nucleation Barrier and Steady State Nucleation Rate

It is convenient to reformulate the nucleation problem in terms of the grand potential Ω. In CNT the system is considered to be at constant pressure p^v, number of molecules N and temperature T and therefore the natural potential is the Gibbs free energy G. The latter is related to Ω through the Legendre transformation

$$G = \Omega + p^v V + \mu^v N \tag{5.37}$$

where V is the volume of the system "droplet + vapor" and $\mu^v(p^v, T)$ is the chemical potential of a vapor molecule. In the absence of a droplet the Gibbs free energy and the grand potential are

$$G_0 = \mu^v N, \quad \text{and} \quad \Omega_0 = -p^v V$$

Then from (5.37) the change in the Gibbs energy due to the droplet formation is

$$\Delta G = \Omega - \Omega_0 \equiv \Delta\Omega \tag{5.38}$$

and therefore the energy barrier to nucleation can be calculated in the grand ensemble. The grand potential functional is related to the intrinsic free energy (in the absence of external field) via

$$\Omega[\rho(r)] = \mathscr{F}_{int}[\rho(r)] - \mu \int dr \rho(r) \tag{5.39}$$

For $\mathscr{F}_{int}[\rho(r)]$ we take the mean field form (5.25) which consists of the local density approximation for the (effective) hard-sphere part of the interaction potential and the random phase approximation for the attractive part $u_1(r)$ considered as a perturbation:

$$\mathscr{F}_{int}[\rho] = \int d\mathbf{r} \psi_d(\rho(\mathbf{r})) + \frac{1}{2} \int d\mathbf{r}_1 d\mathbf{r}_2 \rho(\mathbf{r}_1)\rho(\mathbf{r}_2)u_1(r_{12}) \tag{5.40}$$

The uniform hard-sphere system is described by the Carnahan-Starling approximation. Hence, $\Omega[\rho]$ reads:

$$\Omega[\rho(r)] = \int d\mathbf{r} \psi_d(\rho(\mathbf{r})) + \frac{1}{2} \int d\mathbf{r}_1 d\mathbf{r}_2 \rho(\mathbf{r}_1)\rho(\mathbf{r}_2)u_1(r_{12}) - \mu \int d\mathbf{r}\rho(\mathbf{r}) \tag{5.41}$$

Critical droplet refers to a *metastable* state of the system "droplet + supersaturated vapor". The chemical potential μ in (5.41) is away from its coexistence value μ_{sat}.

This implies that the DFT equation

$$\frac{\delta\Omega}{\delta\rho(r)} = 0$$

resulting in:

$$\mu_d(\rho(\mathbf{r})) = \mu - \int d\mathbf{r}'\rho(\mathbf{r}')u_1(|\mathbf{r}-\mathbf{r}'|) \tag{5.42}$$

refers to a *local* (rather than the global) minimum of the free energy. Still there is a nontrivial solution of Eq. (5.42) which corresponds to a *saddle point* of the functional $\Omega[\rho]$ in the functional space. This solution describes a critical nucleus. The iteration process is now unstable. Nevertheless the solution can be found once an *appropriate initial guess* is chosen. As an initial guess for a radial droplet profile a step-function can be taken with a range parameter R_{init}. If R_{init} is small enough the droplet will shrink in the process of iteration giving rise to a metastable vapor density solution. If R_{init} is large the droplet will grow into a stable liquid. There exist an intermediate value R_{init}^* which in the process of iteration will give rise to the critical droplet neither growing, nor shrinking over a large number of iteration steps n. So the function $\Omega(n)$ will exhibit a long plateau staying at a constant value Ω^*. The energy barrier for nucleation is given by

$$\Delta\Omega^* = \Omega^* - \Omega_0$$

Finally, the steady state nucleation rate can be written as

$$J = J_0 e^{-\beta\Delta\Omega^*} \tag{5.43}$$

The pre-exponential factor can be taken from CNT (see (3.55))

$$J_0 = \frac{(\rho^v)^2}{\rho^l}\sqrt{\frac{2\gamma_\infty}{\pi m_1}} \tag{5.44}$$

since the nucleation rate is far less sensitive to J_0 than to the value of the energy barrier.

The great advantage of the DFT over the purely phenomenological models is that in the DFT one does not have to invoke the macroscopic equilibrium properties and equation of state. Yet, calculations of nucleation behavior are much faster than direct *computer simulations* using Monte Carlo or Molecular Dynamics methods (discussed in Chap. 8).

5.3.2 Results

Oxtoby and Evans [6] calculated density profiles for a critical droplet with a Yukawa attractive potential

$$u_1(r) = -\alpha\kappa^3 \frac{e^{-\kappa r}}{4\pi\lambda r}$$

According to CNT one expects that the density in the center of the droplet is equal to the liquid density at coexistence $\rho_{sat}^l(T)$. However, the results of [6] show that the density in the center of the droplet is lower than ρ_{sat}^l and is lower than the density of the bulk liquid at the same chemical potential as the supersaturated vapor $\rho^l(\mu^v)$. Another important feature of this approach is that the barrier to nucleation, $\Delta\Omega^*$, vanishes as spinodal is approached whereas in the classical theory it remains finite. Zeng and Oxtoby [9] applied the DFT to predict nucleation rates for a Lennard-Jones fluid. Following the preceding discussion the free energy functional is written using the hard-sphere perturbation analysis based on the WCA decomposition of the non-truncated Lennard-Jones potential.

Figure 5.5 based on the results of Ref. [9] shows the comparison of the nucleation rates predicted by the DFT and the CNT. To make such a comparison consistent the macroscopic surface tension used in the CNT was obtained by means of Eq. (5.35). The results in Fig. 5.5 correspond to the fixed classical nucleation rate $J_{CNT} = 1 \, \text{cm}^{-3}\text{s}^{-1}$. At each temperature this condition determines the chemical potential difference (and consequently the supersaturation) for which the DFT calculation is carried out. The results of Ref. [9] demonstrate that CNT and DFT predict the same dependence of nucleation rate on the supersaturation but show the essentially different temperature dependence. Due to the latter the disagreement between the two approaches is up to 5 orders of magnitude in J. The difference between two theories becomes pronounced when the nucleation temperature is away from $k_B T/\varepsilon \approx 1.1$. As one can see

Fig. 5.5 Ratio of the nucleation rates (CNT to DFT) for a Lennard-Jones fluid [9]. DFT calculations at each temperature are carried out for the supersaturation corresponding to the fixed classical nucleation rate $J_{CNT} = 1 \, \text{cm}^{-3}\text{s}^{-1}$. The predictions of both models become equal at $k_B T/\varepsilon \approx 1.08$. The *solid line* is shown to guide the eye

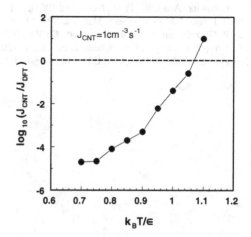

from Fig. 5.5, at lower temperatures CNT underestimates the nucleation rate (compared to DFT) while at higher temperatures it overestimates it. The same trend is demonstrated by the CNT when it is compared to *experimental* nucleation rates: e.g. for water CNT underestimates experimental rates at $T < 230$ K, and overestimates experimental rates at $T > 230$ K [23, 24].

References

1. C. Ebner, W.F. Saam, D. Stroud, Phys. Rev. A **14**, 2264 (1976)
2. R. Evans, Adv. Phys. **28**, 143 (1979)
3. R. Evans, Density functionals in the theory of nonuniform fluids. in *Fundamentals of Inhomogeneous Fluids*, ed. by D. Henderson (Marcel Dekker, New York 1992), p. 85
4. P. Hohenberg, W. Kohn, Phys. Rev. **136**, B864 (1964)
5. W. Kohn, L.J. Sham, Phys. Rev. **140**, A1133 (1965)
6. D.W. Oxtoby, R. Evans, J. Chem. Phys. **89**, 7521 (1988)
7. D.W. Oxtoby, in *Fundamentals of Inhomogeneous Fluids*, ed. by D. Henderson (Marcel Dekker, New York, 1992), Chap. 10
8. D.W. Oxtoby, J. Phys. Cond. Matt. **4**, 7627 (1992)
9. X.C. Zeng, D.W. Oxtoby, J. Chem. Phys. **94**, 4472 (1991)
10. V. Talanquer, D.W. Oxtoby, J. Chem. Phys. **99**, 4670 (1993)
11. V. Talanquer, D.W. Oxtoby, J. Chem. Phys. **100**, 5190 (1994)
12. A.R. Denton, N.W. Ashcroft, Phys. Rev. A **39**, 4701 (1989)
13. V.I. Kalikmanov, *Statistical Physics of Fluids. Basic Concepts and Applications* (Springer, Berlin, 2001)
14. J.S. Rowlinson, B. Widom, *Molecular Theory of Capillarity* (Clarendon Press, Oxford, 1982)
15. W.A. Curtin, N.W. Ashcroft, Phys. Rev. A **32**, 2909 (1985)
16. D. Weeks, D. Chandler, H.C. Andersen, J. Chem. Phys. **54**, 5237 (1971)
17. N.F. Carnahan, K.E. Starling, J. Chem. Phys. **51**, 635 (1969)
18. C.C.M. Luijten, Ph.D. Thesis, Eindhoven University, 1999
19. A.G. Chapela, G. Saville, S.M. Thompson, J.S. Rowlinson, J. Chem. Soc. Faraday Trans. II **73**, 1133 (1977)
20. S. Dietrich, in *Phase Transitions and Critical Phenomena*, vol. 12, ed. by C. Domb, J.L. Lebowitz (Academic Press, New York 1988), p. 1
21. R. Evans, J. Phys. Condens. Matter **2**, 8989 (1990)
22. B. Götzelmann et al., Europhys. Lett. **47**, 398 (1999)
23. J. Wölk, R. Strey, J. Phys. Chem. B **105**, 11683 (2001)
24. D.G. Labetski, V. Holten, M.E.H. van Dongen, J. Chem. Phys. **120**, 6314 (2004)

Chapter 6
Extended Modified Liquid Drop Model and Dynamic Nucleation Theory

In the classical theory and its modifications an arbitrary cluster is characterized by one parameter—the number of molecules in it. In a series of papers [1–3] Reiss and co-workers discussed an alternative form of cluster characterization. It was suggested that a cluster should be characterized not only by the particle number, i, but also by its volume v. As a result dynamics of such an i, v-*cluster* becomes two-dimensional (as opposed to the CNT, where it is one-dimensional) resembling nucleation in binary systems. Using these arguments Weakliem and Reiss [4] put forward the *modified liquid drop model* and performed extensive Monte Carlo simulations to calculate free energy of the i, v-clusters. Based on these ideas Reguera et al. [5] put forward the "extended modified liquid drop" model (EMLD), taking into account the effect of fluctuations which are important for the formation of tiny droplets in a small NVT-system. More recently Reguera and Reiss [6] combined EMLD with the Dynamic Nucleation Theory (DNT) of Shenter et al. [7, 8]. The new model, called "*Extended Modified Liquid Drop Model-Dynamical Nucleation Theory*" (EMLD-DNT), is discussed in the next sections.

6.1 Modified Liquid Drop Model

The CNT studies formation of droplets in an open system. The main feature of the modified liquid drop model of Ref. [4] is that it considers the *closed system* containing N molecules confined within a small spherical volume V at a temperature T. This small NVT system is called an *EMLD-cluster*. Within the volume V various *sharp n-clusters* can form, $n = 1, \ldots, N$. The important difference between the closed an open system is that in the system with the fixed total amount of molecules N the formation and growth of droplets is accompanied by the depletion of the vapor—the effect neglected in CNT, which assumes the existence of an infinite source of vapor molecules. The depletion of the vapor molecules in EMLD results in the decrease of supersaturation so that the droplet can not become arbitrarily large. Following

V. I. Kalikmanov, *Nucleation Theory*, Lecture Notes in Physics 860,
DOI: 10.1007/978-90-481-3643-8_6, © Springer Science+Business Media Dordrecht 2013

Fig. 6.1 A schematic representation of the EMLD-
cluster: a closed system of N
molecules confined inside the
volume V of the radius R.
n out of N molecules form
a liquid drop of the radius r,
while the rest $N - n$ molecules
remain in the vapor phase

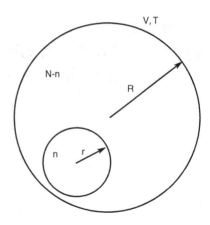

Ref. [6] we analyze the properties of the EMLD-cluster using purely thermodynamic
considerations.

The spherical volume V of the radius R is assumed to have impermeable hard walls.
Under certain conditions a liquid drop with n molecules, $n \leq N - 1$, can be formed
inside V. The rest $N - n$ molecules remain in the vapor phase, occupying the volume
$V - nv^l$, where v^l is the volume per molecule in the bulk liquid phase (see Fig. 6.1).
This vapor has the pressure described within the ideal gas approximation

$$p_1 = \frac{(N - n)\,k_B T}{V - nv^l} \tag{6.1}$$

The sharp n-cluster is treated within the capillarity approximation, i.e. it is assumed
that it has a sharp interface, characterized by the macroscopic surface tension γ_∞,
and the bulk liquid properties inside it. We stress that within this model the *entire*
NVT-system is considered as the EMLD-cluster, and not just one (sharp) n-droplet.
Since we discuss the closed system, the appropriate thermodynamic potential is the
Helmholtz free energy

$$\mathscr{F} = U - T\mathscr{S}$$

where U is the internal energy and $\mathscr{S}$ is the entropy of the EMLD-cluster. Denoting
the vapor and liquid subsystems inside the EMLD-cluster by subscripts 1 and 2,
respectively, we write the differentials of U_1 and U_2 as:

$$dU_1 = T dS_1 - p_1 dV_1 + \mu_1 dN_1 \tag{6.2}$$
$$dU_2 = T dS_2 - p_2 dV_2 + \mu_2 dN_2 + \gamma_\infty dA \tag{6.3}$$

where $A = 4\pi r^2$ is the surface area of the n-cluster. The free energy change associated
with the formation of a sharp n-droplet inside the EMLD-cluster is:

$$(d\mathscr{F})_{NVT} = dU_1 + dU_2 - T dS_1 - T dS_2$$

For the closed system

$$dV_1 = -dV_2, \quad dN_1 = -dN_2 = -n$$

which using (6.2)–(6.3) gives

$$(d\mathscr{F})_{NVT} = -\left(p_2 - p_1 - \frac{2\gamma_\infty}{r}\right) dV_2 + (\mu_2 - \mu_1) dN_2 \tag{6.4}$$

Equilibrium of the sharp n-droplet with the surrounding vapor corresponds to $(dF)_{NVT} = 0$ resulting in

$$\mu_2 = \mu_1 \tag{6.5}$$

$$p_2 - p_1 = \frac{2\gamma_\infty}{r} \tag{6.6}$$

The second equality is the Laplace equation. We can rewrite these results applying the ideal gas approximation for the vapor and considering liquid to be incompressible. Using the thermodynamic relationship

$$(d\mu)_T = \frac{1}{\rho} dp$$

for the vapor and liquid phases, we relate μ_i to the bulk vapor-liquid equilibrium properties

$$\mu_1(p_1) - \mu_{sat} = k_B T \ln \frac{p_1}{p_{sat}} \tag{6.7}$$

$$\mu_2(p_2) - \mu_{sat} = v^l(p_2 - p_{sat}) \tag{6.8}$$

Then, the equilibrium conditions (6.5)–(6.6) result in

$$k_B T \ln \frac{p_1}{p_{sat}} = \frac{2\gamma_\infty}{r} v^l + v^l(p_1 - p_{sat}) \tag{6.9}$$

The last term is usually very small and can be neglected. The result is the classical Kelvin equation (3.61) relating the pressure inside the n-droplet to its radius

$$p_1 = p_{sat} \exp\left(\frac{2\gamma_\infty v^l}{r k_B T}\right) \tag{6.10}$$

Using Eq. (6.1) for the vapor pressure inside the EMLD-cluster, we can solve (6.10) for the size of the coexisting droplet. In CNT, dealing with the open μVT system, the solution of the Kelvin equation determines the critical cluster radius at the given temperature and supersaturation. In the closed system the supersaturation p_1/p_{sat} is not fixed but depends on the amount of molecules in the n-cluster since the total amount N is conserved. Another difference is that while in the open system the solution of the coexistence equations is unique and corresponds to the critical cluster, in the closed system Eqs. (6.5)–(6.6) have two solutions: one corresponding to the critical cluster, i.e. the cluster in the *metastable* equilibrium with the vapor, and the other one—corresponding to the *stable* cluster.

For an arbitrary n-cluster (i.e. not necessarily a critical one), which is *not in equilibrium* with the surrounding vapor, the free energy change $(\mathrm{d}\mathscr{F})_{NVT} \neq 0$ and the chemical potential difference $\mu_2(p_2) - \mu_1(p_1)$ reads:

$$\mu_2(p_2) - \mu_1(p_1) = v^{\mathrm{l}}(p_2 - p_{\mathrm{sat}}) - k_{\mathrm{B}}T \ln \frac{p_1}{p_{\mathrm{sat}}} \tag{6.11}$$

Substituting (6.11) into (6.4) and using the incompressibility of the liquid phase $(\mathrm{d}V_2 = v^{\mathrm{l}}\,\mathrm{d}n)$ we obtain:

$$(\mathrm{d}\mathscr{F})_{NVT} = \left[\frac{2\gamma_\infty}{r(n)} v^{\mathrm{l}} + v^{\mathrm{l}}(p_1(n) - p_{\mathrm{sat}}) - k_{\mathrm{B}}T \ln \frac{p_1(n)}{p_{\mathrm{sat}}} \right] \mathrm{d}n \tag{6.12}$$

We can perform thermodynamic integration of this equation from $n = 0$, corresponding to the state of pure vapor, to an arbitrary n. Then

$$\int_0^n (\mathrm{d}\mathscr{F})_{NVT} = \mathscr{F}(n) - \mathscr{F}(0) \equiv \Delta\mathscr{F}(n)$$

is the free energy difference between the state in which EMLD-cluster contains the n-droplet and the state of pure vapor; thus, $\Delta F(n)$ is the Helmholtz free energy of the n-droplet formation. Taking into account that $r(n) = r^{\mathrm{l}}n^{1/3}$ (where $r^{\mathrm{l}} = (3v^{\mathrm{l}}/4\pi)^{1/3}$) and using (6.1) we find after the integration

$$\Delta\mathscr{F}(n) = -n\,k_{\mathrm{B}}T \ln\left(\frac{p_1}{p_{\mathrm{sat}}}\right) + \gamma_\infty A + n\,k_{\mathrm{B}}T \left(1 - \frac{v^{\mathrm{l}}p_{\mathrm{sat}}}{k_{\mathrm{B}}T}\right) + N k_{\mathrm{B}}T \ln\left(\frac{p_1}{p_0}\right) \tag{6.13}$$

where

$$p_0 = \frac{N k_{\mathrm{B}}T}{V}$$

is the pressure corresponding to the pure vapor. In Eq. (6.13) the first and the second terms are, respectively, the standard bulk and surface contributions to the free energy of cluster formation (recall, that here the supersaturation is $S(n) = p_1(n)/p_{\mathrm{sat}}$); the third term is the volume work (usually small and is commonly neglected); the last term originates from depletion of the vapor molecules in the EMLD-cluster when an n-droplet is formed. In the thermodynamic limit $p_1 = p_0$ recovering the CNT result for the free energy of the cluster formation. Note, that in this model the state of pure vapor corresponds to $n = 0$ and not to $n = 1$; the value $n = 1$ corresponds to a *hypothetical liquid cluster of size 1*, which is not the same as a molecule of the vapor.

The EMLD-cluster can contain sharp droplets of various sizes n. The total free energy of the cluster $\Delta\mathscr{F}_{tot}$ is found by accounting of all possible fluctuations of the droplet size. Each such fluctuation enters the configuration integral Q of the cluster with the Boltzmann factor $e^{-\beta\Delta\mathscr{F}(n)}$ resulting in

$$Q = \sum_{n=0}^{N} e^{-\beta\Delta\mathscr{F}(n)}$$

Then,

$$\Delta\mathscr{F}_{tot} = -k_B T \ln Q = -k_B T \ln \sum_{n=0}^{N} e^{-\beta\Delta\mathscr{F}(n)} \tag{6.14}$$

The probability of finding a sharp cluster with *exactly* n molecules inside is

$$f(n) = \frac{e^{-\beta\Delta\mathscr{F}(n)}}{Q}, \qquad \sum_{n=0}^{N} f(n) = 1$$

The total pressure P_{tot} of the EMLD-cluster is the weighted sum of the vapor pressure in the confinement sphere over all possible n. The vapor pressure consists of p_1 and the pressure exerted by the n-drop (for $n \neq 0$), modelled as a single ideal gas molecule moving within the container

$$P_{tot} = \sum_{n=0}^{N} f(n) \left[p_1 + \frac{k_B T}{V_c} \, \Xi(n) \right]$$

Here $V_c(n) = 4\pi(R - r(n))^3/3$ is the volume accessible for the center of mass of the n-droplet with the radius $r(n)$, R is the radius of the EMLD-cluster

$$R = \left(\frac{3V}{4\pi} \right)^{1/3}$$

and $\Xi(n)$ is the unit step-function.

6.2 Dynamic Nucleation Theory and Definition of the Cluster Volume

Considerations put forward in the previous section, have not yet answered the question: how to choose the volume V of the EMLD-cluster which would be physically relevant for nucleation at the given external conditions? To address it we notice that since nucleation is a kinetic nonequilibrium process, it is plausible to search for the suitable criterion for V in the theory of rate processes. An important development in this direction is the Dynamic Nucleation Theory (DNT) formulated by Shenter et al. [7]. The key issue of DNT is the determination of the evaporation rate for an isolated cluster.

DNT is a molecular based theory: for a given intermolecular interaction potential one derives the Helmholtz free energy of the cluster using an appropriate statistical mechanical sampling. The detailed balance condition is used to derive the relationship between the condensation and evaporation rates which is expressed in terms of the differences in Helmholtz free energies between the N- and $(N + 1)$-clusters. An important quantity in these calculations is the volume V of the N-cluster, or, equivalently, the size of the configurational space of the cluster. In the DNT the ambiguity in the choice of V is removed by means of the *variational transition state theory* [9–11]. At each stage of the nucleation process the DNT defines the dividing surface of the radius r_{cut} in the phase space that separates reactant states from the producing states. From this dividing surface an unambiguous cluster definition emerges which is consistent with the detailed balance between the condensation and evaporation rates [8]. This dividing surface leads to the evaporation rate which is proportional to the Helmholtz free energy of the cluster. Using the variational transition state theory Shenter et al. [12] showed that the proper *kinetic* definition of V is the one that minimizes the evaporation rate. From the above discussion it follows that the proper V corresponds to the minimum of the free energy change of the cluster with respect to its volume, or in other words the physically relevant cluster volume V_m corresponds to the minimum of the pressure:

$$\frac{\partial P_{\text{tot}}}{\partial V}\bigg|_{V_m} = 0$$

The combined (EMLD and DNT) model is called the *EMLD-DNT* theory.

6.3 Nucleation Barrier

In the open $(\mu V T)$-system the work of formation of the physical (N, V_m)-cluster at a temperature T is given by

$$\Delta G = \Delta \Omega = \Delta \mathscr{F}(N, V_m) - V_m(p_0 - P_{\text{tot}}) + N\Delta\mu_0 \qquad (6.15)$$

where $\Delta\mu_0 = \mu_0 - \mu$. The role of the cluster size in this expression is played by N. By construction the EMLD-DNT cluster represents the diffuse interface system. That is why N is not the molecular excess quantity determined by the nucleation theorem (discussed in Chap. 4). The nucleation barrier corresponds to the maximum of ΔG:

$$\left.\frac{\partial \Delta G}{\partial N}\right|_{V_m,T} = 0 \qquad (6.16)$$

Having determined the critical cluster N^* from this equation, we substitute it into Eq. (6.15) to obtain the nucleation barrier

$$\Delta G^*_{\text{EMLD-DNT}} = \Delta \mathscr{F}(N^*, V_m) - V_m(p_0 - P_{\text{tot}}) + N^* k_B T \ln(p_0/P_{\text{tot}}) \qquad (6.17)$$

In the thermodynamic limit p_0 coincides with the actual vapor pressure p^v. The EMLD-DNT nucleation rate reads:

$$J_{\text{EMLD-DNT}} = J_0 \exp\left(-\frac{\Delta G^*_{\text{EMLD-DNT}}}{k_B T}\right) \qquad (6.18)$$

with the CNT kinetic prefactor, J_0.

The advantage of EMLD-DNT is that it does not require information about the microscopic interactions (as DNT or DFT). It uses the same set of the macroscopic parameters as the CNT: $p_{\text{sat}}(T)$, $\rho^l(T)$, $\gamma_\infty(T)$. At the same time the model is able to predict the vanishing of the nucleation barrier at some finite S which signals the thermodynamic spinodal; recall that in the CNT the nucleation barrier remains finite for all values of S. The important conceptual feature of EMLD-DNT is that instead of a sharp density profile it allows for a diffusive cluster. Within this approach *every* sharp n-droplet inside the EMLD-cluster enters the free energy with its Boltzmann weight. Note, that describing a sharp droplet of any size $1 \leq n \leq N - 1$ this model uses the capillarity approximation with the planar surface tension γ_∞ which becomes dubious for small n. The latter, however, is smeared out between all admissible cluster sizes. In Chap. 10 we show the predictions of EMLD-DNT for argon nucleation and compare them to experiments and other theoretical models as well as to the DFT and computer simulations.

References

1. H. Reiss, A. Tabazadeh, J. Talbot, J. Chem. Phys. **92**, 1266 (1990)
2. H.M. Ellerby, C.L. Weakliem, H. Reiss, J. Chem. Phys. **95**, 9209 (1991)
3. H.M. Ellerby, H. Reiss, J. Chem. Phys. **97**, 5766 (1992)
4. C.L. Weakliem, H. Reiss, J. Chem. Phys. **99**, 5374 (1993)
5. D. Reguera et al., J. Chem. Phys. **118**, 340 (2003)
6. D. Reguera, H. Reiss, Phys. Rev. Lett. **93**, 165701 (2004)
7. G.K. Shenter, S.M. Kathmann, B.C. Garrett, Phys. Rev. Lett. **82**, 3484 (1999)
8. S.M. Kathmann, G.K. Shenter, B.C. Garrett, J. Chem. Phys. **116**, 5046 (2002)

 9. E. Wigner, Trans. Faraday Soc. **34**, 29 (1938)
10. J.C. Keck, J. Chem. Phys. **32**, 1035 (1960)
11. J.C. Keck, Adv. Chem. Phys. **13**, 85 (1967)
12. G.K. Shenter, S.M. Kathmann, B.C. Garrett, J. Chem. Phys. **110**, 7951 (1999)

Chapter 7
Mean-Field Kinetic Nucleation Theory

7.1 Semi-Phenomenological Approach to Nucleation

On the microscopic level nucleation behavior is determined by intermolecular interactions in the substance. This is clearly demonstrated by the Density Functional Theory. Its applicability, however, is limited by relatively simple types of interactions. For the substances with highly nonsymmetric molecules the applicability of the standard DFT scheme becomes increasingly difficult. It is therefore desirable to propose a compromise between the microscopic and phenomenological descriptions. One can classify it as a *semi-phenomenological* approach to nucleation. It was pioneered by Dillmann and Meier [1] and developed by Ford, Laaksonen and Kulmala [2], Delale and Meier [3] and Kalikmanov and van Dongen [4]. The main idea of the semi-phenomenological approach is a combination of statistical thermodynamics of clusters with available data on the equilibrium material properties.

The statistical thermodynamic part of all these models is based on the seminal Fisher droplet model of condensation [5], in which the real gas is considered as a system of noninteracting clusters and the cluster distribution function is expressed in terms of the cluster configuration integral. The latter in the Fisher theory contains an undetermined quantity—the Helmholtz free energy per unit surface of the cluster- termed in [5] a *microscopic surface tension*. A common feature of the above-mentioned models is that Fisher's microscopic surface tension is presented in the form of the expansion of the cluster surface tension in powers of the curvature. The coefficient at the first order term is known as the Tolman length (cf. Sect. 2.2.2); the expansion itself is frequently called the Tolman expansion. In [2–4] the Tolman expansion is truncated at the first or second order term.

By doing so one has to realize that the range of validity of this expansion is a matter of great importance. Clearly, since the expansion is in the cluster curvature, it is applicable for sufficiently big clusters. In a large number of experiments, however, the clusters that dominate nucleation behavior (those close to the critical size), are relatively small, containing tens or hundreds of molecules, for them the cluster curvature is not

V. I. Kalikmanov, *Nucleation Theory*, Lecture Notes in Physics 860,
DOI: 10.1007/978-90-481-3643-8_7, © Springer Science+Business Media Dordrecht 2013

a small parameter implying that the Tolman expansion becomes dubious. Therefore, it is desirable to formulate a *nonperturbative* semi-phenomenological approach valid for all cluster sizes. In this chapter we discuss such a model—a *Mean-field Kinetic Nucleation Theory* (MKNT) of Ref. [6].

7.2 Kinetics

Kinetics of cluster formation is governed by the standard CNT assumptions:

- cluster growth and decay are dominated by monomer addition and monomer extraction;
- if a monomer collides a cluster it sticks to it with probability unity (the sticking coefficient is unity); and
- there is no correlation between successive events that change the number of particles in a cluster.

Using Katz's "kinetic approach" of Sect. 3.5, we write the steady state nucleation rate in the form given by Eq. (3.68):

$$
J = \left[\sum_{n=1}^{\infty} \frac{1}{f(n)\, S^n\, \rho_{\text{sat}}(n)} \right]^{-1}
\tag{7.1}
$$

where S is the supersaturation, $f(n)$ is the forward rate of n-cluster formation in the supersaturated vapor, and $\rho_{\text{sat}}(n)$ is the equilibrium cluster distribution at saturation (corresponding to $S = 1$). The purely phenomenological considerations adopted in the CNT and in Katz's kinetic version of the CNT are valid when the major contribution to the nucleation rate comes from big clusters (the notion of a "big cluster" will be specified below). In this case the free energy of cluster formation reads

$$
\Delta G_{\text{sat}}^{\text{CNT}}(n) = \gamma_{\infty}\, s_1\, n^{2/3}
$$

yielding for the cluster distribution function at saturation

$$
\rho_{\text{sat}}(n) \sim e^{-\beta\, \gamma_{\infty}\, s_1\, n^{2/3}}
$$

The formal extension of these expressions to *all* cluster sizes does not pose the problem since the contribution of small clusters to J is negligible.

On the other hand, if nucleation behavior is primarily driven by the formation of *small* clusters, one has to apply microscopic considerations. In this regime the very notion of a surface tension of a small cluster looses its physical meaning. In the next section we consider the model for $\rho_{\text{sat}}(n)$ which is valid for arbitrary n.

7.3 Statistical Thermodynamics of Clusters

A typical interaction between gas molecules consists of a harshly repulsive core and a short-range attraction. The most probable configurations of the gas at low densities and temperatures will be isolated clusters of $n = 1, 2, 3, \ldots$ molecules. Hence, to a reasonable approximation one can describe a real gas as a system of noninteracting clusters.[1] At the same time *intracluster* interactions are important—they are responsible for the formation of a cluster All clusters are in statistical equilibrium, associating and dissociating. Even large clusters have a certain probability of appearing. The partition function of an n-cluster at a temperature T is:

$$Z_n = \frac{1}{\Lambda^{3n}} q_n \tag{7.2}$$

where Λ is the thermal de Broglie wavelength of a particle (being an atom or a molecule); q_n is the configuration integral of the n-cluster in a physical domain of volume V:

$$q_n(T) = \frac{1}{n!} \oint_{\text{cl}} d\mathbf{r}^n \, e^{-\beta U_n}, \tag{7.3}$$

U_n is the potential energy of the n-particle configuration in the cluster; the factor $\frac{1}{n!}$ takes into account the indistinguishability of particles inside the cluster. The symbol $\oint_{\text{cl}}$ indicates that integration is performed only over those atomic configurations that belong to the cluster. At this point it is important to emphasize the difference between the *n-particle* configuration integral Q_n and the *n-cluster* configuration integral q_n. The latter includes only those configurations in the volume V that form the n-cluster, while Q_n contains *all* different configurations of n particles in the volume V; therefore $Q_n \geq q_n$. The cluster *as a whole* can move through the entire volume V of the system, while the particles inside the cluster are restricted to the configurations about cluster's center of mass that are consistent with a chosen definition of the cluster. For that one can adopt, e.g., Stillinger cluster [7]: an atom belongs to a cluster if there exists at least one atom of the same cluster separated from the given one by a distance $r < r_b$, where r_b is some characteristic distance describing the range of interparticle interactions. In other words, an atom belongs to the cluster if inside a sphere of radius r_b there is at least one atom belonging to the same cluster. Early Monte Carlo studies of Lee et al. [8] showed that a cluster's free energy is almost independent of a cluster definition provided that the definition is reasonable and the temperature is sufficiently low. For the present model a particular type of a cluster definition is not important. What matters is that a cluster is a compact object around its center of mass.

---

[1] Note that at high temperatures interactions between clusters can not be neglected.

The partition function $Z^{(n)}$ of the *gas* of N_n noninteracting n-clusters in the volume V at the temperature T is factorized:

$$Z^{(n)} = \frac{1}{N_n!} Z_n^{N_n} \tag{7.4}$$

where the prefactor $1/N_n!$ takes into account the indistinguishability of *clusters* (recall that indistinguishability of atoms *inside* the cluster is taken into account in q_n). The Helmholtz free energy of the *gas of n-clusters* is: $\mathscr{F}^{(n)} = -k_B T \ln Z^{(n)}$, which using Stirling's formula becomes:

$$\mathscr{F}^{(n)} = N_n k_B T \ln \left(\frac{N_n}{Z_n \, \mathrm{e}} \right)$$

The chemical potential of the n-cluster in the gas is

$$\mu_n = \frac{\partial \mathscr{F}^{(n)}}{\partial N_n} = k_B T \ln \left(\frac{N_n}{Z_n} \right) \tag{7.5}$$

Introducing the number density of n-clusters $\rho(n) = N_n/V$ and substituting (7.2) into (7.5), we obtain

$$\mu_n = k_B T \ln \left(\rho(n) \frac{V \Lambda^{3n}}{q_n} \right) \tag{7.6}$$

Equilibrium between the cluster and surrounding vapor molecules requires

$$\mu_n = n \mu^{\mathrm{v}} \tag{7.7}$$

where μ^{v} is the chemical potential of a molecule in the vapor phase. Combining (7.6) and (7.7) we find:

$$\rho(n) = \left(\frac{q_n}{V} \right) z^n \tag{7.8}$$

where

$$z = \mathrm{e}^{\beta \mu^{\mathrm{v}}} / \Lambda^3 \tag{7.9}$$

is the fugacity of a vapor molecule. From the definition of q_n it is clear that the quantity q_n/V involves only the degrees of freedom relative to the center of mass of the cluster and remains finite in thermodynamic limit ($V \to \infty$). The pressure equation of state for the vapor is given by Dalton's law

$$\frac{p^{\mathrm{v}}(\mu^{\mathrm{v}}, T)}{k_B T} = \sum_{n=1}^{\infty} \rho(n)_{\mu^{\mathrm{v}}, T} \tag{7.10}$$

and the overall number density of the gas is:

$$\rho^{\text{v}} = \sum_{n=1}^{\infty} n\rho(n) \tag{7.11}$$

Equation (7.8) holds for every point of the gaseous isotherm. In particular for the saturation point it reads

$$\rho_{\text{sat}}(n) = \left(\frac{q_n}{V}\right) z_{\text{sat}}^n, \qquad z_{\text{sat}} = \frac{e^{\beta \mu_{\text{sat}}}}{\Lambda^3} \tag{7.12}$$

where the chemical potential at saturation $\mu_{\text{sat}}(T)$ can be found from the suitable equation of state. The problem of finding the equilibrium cluster distribution is reduced to the determination of the cluster configuration integral. Up to this point all results were exact. To proceed with calculation of q_n it is necessary to introduce approximations.

7.4 Configuration Integral of a Cluster: Mean-Field Approximation

An n-cluster is an object containing n particles satisfying a certain cluster definition. A geometrical form of the cluster can be quite different. Big clusters tend to a compact spherical shape with a well defined surface area scaling with the cluster size as $n^{2/3}$. This is definitely not true for small clusters: they look more like fractal objects—one can think here about binary-, ternary clusters, etc. It is not clear how to define the surface area of such an object. It is therefore reasonable to replace the concept of the surface *area* of an arbitrary cluster by the number of suitably defined *surface particles*.

With this in mind, following Zhukhovitskii [9], we decompose n into two groups: the core n^{core} and the surface particles n^s

$$n = n^{\text{core}} + n^s \tag{7.13}$$

The physical idea behind this distinction is that the core of the cluster, *if present*, should possess the *liquid-like structure* which can be characterized by a certain property typical for the liquid phase, e.g. by the liquid coordination number N_1. The surface molecules can then be viewed as an adsorption layer covering the core.[2] By definition the integer numbers n^{core} and n^s satisfy

$$n^{\text{core}} \geq 0, \qquad n^s \geq 1$$

---

[2] This decomposition should not be confused with the Gibbs construction involving a dividing surface discussed in Sect. 2.2.

Both quantities can fluctuate around their mean values

$$n^{\text{core}} = \overline{n^{\text{core}}}(n; T) + \delta n^{\text{core}}, \quad n^s = \overline{n^s}(n; T) + \delta n^s$$

so that $\delta n^{\text{core}} + \delta n^s = 0$. For the present purposes it is not necessary to specify the form of these quantities; we postpone this discussion till Sect. 7.5.

Similar to the seminal Fisher droplet model of condensation [5], we write the internal potential energy of an n-cluster as a sum of the bulk and surface contributions

$$U_n = -n\,E_0 + W_n \tag{7.14}$$

Here $-E_0$ ($E_0 > 0$) is the binding energy per particle related to the depth of inter-particle attraction; W_n is the surface energy of the n-cluster. In the Fisher model W_n has the form $W_n = w\,A(n)$, with w being the energy *per unit surface* and $A(n)$ is the cluster surface area. In view of the previous discussion, we present W_n in a different way making use of the concept of surface particles:

$$W_n = w_1\,n^s, \quad w_1 > 0 \tag{7.15}$$

where w_1 is the surface energy *per surface particle*. Both E_0 and w_1 are material constants independent of temperature. The difference between the two models becomes increasingly important for small clusters, for which the surface area of a cluster can not be properly defined.

Let us place the origin of the coordinate system in Eq. (7.3) into the center of mass of the cluster. Then the configuration integral can be written as

$$q_n = V\,e^{n\beta E_0}\,G_n(\beta) \tag{7.16}$$

where

$$G_n(\beta) = \frac{1}{n!}\oint_{\text{cl}} d\mathbf{r}^{n-1}\,e^{-\beta w_1 n^s(\mathbf{r}^{n-1})} \tag{7.17}$$

describes the "surface part" of q_n and has the dimensionality of (volume)$^{n-1}$. The notation $n^s(\mathbf{r}^{n-1})$ indicates that n^s depends on a particular configuration of cluster particles. The configurational Helmholtz free energy of the cluster reads

$$\mathscr{F}_n^{\text{conf}} = -k_{\text{B}}T\ln q_n$$

from which the cluster *configurational entropy* is:

$$\mathscr{S}_n^{\text{conf}} = -\frac{\partial\mathscr{F}_n^{\text{conf}}}{\partial T} = k_{\text{B}}\left[\ln q_n - \beta\,\frac{\partial\ln q_n}{\partial\beta}\right] \tag{7.18}$$

From (7.16)

$$\ln q_n = \ln V + n\beta E_0 + \ln G_n(\beta) \qquad (7.19)$$

yielding

$$\frac{\partial \ln q_n}{\partial \beta} = n E_0 + \frac{1}{G_n} \frac{\partial G_n}{\partial \beta}$$

Let us discuss the last term of this expression. From (7.17):

$$\frac{1}{G_n} \frac{\partial G_n}{\partial \beta} = \frac{\oint_{\text{cl}} d\mathbf{r}^{n-1} \left[-w_1 n^s(\mathbf{r}^{n-1})\right] e^{-\beta w_1 n^s(\mathbf{r}^{n-1})}}{\oint_{\text{cl}} d\mathbf{r}^{n-1} e^{-\beta w_1 n^s(\mathbf{r}^{n-1})}} \equiv -w_1 \overline{n^s}(\beta) \qquad (7.20)$$

where $\overline{n^s}(\beta)$ is the thermal average of n^s. Substituting (7.19) and (7.20) into (7.18) we find

$$\mathscr{S}_n^{\text{conf}}(\beta) = k_B[\ln G_n + \beta w_1 \overline{n^s} + \ln V]$$

The bulk entropy *per molecule* can be identified with the entropy per molecule in the bulk liquid, or equivalently, in the infinitely large cluster

$$\mathscr{S}_0(\beta) = \lim_{n \to \infty} \frac{\mathscr{S}_n^{\text{conf}}}{n}$$

When $n \to \infty$ most of the particles belong to the core while the relative number of surface molecules vanishes:

$$\frac{\overline{n^s}}{n} \to 0 \quad \text{as} \quad n \to \infty$$

(the rigorous proof of this statement is presented in Sect. 7.5). Then

$$\mathscr{S}_0(\beta) = k_B \left[\lim_{n \to \infty} \frac{1}{n} \ln G_n(\beta) \right] \qquad (7.21)$$

By virtue of the cluster definition, mutual distances between molecules in the cluster can not exceed some maximum value, therefore the integral (7.17) remains finite. Let us introduce a temperature dependent parameter v_0 with the dimensionality of volume which can be understood as an average volume per molecule in the cluster. Scaling all distances with $v_0^{1/3}$, we rewrite G_n as

$$G_n(\beta) = v_0^{n-1} \left[\frac{1}{n!} \oint_{\text{cl}} d\tilde{\mathbf{r}}^{n-1} e^{-\beta w_1 n^s(\tilde{\mathbf{r}}^{n-1})} \right]$$

where $\tilde{\mathbf{r}}_i$ are the dimensionless positions of the cluster molecules; the integral in the square brackets is now dimensionless (and finite). The number of surface molecules in the n-cluster depends on a particular configuration, but can have the values in the range $1 \leq n^s \leq n$. Replacing integration over $3(n-1)$ configuration space by summation over all possible values of n^s, we obtain

$$G_n(\beta) = v_0^{n-1} \sum_{1 \leq n^s \leq n} g(n, n^s) \, e^{-\beta w_1 n^s} \tag{7.22}$$

where the degeneracy factor $g(n, n^s)$ gives the number of different molecular configurations in the n-cluster having the same number n^s of the surface molecules.

We calculate the positive definite series $G_n(\beta)$ using the *mean-field approximation*. We assume that the sum in Eq. (7.22) is dominated by its largest term to the extent that it is possible replace the entire sum by this largest term while completely neglecting the others:

$$G_n(\beta) \approx v_0^{n-1} \max_{n^s} \left\{ g(n, n^s) \, e^{-\beta w_1 n^s} \right\} \tag{7.23}$$

The similar approximation is used in the theory of phase transitions giving rise to the Landau theory [10]. The maximum in (7.23) is attained at the most probable number of surface particle in the n-cluster, $\overline{n^s}(n; \beta)$, corresponding to the particle configuration with the maximum statistical weight. Thus, within the mean-field approximation

$$G_n(\beta) = v_0^{n-1} g(n, \overline{n^s}) \, e^{-\beta w_1 \overline{n^s}} \tag{7.24}$$

Recalling the physical meaning of $g(n, n^s)$, one can expect that $\ln g(n, \overline{n^s})$ is related to the configurational entropy of the n-cluster. To verify this conjecture we use Eqs. (7.21) and (7.24) to obtain:

$$\mathscr{S}_0 = k_B \left[\ln v_0 + \lim_{n \to \infty} \frac{1}{n} \ln g(n, \overline{n^s}) \right] \tag{7.25}$$

Similar to the decomposition of U_n, we decompose the configurational entropy of the cluster as

$$\mathscr{S}_n^{\text{conf}} = n \, \mathscr{S}_0 + v_1 \overline{n^s} \tag{7.26}$$

where the temperature independent material parameter v_1 is the configurational *surface entropy per particle*; the term $v_1 \overline{n^s}$ characterizes the number of distinct cluster configurations of $\overline{n^s}$ surface particles.[3] Comparing Eqs. (7.25) and (7.26) it is plausible to assume that the surface entropy satisfies

$$v_1 \overline{n^s} = k_B \, n \, \ln v_0 + k_B \ln g(n, \overline{n^s}(\beta)) - n \mathscr{S}_0 \tag{7.27}$$

---

[3] Note that as a thermodynamic quantity $\mathscr{S}_n^{\text{conf}}$ depends on the *average* number of surface molecules $\overline{n^s}$, whereas U_n as a *microscopic* quantity depends on n^s itself.

yielding:

$$g(n, \overline{n^s}) = v_0^{-n} \exp\left[\frac{n\mathscr{S}_0}{k_B} + \frac{v_1 \overline{n^s}}{k_B}\right] \tag{7.28}$$

Combining (7.28), (7.16) and (7.24), we obtain

$$\frac{q_n}{V} = C \left\{\exp\left[\beta E_0 + \frac{\mathscr{S}_0}{k_B}\right]\right\}^n \{\exp\left[-\beta(w_1 - v_1 T)\right]\}^{\overline{n^s}} \tag{7.29}$$

where

$$C = v_0^{-1} \tag{7.30}$$

Substituting (7.29) into (7.8), the number density of n-clusters takes the form

$$\rho(n) = C\, y^n\, x^{\overline{n^s}} \tag{7.31}$$

where for convenience we introduced the following notations

$$y \equiv z \exp\left[\beta E_0 + \frac{\mathscr{S}_0}{k_B}\right] \tag{7.32}$$

$$x \equiv \exp\left[-\left(\frac{w_1 - v_1 T}{k_B T}\right)\right] \tag{7.33}$$

The quantity $x > 0$ measures the temperature; at low temperatures x is small. The quantity y measures the fugacity, or, equivalently, the chemical potential of the vapor.

A big cluster can be associated with a liquid droplet in the vapor. The growth of a macroscopic droplet corresponds in this picture to condensation. Following [5], let us discuss the probability of finding an n-cluster in the vapor at the temperature T and the chemical potential μ^v. This probability is proportional to $\rho(n)$.

If in (7.31) $y < 1$, which corresponds to a small z, or equivalently to a large and negative μ^v, then $\rho(n)$ exponentially decays as $\exp[-\text{const} \times n]$. As y approaches unity this decrease becomes slower. When $y = 1$, $\rho(n)$ still decays but only as $\exp[-\text{const} \times \overline{n^s}(n)]$. Finally, if y slightly exceeds unity, then $\rho(n)$ first decreases, reaching a minimum at some $n = n_0$, and then increases without bounds (see Fig. 7.1). The large (divergent) probability of finding a very large cluster signals the condensation. Thus, we identify

$$y_{sat} = 1 \tag{7.34}$$

with the saturation point. Applying Eq. (7.32) to saturation and using (7.34) we find

$$\exp\left[\beta E_0 + \frac{\mathscr{S}_0}{k_B}\right] = \frac{1}{z_{sat}} \tag{7.35}$$

Fig. 7.1 The number density of n-clusters $\rho(n)$ for various values of the fugacity, or equivalently, parameter y. For $y > 1$ $\rho(n)$ attains a minimum at $n = n_0$ and for $n > n_0$ it diverges

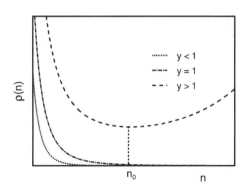

where $z_{sat}(T)$ is the fugacity at saturation. From (7.31) and (7.34) the cluster distribution at saturation reads:

$$\rho_{sat}(n) = C \, \exp\left[- \frac{(w_1 - v_1 T)\,\overline{n^s}}{k_B T} \right]$$

The quantity

$$\gamma_{micro} = w_1 - v_1 T \tag{7.36}$$

is the Helmholtz *free energy per surface particle* of the cluster, it includes both energy and entropy contributions and depends on the temperature but *not on the cluster size*. The size-dependence of the surface energy is contained in $\overline{n^s}(n)$. By analogy with the fluctuation theory γ_{micro} can be termed a *microscopic* surface tension *per particle* (the combination of the terms "microscopic" and "surface tension" is purely terminological and should not cause confusion). It is convenient to introduce the dimensionless quantity

$$\theta_{micro} = \frac{\gamma_{micro}}{k_B T} \tag{7.37}$$

which we term the "reduced microscopic surface tension", being the Helmholtz free energy per surface particle in $k_B T$ units. From (7.29), (7.35) and (7.37) the configuration integral of the n-cluster takes the form

$$\frac{q_n}{V} = C \, z_{sat}^{-n} \, e^{-\theta_{micro}\,\overline{n^s}(n)} \tag{7.38}$$

This is the central result of the model. Substitution of (7.38) into (7.8) yields the cluster distribution function in supersaturated vapor

$$\rho(n) = C \, e^{n\beta\,(\mu^v - \mu_{sat})} \, e^{-\theta_{micro}\,\overline{n^s}(n)}, \quad 1 \leq n < \infty \tag{7.39}$$

At saturation $\mu^{v} = \mu_{sat}(T)$ yielding

$$\rho_{sat}(n) = C\,e^{-\theta_{micro}\,\overline{n^s}(n)} \tag{7.40}$$

The unknown parameter C can be found from (7.11) and (7.40):

$$C = \frac{\rho^{v}_{sat}}{\sum_{n=1}^{\infty} n\,h^{-\overline{n^s}(n)}}, \quad \text{where } h \equiv e^{\theta_{micro}} \tag{7.41}$$

Without going into details of the behavior of $\overline{n^s}(n)$ for *arbitrary* n we make use of the physically obvious fact that *small* clusters (with $n \leq N_1$) do not have the liquid core implying that:

$$\overline{n^s}(n) = n, \quad \text{for } n \leq N_1 \tag{7.42}$$

At low temperatures we expect that

$$h \gg 1 \tag{7.43}$$

or equivalently:

$$\theta_{micro}(T) > 2 \tag{7.44}$$

The above constraint is the domain of validity of the model. From (7.41) to the leading order in $1/h$

$$C = \rho^{v}_{sat}\,e^{\theta_{micro}} \tag{7.45}$$

Thus, the cluster distribution at saturation (7.40) reads:

$$\rho_{sat}(n) = \rho^{v}_{sat}\,e^{-\theta_{micro}\left[\overline{n^s}(n)-1\right]} \tag{7.46}$$

Recalling that $C = 1/v_0$ we conclude from (7.45) that

$$v_0 = v^{v}_{sat}\,e^{-\theta_{micro}} \tag{7.47}$$

where $v^{v}_{sat} = 1/\rho^{v}_{sat}$ is the volume per molecule in the bulk vapor at saturation. Equation (7.47) is the manifestation of the fact that molecules in the cluster are on average more densely packed than in the vapor at the same temperature. To accomplish the theory it is necessary to present approximations for θ_{micro} and $\overline{n^s}(n)$. Consider the vapor compressibility factor at saturation

$$Z^{v}_{sat} = \frac{p_{sat}}{\rho^{v}_{sat}k_B T}$$

Using Eqs. (7.11), (7.10) and (7.40) it can be written as

$$Z_{\text{sat}}^{v} = \frac{\sum_{n=1}^{\infty} \rho_{\text{sat}}(n)}{\sum_{n=1}^{\infty} n \rho_{\text{sat}}(n)} = \frac{\sum_{n=1}^{\infty} h^{-\overline{n^s}}(n)}{\sum_{n=1}^{\infty} n \, h^{-\overline{n^s}}(n)}$$

Truncating both series at N_1 and using (7.42) we have

$$Z_{\text{sat}}^{v} = \frac{\sum_{n=1}^{N_1} h^{-n}}{\sum_{n=1}^{N_1} n \, h^{-n}} = -\frac{h^{-N_1}(-1+h)(-1+h^{N_1})}{-h+h^{-N_1}(h-N_1+hN_1)}$$

Expanding the right-hand side in $1/h$, we obtain:

$$Z_{\text{sat}}^{v} = 1 - \frac{1}{h} + N_1 \left(\frac{1}{h}\right)^{N_1} + O\left(\frac{1}{h}\right)^{N_1+1}$$

With the high degree of accuracy $\sim O\left(h^{-N_1}\right)$ we can set

$$Z_{\text{sat}}^{v} = 1 - \frac{1}{h} \tag{7.48}$$

On the other hand,

$$Z_{\text{sat}}^{v} = 1 + Z_{\text{exc,sat}}^{v} \tag{7.49}$$

where the first term is the ideal gas part and $Z_{\text{exc,sat}}^{v}$ is the excess (over ideal) contribution. Comparing (7.48) and (7.49) we find

$$h = -\frac{1}{Z_{\text{exc,sat}}^{v}} \tag{7.50}$$

This result shows that the microscopic surface tension originates from the nonideality of the vapor and can be determined from a suitable equation of state; its simplest form is the second order virial expansion [11]:

$$Z_{\text{sat}}^{v} = 1 + \underbrace{\frac{B_2 \, p_{\text{sat}}}{k_B T}}_{Z_{\text{exc,sat}}^{v}} \tag{7.51}$$

where $B_2(T)$ is the second virial coefficient. From (7.50) and (7.51)

$$\theta_{\text{micro}} = -\ln\left[\frac{(-B_2) \, p_{\text{sat}}}{k_B T}\right] \tag{7.52}$$

This feature of the model makes it especially attractive for applications: in order to find θ_{micro} one does not need to solve the two-phase equilibrium equations but can

Fig. 7.2 Vapor and liquid compressibility factors at saturation. The vapor compressibility factor $Z_{\text{sat}}^{\text{v}}(T)$ decreases with T, while liquid compressibility factor $Z_{\text{sat}}^{\text{l}}(T)$ increases with T. Both curves meet at the critical point T_c

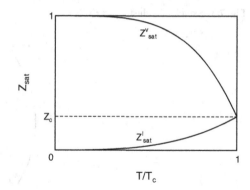

use the experimental or tabulated data on $B_2(T)$ and $p_{\text{sat}}(T)$ which are available for a large amount of substances in a broad temperature range (see [11]). From (7.43) and (7.52) the range of validity of the theory is given by:

$$\left| \frac{B_2(T)\, p_{\text{sat}}(T)}{k_B T} \right| \ll 1 \tag{7.53}$$

At low temperatures $Z_{\text{sat}}^{\text{v}}$ is close to unity, so that $Z_{\text{exc,sat}}^{\text{v}} \approx 10^{-8} \div 10^{-5}$. At higher temperatures $Z_{\text{sat}}^{\text{v}}$ decreases (see Fig. 7.2). If we formally apply (7.50) at the critical point we would find

$$h(T_c) = \frac{1}{1 - Z_c}$$

The critical compressibility factor lies in the limits $Z_c \approx 0.2 \div 0.4$ [11] indicating that the constraint (7.43) is violated. For example, for van der Waals fluids $Z_c = 3/8$ [12] yielding

$$h_c^{\text{vdW}} = \frac{8}{5}$$

The failure of the model at high temperatures manifests the fact that in this domain one has to take into account intercluster interactions which are completely neglected in the present model. Besides, in the close vicinity of T_c fluctuations become increasingly important and the mean-field approach can be in error.

7.5 Structure of a Cluster: Core and Surface Particles

Equation (7.13) written for the average quantities is

$$n = \overline{n^{\text{core}}} + \overline{n^s} \tag{7.54}$$

Fig. 7.3 Sketch of a cluster

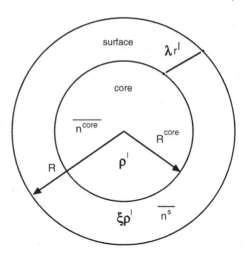

By definition the core possesses the liquid-like structure characterized by the coordination number N_1 in the liquid phase; N_1 gives the average number of nearest neighbors for a molecule in the bulk liquid. This quantity influences a number of physical properties: density, viscosity, diffusivity, etc. The determination of N_1 is a nontrivial problem in its own right (it is discussed in Sect. 7.6). In the present section we assume N_1 to be known. By construction the clusters with $n \leq N_1$ do not have the core—all their particles belong to the surface:

$$\overline{n^s}(n) = n, \quad \overline{n^{\mathrm{core}}}(n) = 0, \qquad \text{for} \quad n \leq N_1 \tag{7.55}$$

Consider now a cluster with $n \geq N_1 + 1$. It has core, which contains on average $\overline{n^{\mathrm{core}}}$ particles and can be characterized by a radius R^{core} and the number density ρ^l (we discuss spherical clusters for simplicity). Following Zhukhovitskii [9] we characterize the surface layer, containing $\overline{n^s}$ particles, by a thickness λr^l, $\lambda \geq 1$ and the constant density $\xi \rho^l$, lower than the bulk liquid density: $\xi < 1$; r^l is the average intermolecular distance in the bulk liquid (see Fig. 7.3). One can view the surface molecules as an "adsorption layer for the core" separating it from the bulk vapor surrounding the cluster. This analogy suggests that one can expect λ to vary in a narrow range: $1 \leq \lambda < 2$, with the left boundary corresponding to a monolayer and the right boundary—to a double layer of surface molecules. We stress that this division is purely schematic and serves the purposes of the model.

Physically a cluster can be viewed as a density fluctuation in the vapor, characterized by a smooth profile $\rho_m(r)$ asymptotically tending to the bulk vapor density at $r \to \infty$. It is convenient to define a cluster radius R at the location of the equimolar dividing surface R_e which by definition is characterized by the zeroth (*physical*) adsorption [13]. Outside R_e The model construction illustrated in Fig. 7.3 replaces the smooth profile $\rho_m(r)$ by a *two*-step function $\rho(r)$ with the width of the middle step

Fig. 7.4 Two-step density profile $\rho(r)$ discussed in the model. Also shown is the *true* smooth thermodynamic profile $\rho_m(r)$ (Reprinted with permission from Ref. [6], copyright (2006), American Institute of Physics.)

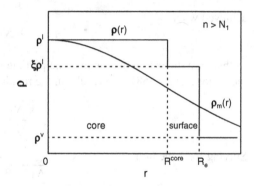

$\lambda r^1 = R_e - R^{\text{core}}$; outside R_e: $\rho(r) = \rho^{\text{v}}$, as shown in Fig. 7.4. The core radius is then

$$R^{\text{core}} = R_e - \lambda r^1, \qquad n \geq N_1 + 1$$

Recall that CNT replaces $\rho_m(r)$ by a *single*-step function

$$\rho_{\text{CNT}}(r) = \rho^1 (1 - \varXi(r - R_e)) + \rho^{\text{v}} \varXi(r - R_e)$$

where $\varXi(x)$ is the Heaviside unit step function. Clearly, for large clusters the relative width of the adsorption layer $\lambda r^1 / R^{\text{core}} \to 0$ and both approaches become asymptotically identical.

Within the two-step approximation Eq. (7.54) reads:

$$n = \underbrace{\rho^1 \frac{4\pi}{3} (R^{\text{core}})^3}_{\overline{n^{\text{core}}}} + \underbrace{\xi\rho^1 \frac{4\pi}{3} [(R^{\text{core}} + \lambda r^1)^3 - (R^{\text{core}})^3]}_{\overline{n^s}} \qquad (7.56)$$

Let us introduce a dimensionless core radius

$$X = \frac{R^{\text{core}}}{r^1}$$

From (3.23)

$$X^3 = \overline{n^{\text{core}}} \qquad (7.57)$$

It is convenient to use the pair of material parameters λ and $\omega \equiv \xi \lambda$ instead of the pair (λ, ξ). Then Eq. (7.56) reads:

$$X^3 = -3\omega X^2 - 3\omega\lambda X + (n - \omega\lambda^2) \qquad (7.58)$$

This cubic equation for $X(n)$ has the unique real positive root if $n - \omega\lambda^2 > 0$. For $n \geq N_1 + 1$: $\overline{n^{\text{core}}} \geq 1$, implying that $X \geq 1$. The minimum value $X = 1$ is achieved when

$$n = 1 + N_1$$

In this case the core contains just one particle while the rest N_1 particles belong to the surface. Eq. (7.58) then results in the relation between ω and λ:

$$\omega = \frac{N_1}{3 + 3\lambda + \lambda^2}$$

Solving for λ we find:

$$\lambda = \sqrt{\frac{N_1}{\omega} - \frac{3}{4}} - \frac{3}{2} \tag{7.59}$$

Since $\lambda > 1$, ω should lie in the limits

$$0 < \omega < N_1/7$$

The average number of surface particles is found from (7.57)

$$\overline{n^s}(n) = n - [X(n)]^3$$

where $X(n)$ is the solution of Eq. (7.58). It is convenient to introduce two dimensionless quantities

$$\alpha = \frac{\overline{n^s}}{n} = 1 - \frac{X^3}{n} \quad \text{and} \tag{7.60}$$

$$\zeta = n^{-1/3} \tag{7.61}$$

Then

$$X = \zeta^{-1}(1 - \alpha)^{1/3}$$

In these variables (7.58) becomes the equation for $\alpha(\zeta)$:

$$\alpha = 3\omega\zeta(1 - \alpha)^{2/3} + 3\omega\lambda\zeta^2(1 - \alpha)^{1/3} + \zeta^3\omega\lambda^2 \tag{7.62}$$

Equation (7.62) can be used to determine the parameter ω. To this end let us consider the behavior of the model for large clusters: $n \to \infty$. In this case almost all cluster particles belong to the core:

$$\frac{\overline{n^{\text{core}}}}{n} \to 1, \quad \frac{\overline{n^s}}{n} \to 0 \quad \text{for} \quad n \to \infty$$

The problem has two small parameters:

$$0 < \alpha \ll 1, \quad 0 < \zeta \ll 1$$

Keeping in Eq. (7.62) the first order terms in α and ζ, we obtain:

$$\alpha = 3\omega\zeta$$

Then from (7.60)–(7.61)

$$\overline{n^s} = 3\omega\, n^{2/3}, \qquad n \to \infty \tag{7.63}$$

This result comes as no surprise, since at large n the droplet is a compact spherical object, its radius scales as $n^{1/3}$ and the number of surface atoms scales as the surface area $\sim n^{2/3}$. From (7.46) to the same order in n

$$\rho_{sat}(n) \sim \exp[-3\omega\, \theta_{micro}\, n^{2/3}], \qquad n \to \infty \tag{7.64}$$

On the other hand, for big clusters the CNT description is valid:

$$\rho_{sat}(n) \sim \exp[-\theta_{\infty}\, n^{2/3}], \qquad n \to \infty \tag{7.65}$$

Comparing (7.64) and (7.65) we find

$$\omega = \frac{1}{3}\frac{\theta_{\infty}}{\theta_{micro}} \tag{7.66}$$

7.6 Coordination Number in the Liquid Phase

The remaining unknown parameter of the model—the coordination number in the liquid phase N_1—can be derived from X-ray diffraction experiments. In these experiments one measures the static structure factor whose Fourier transform gives the pair correlation function $g(r; \rho, T)$ [12]. N_1 can be obtained by integration of the area under the first peak of $g(r)$. However, there is an ambiguity in the way this integration is carried out, which sometimes results in a substantial discrepancy in the values of N_1. Moreover, an error in numerical integration may lead to the values $N_1 > 12$ which is impossible since $N_1 = 12$ corresponds to the closed packing $\eta = 0.74$ (face-centered cubic structure). Typically for various liquids close to the melting point: $N_1(\rho, T) = 4 \div 8$.

Cahoon [14] proposed a simple alternative method for calculation of N_1 for a liquid. It is based on the observation that *in a solid* there is one-to-one correspondence between the coordination number and the packing fraction

$$\eta \equiv \frac{\pi d^3}{6v_a} \tag{7.67}$$

Table 7.1 Coordination numbers and volume fractions for different cubic crystal structures

Unit cell	N_1	v_a	η
Diamond cubic	4	$8d^3/3\sqrt{3}$	0.34009
Simple cubic	6	d^3	0.52360
Body-centered cubic	8	$4d^3/3\sqrt{3}$	0.68018
Face-centered cubic	12	$d^3/\sqrt{2}$	0.74048

characterizing a given crystal structure—face-centered cubic, body-centered cubic, simple cubic, diamond cubic. In (7.67) d is the atomic diameter and v_a is the atomic specific volume. The relation between N_1 and η in a solid is given in Table 7.1.

The main idea of Ref. [14] is that for any pure isotropic liquid (or amorphous) material the function relating N_1 to η will be similar to that for the isotropic crystal solid with the exception that *noninteger* values are permissible. Using Table 7.1 the dependence $N_1(\eta)$ for $4 \leq N_1 \leq 8$ can be well approximated by

$$N_1 = 5.5116\eta^2 + 6.1383\eta + 1.275 \qquad (7.68)$$

For the case of a liquid the atomic volume v_a should be replaced by $1/\rho^l$ and the atomic diameter d should be replaced by the position d_g of the first peak of the pair correlation function $g(r;\ \rho^l, T)$. Thus,

$$\eta = \frac{\pi}{6}\rho^l d_g^3 \qquad (7.69)$$

If the diffraction data is not available one can find d_g using the ideas of perturbation approach in the theory of liquids applying the Weeks–Chandler–Anderson decomposition scheme of the interaction potential given by Eqs. (5.20), (5.21) and shown in Fig. 5.2. Within the WCA d_g is approximated by the effective hard-sphere diameter

$$d_{\text{hs}}^3 = 3\int_0^{r_m}\left[1 - e^{-\beta u_0(r)}\right]r^2\,dr \qquad (7.70)$$

where $u_0(r)$ is the WCA reference potential (5.20).[4]

7.7 Steady State Nucleation Rate

Combining the results of the previous sections the steady state nucleation rate is

$$J = K_0\left[\sum_{n=1}^{\infty}e^{-H(n)}\right]^{-1} \qquad (7.71)$$

---

[4] Equation (7.70) is the mean-field approximation to the original WCA expression, where the cavity function of the hard sphere system is set to unity (for details see e.g. [12] Chap. 5).

where

$$K_0 = \rho_{sat}^v \, f_{1,sat} \, S, \quad f_{1,sat} = \frac{p_{sat} \, s_1}{\sqrt{2\pi m_1 k_B T}} \tag{7.72}$$

and

$$H(n) = \frac{2}{3} \ln n + n \ln S - \theta_{micro} \left[\overline{n^s}(n) - 1 \right] \tag{7.73}$$

Here θ_{micro} is given by Eq. (7.52); $\overline{n^s}(n) = n$ for $n \le N_1$, and

$$\overline{n^s}(n) = n - [X(n)]^3, \quad \text{for} \quad n \ge N_1 + 1 \tag{7.74}$$

$X(n)$ is the real positive root of Eq. (7.58) in which the parameters λ and ω are found from

$$\omega = \frac{1}{3} \frac{\theta_\infty}{\theta_{micro}}, \quad \lambda = \sqrt{\frac{N_1}{\omega} - \frac{3}{4}} - \frac{3}{2} \tag{7.75}$$

The coordination number N_1 is expressed in terms of the molecular packing fraction in the liquid phase

$$\eta = (\pi/6) \rho^l \, d_{hs}^3 \tag{7.76}$$

(where $d_{hs}(T)$ is the effective hard sphere diameter in the theory of liquids) by means of Eq. (7.68). We refer to this model as a *Mean-field Kinetic Nucleation Theory* (MKNT) [6].

It is easy to see that $-H(n)$ is the free energy of the cluster formation in $k_B T$ units

$$-H(n) = \beta \Delta G(n)$$

Apart from the small logarithmic corrections (which can safely be set to a constant $\frac{2}{3} \ln n_c$, n_c being the critical cluster) the free energy reads

$$\beta \Delta G(n) = -n \ln S + \theta_{micro} \left[\overline{n^s}(n) - 1 \right] \tag{7.77}$$

At small n the surface part of $\Delta G(n)$ is

$$\beta \Delta G^{surf}(n) = \theta_{micro} (n - 1), \quad n \le N_1 \tag{7.78}$$

yielding $\Delta G^{surf}(n = 1) = 0$. Equation (7.78) implies that the limiting consistency (cf. Sect. 3.6) is an intrinsic property of MKNT. By virtue of the kinetic approach MKNT satisfies also the law of mass action.

Let us define a critical cluster as the one that makes the major contribution to the series in (7.71). The latter corresponds to the minimum of $H(n)$, or equivalently—to the maximum of $\Delta G(n)$. A close inspection of Eq. (7.77) shows that the function $\Delta G(n)$ has *two* maxima. For small n the Gibbs energy is an increasing linear function of n

$$\beta \Delta G(n) = (\theta_{\text{micro}} - \ln S)\, n - \theta_{\text{micro}}, \quad n \le N_1$$

Expression in the round brackets is positive: the supersaturation can not exceed some maximum value given by the *pseudospinodal* corresponding to the nucleation barrier $\approx k_{\text{B}} T$. From the pseudospinodal condition, discussed in Chap. 9, it follows that

$$\ln S < \theta_{\text{micro}}$$

Hence, $\Delta G(N_1 - 1) < \Delta G(N_1)$. Due to the model construction when the cluster size is increased from N_1 to $N_1 + 1$, the number of surface particles does not change

$$\overline{n^s}(N_1) = \overline{n^s}(N_1 + 1) = N_1$$

leading to $\Delta G(N_1) > \Delta G(N_1 + 1)$. Thus, $\Delta G(n)$ has a maximum at $n = N_1$ which is an artifact of the model and has to be ignored. The *second* maximum corresponds to the critical cluster n_c:

$$\Delta G' \equiv \frac{\mathrm{d}\Delta G}{\mathrm{d}n} = -\ln S + \theta_{\text{micro}} \frac{\mathrm{d}\overline{n^s}}{\mathrm{d}n} = 0 \qquad (7.79)$$

Expanding $\Delta G(n)$ to the second order around n_c

$$\Delta G(n) \approx \Delta G^* + \frac{1}{2} \Delta G''(n_c)\,(n - n_c)^2, \quad \Delta G^* \equiv \Delta G(n_c)$$

we have:

$$\sum_{n=1}^{\infty} e^{\beta \Delta G(n)} \approx e^{\beta \Delta G^*} \int_{-\infty}^{\infty} \mathrm{d}x \, \exp\left[\frac{1}{2}\beta \Delta G''(n_c)\, x^2\right] = \frac{e^{\beta \Delta G^*}}{\mathscr{Z}}$$

where

$$\mathscr{Z} = \sqrt{\frac{-\beta \Delta G''(n_c)}{2\pi}}$$

is the Zeldovich factor. The steady-state nucleation rate reads:

$$J = J_0\, e^{-\beta \Delta G^*} \qquad (7.80)$$

where

$$J_0 = \mathscr{Z}\, \rho_{\text{sat}}^{\text{v}}\, f(n_c) \qquad (7.81)$$

is the kinetic prefactor, $f(n_c) = f_{1,\text{sat}}\, S\, n_c^{2/3}$, and

$$\beta \Delta G^* = -n_c \ln S + \theta_{\text{micro}} [\overline{n^s}(n_c) - 1] \qquad (7.82)$$

is the nucleation barrier. Apparently, the notion of a critical cluster is a convenient concept but not a necessity: Eqs. (7.81)–(7.82) are the approximation to the exact result (7.71)–(7.73).

7.8 Comparison with Experiment

7.8.1 Water

In Chap. 3 we discussed nucleation of water vapor comparing predictions of CNT with various experimental data available in the literature [15–17]—see Fig. 3.4. Now we can supplement this comparison by adding the predictions of MKNT for the same experimental conditions. Figure 7.5 shows the relative nucleation rate

$$J_{rel} = J_{exp}/J_{th}$$

with the closed symbols referring to $J_{th} = J_{MKNT}$ and open symbols referring to $J_{th} = J_{CNT}$. Circles (open and closed) correspond to the experiment of Wölk

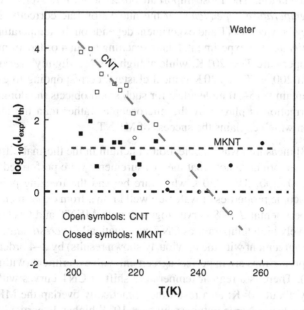

Fig. 7.5 Relative nucleation rate $J_{rel} = J_{exp}/J_{th}$ for water; $Log(J_{rel}) \equiv \log_{10} J_{rel}$. *Closed symbols*: MKNT, *open symbols*: CNT. *Circles*: experiment of Wölk et al. [15]; *squares* experiment of Labetski et al. [17]. The lines labelled 'CNT' and 'MKNT', shown to guide the eye, illustrate the temperature dependence of the relative nucleation rate for the CNT and MKNT, respectively. Also shown is the "ideal line" (*dashed*): $J_{exp} = J_{th}$

Fig. 7.6 Nucleation rate for
water: theory (MKNT, CNT)
versus experiment of Brus
et al. [18]. *Solid lines*: MKNT,
dashed lines: CNT; *closed
symbols*: experiment [18].
Labels: nucleation temperature in K; horizontal labels
refer to the theory, inclined
(italicized) labels refer to
experiment. The CNT line
for 300 K is almost coinciding with the MKNT line for
310 K; and the CNT line for
310 K is almost coinciding
with the MKNT line for 320 K

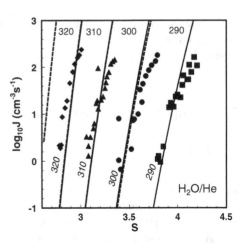

et al. [15]; squares (open and closed)—to the experiment of Peeters et al. [16] and
Labetski et al. [17]. The thermodynamic data for water used in both models are
given in Appendix A. In the whole temperature range the MKNT predictions are
1–2 orders of magnitude off the experimental data, while the CNT demonstrates
much larger deviation. The most important observation, however, is that MKNT
predicts the *temperature dependence* of the nucleation rate correctly. Meanwhile,
the discrepancy between CNT and experiment depends on the temperature: at low T
CNT *underestimates* the experimental data, reaching about 4 orders of magnitude at
the lowest temperature $T = 201$ K, while at high T CNT slightly *overestimates* the
experiment. At $200 < T < 220$ K critical clusters, corresponding to experimental
conditions, contain ≈ 15–20 molecules; for such small objects the dominant role in
the cluster formation is played by the microscopic (rather than the macroscopic)
surface tension which explains the success of MKNT.

Brus et al. [18] measured water nucleation in helium in the thermal diffusion cloud
chamber. It is important to note that the measurements were performed for temperatures $T = 290, 300, 310, 320$ K which are beyond the freezing point implying
that all macroscopic properties of water are well known from experiment. Figure 7.6
shows the experimental $J - S$ curves together with MKNT and CNT predictions.
At these relatively high temperatures CNT systematically *overestimates* experiment
(in qualitative agreement with the previously shown results) by 2–4 orders of magnitude. MKNT predictions are in perfect agreement with experiment (within one order
of magnitude). There is a regular temperature shift of CNT curves with respect to
experiment by about 10° K; as a result they practically overlap the MKNT curves
related to the nucleation temperatures which are 10° K higher. In particular, the 290 K
CNT line overlaps the 300 K MKNT line; the 300 K CNT line overlaps the 310 K
MKNT line, etc.

Fig. 7.7 Relative nucleation rate $\log_{10} J_{rel}$ for nitrogen. *Closed circles*: MKNT, *open diamonds*: CNT. Experiment: [19]; $J_{exp} = 7 \pm 2\,\mathrm{cm}^{-3}\,\mathrm{s}^{-1}$. Also shown is the "ideal line" (*dashed*): $J_{exp} = J_{th}$

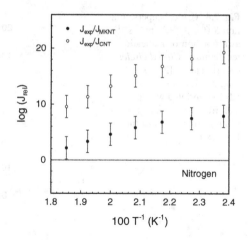

7.8.2 Nitrogen

Figure 7.7 shows the comparison of experimental nucleation rate J_{exp} of Ref. [19] with predictions of the CNT J_{CNT} and MKNT J_{MKNT}. Thermodynamic data used in the analysis is given in Appendix A. As in Fig. 7.5 the relative nucleation rate is: $J_{rel} = J_{exp}/J_{th}$ with $J_{th} = J_{CNT}$ or J_{MKNT}.

The dashed line corresponds to the "ideal case": $J_{exp} = J_{th}$. As one can see, MKNT predictions for J deviate on average from the experimental data by 2–7 orders of magnitude while the CNT predictions are 10–20 orders of magnitude lower then the experimental values.

7.8.3 Mercury

Vapor–liquid nucleation in mercury is a somewhat extreme example of nucleation studies requiring very high supersaturations. The reason for that is that the surface tension of mercury is more than ten times higher than that of molecular fluids, implying that a high supersaturation is necessary to compensate for the energy cost to build the cluster surface. An important feature of mercury, typical for fluid metals, is that its electronic structure strongly depends on the thermodynamic state of the system [20] implying that the interatomic interaction also depends on the thermodynamic state. Mercury vapor is a simple rare-gas system with only van der Waals dispersive interactions. Being combined in clusters, mercury atoms can behave differently: in small clusters interactions between them are purely dispersive (as in the vapor); however, beyond a certain cluster size the energy gap becomes smaller than $k_B T$ resulting in the nonmetal-to-metal transition. This transition was estimated to appear at the cluster size $n \approx 70$ according [21], $n \approx 80$ according to [22], $n \approx 135$

Fig. 7.8 Critical supersatura-
tion S (for $J = 1\,\mathrm{cm}^{-3}\,\mathrm{s}^{-1}$)
as a function of nucleation
temperature. *Closed circles*:
experiment of [25]; *solid
line*: MKNT, *dashed line*:
CNT (Reprinted with permission from Ref. [6], copyright
(2006), American Institute of
Physics.)

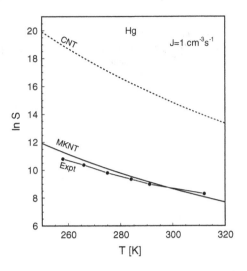

according [23]. Recently the cluster size dependent interaction potential of mercury
was proposed by Moyano et al. [24].

Experimental study of mercury nucleation in helium as a carrier gas was carried out
by Martens et al. [25]. The measurements were made in an upward diffusion cloud
chamber. This technique makes it possible to detect the onset of nucleation rather
than to measure directly the nucleation rates (as e.g. in shock wave tube experiments).
The onset corresponds to the nucleation rate

$$J_{\exp} = 1\ \mathrm{cm}^{-3}\mathrm{s}^{-1} \tag{7.83}$$

The supersaturation giving rise to the onset of nucleation is referred to as the critical
supersaturation. Figure 7.8 shows the experimental results (closed circles) for the
critical supersaturation S as a function of the nucleation temperature. Also shown
are predictions of the MKNT (solid line), CNT (dashed line). The thermodynamic
data for mercury is presented in Appendix A. As stated in [25] the measured values
of S are about 3 orders of magnitude lower than the CNT predictions; the MKNT
results are in good agreement with experiment.

Recalling how sensitive the nucleation rate is to the value of S it is instructive to
illustrate the difference between the two models in terms of J. For that purpose we
choose an experimental point $T = 284\,\mathrm{K}$, $\ln S = 9.35$ and compare experimental
and theoretical results corresponding to these conditions. Experimental rate is given
by the onset condition (7.83) while theoretical predictions are:

$$J_{\mathrm{CNT}}(T = 284\,\mathrm{K},\ \ln S = 9.35) = 4.7 \times 10^{-67}\,\mathrm{cm}^{-3}\mathrm{s}^{-1}$$
$$J_{\mathrm{MKNT}}(T = 284\,\mathrm{K},\ \ln S = 9.35) = 6.1 \times 10^{-3}\,\mathrm{cm}^{-3}\,\mathrm{s}^{-1}$$

The CNT predictions deviate from experiment by about 67 orders of magnitude (!) while MKNT predictions are within 3 orders of magnitude.[5]

7.9 Discussion

7.9.1 Classification of Nucleation Regimes

The general MKNT result for the nucleation rate can be written as

$$
J = K_0 \left[\underbrace{\sum_{n=1}^{N_1} e^{-H(n)}}_{\text{small}} + \underbrace{\sum_{n=N_1}^{N_{\text{class}}} e^{-H(n)}}_{\text{intermediate}} + \underbrace{\sum_{n=N_{\text{class}}}^{\infty} e^{-H(n)}}_{\text{large}} \right]^{-1}
\tag{7.84}
$$

where we divided all clusters into 3 groups: (i) small, containing $1 \leq n \leq N_1$ particles; (ii) intermediate with $N_1 \leq n \leq N_{\text{class}}$; and (iii) large, containing $n \geq N_{\text{class}}$ particles. Using considerations of Sect. 7.5 we can estimate N_{class} from the constraint

$$
\frac{1}{N_{\text{class}}^{1/3}} \sim (1 \div 1.5) \times 10^{-1}
$$

yielding $N_{\text{class}} \simeq (3 \div 10) \times 10^2$.

For *large* clusters the distribution function can be approximated by the classical form yielding:

$$
\underbrace{\sum_{n=N_{\text{class}}}^{\infty} e^{-H(n)}}_{\text{large}} \approx \sum_{n=N_{\text{class}}}^{\infty} \frac{1}{n^{2/3} \, S^n \, e^{-\theta_\infty n^{2/3}}}
\tag{7.85}
$$

If the critical cluster falls into this domain the contribution of the other two groups is negligible and we recover the kinetic CNT result (3.69). This case can be called a *macroscopic nucleation regime*; here the surface part of the free energy is entirely determined by the macroscopic surface tension θ_∞.

---

[5] Note, that the nucleation rate is very sensitive to the surface tension: if γ_∞ is measured within the 10% relative accuracy (common for most of the liquid metals), the accuracy of the predicted nucleation rate for mercury lies within 4 orders of magnitude.

For *small* clusters MKNT gives:

$$\underbrace{\sum_{n=1}^{N_1} e^{-H(n)}}_{\text{small}} = \sum_{n=1}^{N_1} \frac{1}{n^{2/3}\, S^n\, e^{-\theta_{\text{micro}}\,(n-1)}} \tag{7.86}$$

The free energy is determined solely by the microscopic surface tension while the macroscopic surface tension does not play a role.

In majority of nucleation experiments the critical cluster contains $\sim 10 - 10^2$ molecules, falling into the domain of intermediate clusters where the influence of both micro- and macroscopic surface tension is important. This *intermediate nucleation regime* represents a challenging problem for a nucleation theory. MKNT solves it by providing a smooth interpolation between the two limits—of large and small clusters—for which it becomes "exact" by construction. Such interpolation is possible because MKNT is not based on a perturbation in the cluster curvature making it possible to describe all nucleation regimes within one model.

7.9.2 Microscopic Surface Tension: Universal Behavior for Lennard-Jones Systems

The microscopic surface tension of a substance is determined solely by the vapor compressibility factor at coexistence $Z_{\text{sat}}^{\text{v}}$ (cf. Eq. (7.50)). Consider the substances for which the intermolecular interaction potential $u(r)$, where r is the separation between the molecules, has a two-parametric form,

$$u(r) = \varepsilon\, \Phi\left(\frac{r}{\sigma}\right)$$

Here ε describes the depth of the interaction and σ describes the molecular size. An example of such substances are Lennard-Jones fluids characterized by the potential

$$u_{\text{LJ}}(r) = 4\varepsilon \left[\left(\frac{r}{\sigma}\right)^{-12} - \left(\frac{r}{\sigma}\right)^{-6}\right]$$

Scaling the distance with σ and the energy with ε we introduce the dimensionless quantities

$$t_{\text{LJ}} = \frac{k_{\text{B}}T}{\varepsilon}, \quad \rho_{\text{LJ}} = \rho\sigma^3, \quad p_{\text{LJ}} = \frac{p\sigma^3}{\varepsilon}$$

The compressibility factor then reads

$$Z = \frac{p}{\rho k_{\text{B}} T} = \frac{p_{\text{LJ}}}{\rho_{\text{LJ}} t_{\text{LJ}}}$$

At coexistence

$$Z_{\text{sat}}^{\text{v}} = \frac{p_{\text{LJ, sat}}^{\text{v}}}{\rho_{\text{LJ, sat}}^{\text{v}} t_{\text{LJ}}}$$

is a universal function of t_{LJ} which is a manifestation of the law of corresponding states. This implies using (7.37) and (7.50) that

$$\gamma_{\text{micro}} = -\varepsilon \left[t_{\text{LJ}} \ln \left(-Z_{\text{exc,sat}}^{\text{v}} \right) \right] \tag{7.87}$$

The expression in the square brackets is also a universal function of t_{LJ}. Introducing the reduced temperature $t = T/T_c$ we present t_{LJ} as

$$t_{\text{LJ}} = t_{\text{LJ},c}\, t \tag{7.88}$$

where the critical temperature for Lennard-Jones fluids $t_{\text{LJ},c}$ can be determined from Monte Carlo simulations (see e.g. [26]): $t_{\text{LJ},c} = k_B T_c/\varepsilon \approx 1.34$. Dividing both sides of Eq. (7.87) by $k_B T_c$, we find

$$\frac{\gamma_{\text{micro}}}{k_B T_c} = \Psi(t)$$

where $\Psi(t)$ is again a universal function. From Eq. (7.36) we expect that at low temperatures this function is linear

$$\frac{\gamma_{\text{micro}}}{k_B T_c} = \left(\frac{w_1}{k_B T_c} \right) - \left(\frac{v_1}{k_B} \right) t \tag{7.89}$$

implying that for Lennard-Jones fluids at low temperatures the dimensionless surface entropy per particle, v/k_B, and the surface energy per particle, $w_1/k_B T_c$, are universal parameters.

This conjecture can be verified using the mean-field density functional calculations described in Chap. 5. Bulk equilibrium follows Eqs. (5.30)–(5.31) with

$$\mu = \mu_d(\rho) - 2\rho a \tag{7.90}$$
$$p = p_d(\rho) - \rho^2 a \tag{7.91}$$

Here the hard-sphere pressure, p_d, and chemical potential, μ_d, follow the Carnahan-Starling theory [27] and the background interaction parameter for a Lennard-Jones fluids is given by Eq. (5.34):

$$a = \frac{16\pi \sqrt{2}}{9} \varepsilon \sigma^3 \tag{7.92}$$

Figure 7.9 shows the results of the DFT calculations (the dotted line). Indeed, to the high degree of accuracy the temperature dependence of γ_{micro} turns out to be linear,

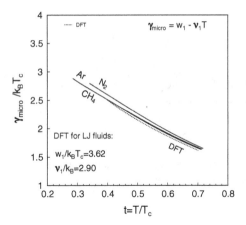

Fig. 7.9 Microscopic surface tension γ_{micro} for Lennard-Jones systems. *Dotted line*: DFT calculation for a Lennard-Jones system; *solid lines*: MKNT results for argon (Ar), nitrogen (N_2) and methane (CH_4) obtained using Eq. (7.52) and empirical fit for the second virial coefficient [11]. The MKNT curves for argon and methane are practically indistinguishable (Reprinted with permission from Ref. [6], copyright (2006), American Institute of Physics.)

supporting the conjecture (7.89), and

$$\frac{w_1}{k_B T_c} \approx 3.62 \tag{7.93}$$

$$\frac{v_1}{k_B} \approx 2.90 \tag{7.94}$$

In Fig. 7.9 we compare the DFT result with predictions of MKNT for various simple fluids—argon, nitrogen, methane— using Eq. (7.52) and empirical fit to the second virial coefficient for nonpolar substances [11] (see Eq. (F.2)). All MKNT curves are close to the DFT line (the curves for argon and methane are practically indistinguishable) confirming the validity of the MKNT conjecture for $\gamma_{\text{micro}}(T)$ and the numerical values of its parameters (7.93)–(7.94).

7.9.3 Tolman's Correction and Beyond

In order to treat small clusters within the classical approach, some nucleation models replace γ_∞ in the CNT free energy of cluster formation by the curvature dependent form, representing the expansion of the surface tension of a cluster in powers of its curvature. To discuss the range of validity of such an expansion we study the general Eq. (7.62), valid for all cluster sizes.

For *large clusters* we solve the cubic equation (7.62) for $\alpha(\zeta)$ keeping the terms up to the 3rd order in ζ, which results in

$$\alpha = 3\omega\zeta \left[1 - 2\left(\omega - \frac{\lambda}{2}\right)\zeta + \left(\frac{\lambda^2}{3} - 3\lambda\omega + 3\omega^2\right)\zeta^2 \right], \quad \zeta \to 0 \qquad (7.95)$$

Multiplying (7.95) by n and using (7.60)–(7.61) we obtain

$$\overline{n^s} = 3\omega n^{2/3} \left[1 - 2\left(\omega - \frac{\lambda}{2}\right)n^{-1/3} + \left(\frac{\lambda^2}{3} - 3\lambda\omega + 3\omega^2\right)n^{-2/3} \right] \qquad (7.96)$$

To the same order of accuracy the cluster distribution function at saturation reads:

$$\rho_{\text{sat}}(n) = \rho_{\text{sat}}^{\text{v}} \exp\left(-\frac{\gamma_n A(n)}{k_B T}\right)$$

where $A(n) = s_1 n^{2/3}$ and

$$\gamma_n = \gamma_\infty \left[1 - 2\left(\omega - \frac{\lambda}{2}\right)n^{-1/3} + \left(\frac{\lambda^2}{3} - 3\lambda\omega + 3\omega^2\right)n^{-2/3} \right], \quad n \to \infty \qquad (7.97)$$

is the curvature dependent surface tension of the cluster. Its radius is given by the radius of the equimolar surface R_e. Within the framework of the Gibbs thermodynamics $\gamma_n = \gamma[R_e] \equiv \gamma_e$ is the surface tension measured at the equimolar surface (see Sect. 2.2.2). Using the relationship $R_e = r^1 n^{1/3}$ we rewrite (7.97) as

$$\gamma_e = \gamma_\infty \left[1 - \frac{2r^1\left(\omega - \frac{\lambda}{2}\right)}{R_e} + \frac{(r^1)^2\left(\frac{\lambda^2}{3} - 3\lambda\omega + 3\omega^2\right)}{R_e^2} \right], \qquad \text{large } R_e \quad (7.98)$$

Recall that for sufficiently large droplets the surface tension *at the surface of tension*, $\gamma_t = \gamma[R_t]$, can be expressed using the Tolman formula (2.41)

$$\gamma_t = \gamma_\infty \left(1 - \frac{2\delta_T}{R_t} + \dots\right) \qquad (7.99)$$

In view of the Ono-Kondo equation (2.35) the surface tension is at minimum at R_t. Hence, to the first order in δ_T Eq. (7.98) reads:

$$\gamma_t = \gamma_\infty \left[1 - \frac{2r^1\left(\omega - \frac{\lambda}{2}\right)}{R_t} + \dots \right] \qquad (7.100)$$

Comparing (7.100) with (7.99) we identify the Tolman length

$$\delta_T = r^1 \left(\omega - \frac{\lambda}{2} \right) \tag{7.101}$$

Since both ω and λ are of order 1, δ_T is of the order of molecular size. It is tempting to apply the Tolman equation (7.100) truncated at the first order term for relatively small n-clusters once the Tolman correction is small

$$\left| 2 \left(\omega - \frac{\lambda}{2} \right) \right| n^{-1/3} \ll 1 \tag{7.102}$$

This condition, however, is not sufficient. By definition (see Eq. (2.37)) the Tolman length is independent of the radius; meanwhile, for sufficiently small droplets $\delta = R_e - R_t$ is a strong function of R_t [28, 29]. Koga et al. [30] obtained $\delta(R_t)$ for simple fluids (Lennard-Jones and Yukawa) using the DFT. They showed that for these systems the Tolman equation becomes valid for clusters containing more than $10^5 \div 10^6$ molecules. This result implies that in contrast with conventional expectation one can not apply the Tolman equation down to droplets containing few tens or hundreds of molecules. For the nucleation theory it means that even in the macroscopic nucleation regime one can use Tolman's correction only for extremely large clusters, containing $n \sim 10^5 \div 10^6 \gg N_{class}$ molecules which are practically irrelevant for experimental conditions (critical clusters contain usually $\sim 10^1 \div 10^2$ molecules).

MKNT allows to find an independent criterion of the applicability of the Tolman equation to nucleation problems by analyzing the next-to-Tolman term in the curvature expansion (7.97). The following condition should be satisfied:

$$\left| \frac{\left(\frac{\lambda^2}{3} - 3\lambda\omega + 3\omega^2 \right)}{2 \left(\omega - \frac{\lambda}{2} \right)} \right| n^{-1/3} \ll 1 \tag{7.103}$$

Combining (7.102) and (7.103) we conclude that the Tolman correction is applicable for the clusters satisfying

$$n^{1/3} \gg \max \left\{ \left| 2 \left(\omega - \frac{\lambda}{2} \right) \right|, \left| \frac{\left(\frac{\lambda^2}{3} - 3\lambda\omega + 3\omega^2 \right)}{2 \left(\omega - \frac{\lambda}{2} \right)} \right| \right\} \tag{7.104}$$

For typical experimental conditions the second requirement (Eq. (7.103)) is much stronger than the first one (Eq. (7.102)). For the cases studied in Sect. 7.8 the Tolman Eq. (7.100) is valid for clusters containing $\sim (4 \div 5) \times 10^4 \div 10^5$ particles which is close to the result of Ref. [30] for simple fluids. Note, that although δ_T is not used

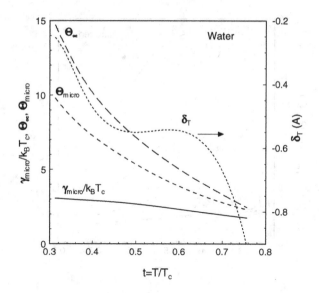

Fig. 7.10 Equilibrium properties of water used in MKNT: θ_∞, θ_{micro}, $\gamma_{micro}/k_B T_c$ (*left y-axis*) and the Tolman length δ_T [Å] according to Eq. (7.101) (*right y-axis*) (Reprinted with permission from Ref. [6], copyright (2006), American Institute of Physics.)

in MKNT, its behavior is important for understanding the asymptotic properties of big clusters.

The temperature dependence of various equilibrium properties of water and nitrogen, used in MKNT analysis, is shown in Figs. 7.10 and 7.11. The temperature range is limited from above by the MKNT validity criterion (7.43). For both substances θ_{micro} is lower than its macroscopic counterpart θ_∞ for all T, however the difference between them decreases with the temperature. The microscopic surface tension γ_{micro} is approximately linear in t supporting the MKNT assumption. The dashed line in Fig. 7.11, labelled "LJ", is the universal form of $\gamma_{micro}/k_B T_c$ for Lennard-Jones fluids given by (7.93)–(7.94).

For water the Tolman length demonstrates a rather unusual behavior. For all temperatures δ_T is negative; at low T it decreases reaching a weak minimum at $T \approx 308$ K. For higher temperatures δ_T slowly increases reaching a maximum at $T \approx 394$ K and then decreases again. For nitrogen at experimental temperatures δ_T is positive and close to zero: $0.04 < \delta_T < 0.06$ Å. At high enough temperatures δ_T becomes negative and monotonously decreases. Such a behavior suggests the divergence of δ_T near T_c. However, the definitive conclusion and the corresponding critical exponent can not be drawn from the present model since the critical region is beyond its range of validity.

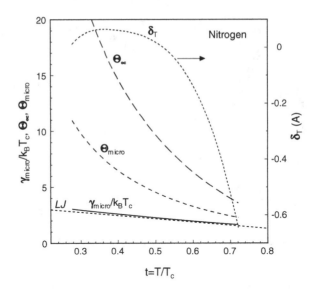

Fig. 7.11 MKNT: equilibrium properties of nitrogen.: θ_∞, θ_{micro}, $\gamma_{micro}/k_B T_c$ (*left y-axis*) and the Tolman length δ_T [Å] according to Eq. (7.101) (*right y-axis*). The *dashed line* labelled "LJ" shows $\gamma_{micro}/k_B T_c$ for Lennard-Jones fluids according to Eq. (7.89) with the universal parameters given by (7.93)–(7.93) (for details see the text)

7.9.4 Small Nucleating Clusters as Virtual Chains

In the limit of small (nano-sized) clusters the behavior of the system (the free energy barrier and the distribution function) as a function of the cluster size n will be essentially different from that given by the phenomenological CNT. It is instructive to consider the physical picture emerging from the MKNT in the case of small n. For $n \leq N_1$ the equilibrium cluster distribution is

$$\rho_{sat}(n) = \rho_{sat}^v \, e^{-\theta_{micro}\,(n-1)}, \quad 1 \leq n \leq N_1 \tag{7.105}$$

Using the relation (7.12) we find for the configuration integral of an n-cluster:

$$q_n = \rho_{sat}^v \, V \, e^{-\theta_{micro}\,(n-1)} \, e^{-n\beta\mu_{sat}} \, \Lambda^{3n} = \left[\rho_{sat}^v \Lambda^3 e^{-\beta\mu_{sat}}\right] V \left(\frac{\Lambda^3 \, e^{-\beta\mu_{sat}}}{h}\right)^{n-1} \tag{7.106}$$

The chemical potential (not close to T_c) can be approximated by

$$\mu_{sat} = k_B T \, \ln(\rho_{sat}^v \Lambda^3)$$

Fig. 7.12 Virtual chain
cluster

which after substitution into (7.106) results in:

$$q_n = V \, K^{n-1}, \qquad 1 \le n \le N_1 \qquad (7.107)$$

where

$$K = \frac{1}{\rho^{\mathrm{v}}_{\mathrm{sat}} \, e^{\theta_{\mathrm{micro}}}} \qquad (7.108)$$

The factorized form (7.107) of the configuration integral suggests that the n-cluster in this case is characterized by short range nearest neighbor interactions between particles and contains $n-1$ bonds, each bond contributing the same quantity K to q_n. This is the minimum possible number of bonds for an n-cluster (a spherical droplet represents the opposite limit of maximum number of bonds). The latter means that a small cluster is not a compact object but rather reminds a polymer chain of atoms with nearest neighbor interactions [31]. Each particle of such cluster is bonded to the two neighboring particles belonging to the same chain with exception of the end-point particles having one neighbor. The simplest example of such a cluster is a linear chain of molecules. A somewhat more complex cluster structure satisfying Eq. (7.107) can have branch points where a particle of the given chain has contacts with particles belonging to another chain; however, loops are prohibited (see Fig. 7.12). This structure was studied in Ref. [32] where it was termed a "system of virtual chains" indicating that the sequence of atoms in the chains is not fixed: they are associating and dissociating.

The value of K depends on the interatomic potential $u(r)$ and temperature. For the nearest neighbor pairwise additive interaction we obtain from the definition of q_n:

$$q_n = V \left(\frac{q_2}{V}\right)^{n-1}, \qquad 1 \le n \le N_1 \qquad (7.109)$$

Comparing (7.107) and (7.109) we identify

$$K = \frac{q_2}{V} = \frac{1}{2} \oint_{\mathrm{cl}} \mathrm{d}r \, e^{-\beta u(r)} \qquad (7.110)$$

From (7.109) and (7.12) it is seen that K represents the *dimer association constant*:

$$K = \frac{\rho_{sat}(2)}{[\rho_{sat}(1)]^2} \tag{7.111}$$

Hence the microscopic surface tension can be related to the dimer association constant:

$$\theta_{micro} = -\ln\left(\rho_{sat}^v\, K\right)$$

References

1. A. Dillmann, G.E.A. Meier, J. Chem. Phys. **94**, 3872 (1991)
2. I.J. Ford, A. Laaksonen, M. Kulmala, J. Chem. Phys. **99**, 764 (1993)
3. C.F. Delale, G.E.A. Meier, J. Chem. Phys. **98**, 9850 (1993)
4. V.I. Kalikmanov, M.E.H. van Dongen, J. Chem. Phys. **103**, 4250 (1995)
5. M.E. Fisher, Physics **3**, 255 (1967)
6. V.I. Kalikmanov, J. Chem. Phys. **124**, 124505 (2006)
7. F.H. Stillinger, J. Chem. Phys. **38**, 1486 (1963)
8. J.K. Lee, J.A. Baker, F.F. Abraham, J. Chem. Phys. **58**, 3166 (1973)
9. D.I. Zhukhovitskii, J. Chem. Phys. **101**, 5076 (1994)
10. J.J. Binney, N.J. Dowrick, A.J. Fisher, M.E.J. Newman, *The Theory of Critical Phenomena* (Clarendon Press, Oxford, 1995)
11. R.C. Reid, J.M. Prausnitz, B.E. Poling, *The Properties of Gases and Liquids*, 4th edn. (McGraw-Hill, New York, 1987)
12. V.I. Kalikmanov, *Statistical Physics of Fluids. Basic Concepts and Applications*(Springer, Berlin, 2001)
13. J.S. Rowlinson, B. Widom, *Molecular Theory of Capillarity* (Clarendon Press, Oxford, 1982)
14. J.R. Cahoon, Can. J. Phys. **82**, 291 (2004)
15. J. Wölk, R. Strey, J. Phys. Chem. B **105**, 11683 (2001)
16. P. Peeters, J.J.H. Giels, M.E.H. van Dongen, J. Chem. Phys. **117**, 5467 (2002)
17. D.G. Labetski, V. Holten, M.E.H. van Dongen, J. Chem. Phys. **120**, 6314 (2004)
18. D. Brus, V. Ždimal, J. Smolik, J. Chem. Phys. **129**, 174501 (2008)
19. K. Iland, Ph.D. Thesis, University of Cologne, 2004
20. F. Hensel, Phil. Trans. R. Soc. London A **356**, 97 (1998)
21. K. Rademann, Z. Phys, D **19**, 161 (1991)
22. P.P. Singh, Phys. Rev. B **49**, 4954 (1994)
23. G.M. Pastor, P. Stampfli, K.H. Benemann, Phys. Scr. **38**, 623 (1988)
24. G. Moyano et al., Phys. Rev. Lett. **89**, 103401 (2002)
25. J. Martens, H. Uchtmann, F. Hensel, J. Phys. Chem. **91**, 2489 (1987)
26. D. Frenkel, B. Smit, *Understanding Molecular Simulaton* (Academic Press, London, 1996)
27. N.F. Carnahan, K.E. Starling, J. Chem. Phys. **51**, 635 (1969)
28. A.H. Falls, L.E. Scriven, H.T. Davis, J. Chem. Phys. **75**, 3986 (1981)
29. V. Talanquer, D.W. Oxtoby, J. Chem. Phys. **99**, 2865 (1995)
30. K. Koga, X.C. Zeng, A.K. Schekin, J. Chem. Phys. **109**, 4063 (1998)
31. U. Gedde, *Polymer Physics* (Chapman & Hall, London, 1995)
32. D.I. Zhukhovitskii, J. Chem. Phys. **110**, 7770 (1999)

Chapter 8
Computer Simulation of Nucleation

8.1 Introduction

Besides the development of analytical theories describing the nucleation process, molecular simulation is a powerful tool for investigating nucleation on a microscopic level. In the best case analytical solutions of theoretical approaches require various approximations to simplify the problem. Often approximations have to be made with respect to the molecular details of a substance. It is not always obvious how such approximations affect the performance of the theory. On the other hand, it is very difficult to get such information on the molecular-level processes from nucleation experiments. In this context molecular simulation is a technique that complements the theoretical and experimental methods. To a certain extent molecular simulation may be regarded as a computer experiment. Interactions between atoms and molecules are mapped on a potential model that determines intermolecular forces. One may distinguish two simulation techniques: molecular dynamics (MD) simulation and molecular Monte Carlo (MC) methods.

MD simulations represent numerical integration of classical equations of motion for the system of interacting particles (atoms, molecules, etc.) while tracking the system in time. Taking the *time averages* of the physical quantities of interest during its evolution and relying on the ergodicity hypothesis, one can state that these averages are equivalent to the *ensemble averages* of these quantities in the micro-canonical (NVE) ensemble.

In Monte Carlo simulations the physical time is not present: the motion of a molecule does not correspond to a real trajectory of a molecule; instead, one simulates the behavior of the system in the *configurational space*. Therefore, the results of MC simulations are *ensemble averages*. MC allows moves that lead fast towards the equilibrium state making it possible to overcome high energy barriers; this would not be possible to achieve within a reasonable computational time using MD methods.

Hence, both approaches and their further developments have advantages that can be employed in a complementary way to address molecular level investigations of

V. I. Kalikmanov, *Nucleation Theory*, Lecture Notes in Physics 860, DOI: 10.1007/978-90-481-3643-8_8, © Springer Science+Business Media Dordrecht 2013

nucleation processes. From the above mentioned general observations it is clear that MD simulations have the advantage of gaining insights into the kinetics of the nucleation process, while MC typically gives information on the thermodynamics of the process. In this chapter we introduce molecular simulation methods that are relevant for modelling of the nucleation process and discuss their application to various nucleation problems.

8.2 Molecular Dynamics Simulation

8.2.1 Basic Concepts and Techniques

Classical molecular dynamics simulation is based on the numerical integration of Newton's equations of motion. A principal scheme of MD can be summarized as follows:

1. One introduces a *potential model* (interaction potential) describing interactions between particles (atoms or molecules) in the system.
2. The net force acting on a specific particle by all other particles is given by the negative gradient of the intermolecular potential with respect to the particle separation.
3. From the Newton law one calculates the acceleration caused by the force acting on the particle.
4. Finally, particles are moved according to the direction and magnitude of the resulting force using a numerical technique.

In the context of MD simulations it is useful to generalize Newton's dynamics by Lagrange dynamics. In classical mechanics, the equations of motion are contained in the fundamental Lagrange equation [1]

$$\frac{d}{dt}\left(\frac{\partial L}{\partial \dot{q}_k}\right) - \left(\frac{\partial L}{\partial q_k}\right) = 0, \quad k = 1, 2, \ldots, s \tag{8.1}$$

where s is the number of degrees of freedom of the system containing N particles; the "upper dot" denotes the time derivative. The *Lagrangian L* of the system is the function of the generalized coordinates $\mathbf{q} = (\mathbf{q_1}, \mathbf{q_2}, \ldots, \mathbf{q_N})$ and their time derivatives $\dot{\mathbf{q}}$ defined in terms of the kinetic energy E_{kin} and potential energy U:

$$L(\mathbf{q}, \dot{\mathbf{q}}) = E_{kin} - U.$$

For a system of N molecules the kinetic energy in Cartesian coordinates reads

$$E_{kin} = \sum_{i=1}^{N} \frac{\mathbf{p}_i^2}{2 m_i}. \tag{8.2}$$

where m_i is the mass of a particle i and $\mathbf{p}_i$ is its momentum. The generalized momenta p_k are given by

$$p_k = \frac{\partial L}{\partial \dot{q}_k}, \quad k = 1, 2, \ldots, s$$

The potential energy may be divided into terms which depend on external potentials, particle interactions of pairs, triplets etc.:

$$U = \sum_{i=1}^{N} u_1(\mathbf{r}_i) + \sum_{i=1}^{N} \sum_{j>i}^{N} u_2(\mathbf{r}_i, \mathbf{r}_j) + \sum_{i=1}^{N} \sum_{j>i}^{N} \sum_{k>j>i}^{N} u_3(\mathbf{r}_i, \mathbf{r}_j, \mathbf{r}_k) + \ldots \quad (8.3)$$

The first term in Eq. (8.3) contains the external potential u_1 acting on the individual molecules. The second term contains the potential u_2 of the interaction between any pair of particles in the system. We could continue to include the interactions between triplets, quadruplets etc. The second term, however, is typically the most important one. Neglecting any external fields, the first term in Eq. (8.3) equals zero. In addition, the contributions from triplet or higher order interactions are typically much smaller and are often neglected, such as for the widely used Lennard-Jones potential. Thus, the total potential of the system reduces to a good accuracy to the sum of all pairwise interactions:

$$U = \sum_{i=1}^{N} \sum_{j>i}^{N} u_2(\mathbf{r}_i, \mathbf{r}_j) \quad (8.4)$$

The second summation $j > i$ provides the exclusion of double counting. If we insert the expressions for potential and kinetic energy into Eq. (8.1), the Lagrange equation reduces to

$$m_i \ddot{\mathbf{r}}_i = \mathbf{f}_i, \quad (8.5)$$

which is the Newton second law in terms of the force $\mathbf{f}_i$ acting on particle i. For time and velocity-independent interaction potentials, the force is given by

$$\mathbf{f}_i = \nabla_{\mathbf{r}_i} L = -\nabla_{\mathbf{r}_i} U. \quad (8.6)$$

Within Lagrangian dynamics equations of motion can be solved while conserving the energy and momentum.

Application of Lagrangian dynamics to the simple problem of one particle with a mass m interacting with the environment by a potential $U(x)$ gives:

$$L = \frac{1}{2} mv^2 - U(x)$$

where v is the velocity of the particle. The Lagrange equation (8.1) now reads:

$$\frac{d}{dt}\underbrace{\left(\frac{\partial L}{\partial \dot{x}}\right)}_{mv} = \underbrace{\left(\frac{\partial L}{\partial x}\right)}_{-\frac{dU}{dx}}$$

On the left-hand side the derivative of the momentum mv with respect to the time t yields ma, where a is the acceleration of the particle, while the right-hand side is equal to the force f. Lagrangian dynamics does not only recover the Newton law but can also be applied to more complicated problems. For example it is useful for the proper derivation of the equations of motion for a system including a thermostat. In this case the Lagrange equation has to be extended by a term representing the energy of the thermostat. The application of the Lagrange differential equation to this approach gives the desired proper equation of motion for the *NVT* ensemble.

Besides Lagrangian dynamics one can also describe the evolution of the system using the Hamiltonian dynamics. Let us introduce the Hamiltonian of the system [1]

$$\mathcal{H}(\mathbf{p}, \mathbf{q}) = \sum_k \dot{q}_k p_k - L(\mathbf{q}, \dot{\mathbf{q}})$$

With its help the equations of motion can be written in the *Hamiltonian form*:

$$\dot{q}_k = \frac{\partial H}{\partial p_k}, \tag{8.7}$$

$$\dot{p}_k = -\frac{\partial H}{\partial q_k}. \tag{8.8}$$

For a time- and velocity-independent interaction potential the Hamiltonian of the system simply reduces to its total energy, which in this case must be conserved:

$$H(\mathbf{p}, \mathbf{q}) = E_{\text{kin}}(\mathbf{p}) + U(\mathbf{q}).$$

Finally, we can write down the Hamiltonian equations (8.7)–(8.8) in Cartesian form:

$$\dot{\mathbf{r}}_i = \mathbf{p}_i/m, \tag{8.9}$$

$$\dot{\mathbf{p}}_i = -\nabla_{r_i} U = \mathbf{f}_i. \tag{8.10}$$

Computing of the trajectories in the phase space involves solving either a system of $3N$ second-order differential equation (8.5) or an equivalent set of $6N$ first-order differential equations (8.9)–(8.10). The classical equations of motion are deterministic and invariant to time reversal. This means that if we change the sign of the velocities, the particles will trace back on exactly the same trajectories. In computer simulations, exact reversibility usually is not observed because of the limited accuracy of the numerical calculations and the chaotic behavior of the dynamics of a many-body system.

The "natural" ensemble for MD simulation is the constant energy (micro-canonical) *NVE* ensemble. For this ensemble the equations of motion can be solved numerically by calculating the motion of the molecules caused by the forces in small time steps Δt. How small these time steps have to be, depends on the type of interaction between the molecules. For a very steep potential a very short time step is required while for a rather flat potential the time step may be larger. A typical time step for MD simulations is typically in the order of a femto-second (10^{-15} s). In order to calculate the position of a molecule at the time $t + \Delta t$ one expands the position of the molecule in a Taylor series around the present position at time t:

$$x(t + \Delta t) = x(t) + \underbrace{\frac{dx}{dt}}_{v(t)} \Delta t + \frac{1}{2} \underbrace{\frac{d^2x}{dt^2}}_{a(t)} \Delta t^2 + \frac{1}{6} \frac{d^3x}{dt^3} \Delta t^3 + O(\Delta t^4)$$

From Newton's law we can calculate the acceleration a at the moment t for the given force $f(t)$:

$$a(t) = \frac{f(t)}{m}$$

yielding

$$x(t + \Delta t) = x(t) + v(t)\,\Delta t + \frac{f(t)}{2m} \Delta t^2 + \frac{1}{6} \frac{d^3x}{dt^3} \Delta t^3 + O(\Delta t^4) \tag{8.11}$$

In order to solve the above equation, we need to know the velocity at time t. However, it can be eliminated by expanding the position of the molecules backward in time. Replacing in (8.11) Δt by $-\Delta t$, we obtain

$$x(t - \Delta t) = x(t) - v(t)\,\Delta t + \frac{f(t)}{2m} \Delta t^2 - \frac{1}{6} \frac{d^3x}{dt^3} \Delta t^3 + O(\Delta t^4) \tag{8.12}$$

Adding expansions (8.11) and (8.12), the odd terms cancel resulting in

$$x(t + \Delta t) = 2x(t) - x(t - \Delta t) + \frac{f(t)}{2m} \Delta t^2 + O(\Delta t^4) \tag{8.13}$$

This way of numerically integrating the equations of motion is called the *Verlet algorithm* [2]. Though it is a rather simple approach, its error is just of the order of $O(\Delta t^4)$. For calculation of positions of the particles the velocity is not needed. Meanwhile, it is required for calculation of the kinetic energy of the system, instantaneous temperature, transport properties, to mention just a few. One can calculate v by subtracting the expansions (8.11) and (8.12):

$$v(t) = \frac{x(t + \Delta t) - x(t - \Delta t)}{2\Delta t} + O(\Delta t^3) \tag{8.14}$$

Thus, velocity at time t is calculated *after* the positions at time $(t + \Delta t)$ and $(t - \Delta t)$ have been already found. The positions in Eq. (8.13) are exact up to errors of the order of $O(\Delta t^4)$, while the velocities are exact up to an error of the order of $O(\Delta t^3)$. The Verlet algorithm is properly centered on the respective positions at $(t \pm \Delta t)$ and, thus, is time reversible. However, the difference between the accuracies in positions and velocities may result in deviations from the classical trajectories. Consequently, most Verlet-type algorithms exhibit a drift in the energy of the system on short and long time-scales, which is strongly influenced by the length of the time-step Δt.

Since calculation of the velocities in the original Verlet method contains a comparatively large error, a more accurate form, known as the *velocity-Verlet* algorithm, was proposed [3]:

$$x(t + \Delta t) = x(t) + \Delta t \, v(t) + \frac{1}{2m} f(t) \, \Delta t^2 \tag{8.15}$$

$$v(t + \Delta t) = v(t) + \frac{\Delta t}{2m} \left[f(t) + f(t + \Delta t) \right] \tag{8.16}$$

Calculation using the velocity-Verlet algorithm typically proceeds through the two steps: first, the new positions $x(t + \Delta t)$ and the velocities are calculated at an intermediate time interval:

$$v\left(t + \frac{1}{2}\Delta t\right) = v(t) + \frac{1}{2m} f(t) \, \Delta t$$

Then, the acceleration at $t + \Delta t$ is calculated and the velocities are updated:

$$v(t + \Delta t) = v\left(t + \frac{1}{2}\Delta t\right) + \frac{1}{2m} f(t + \Delta t) \, \Delta t$$

The original Verlet algorithm may be recovered by eliminating the velocity from Eq. (8.16). In the velocity-Verlet method, the new velocities are obtained with the same accuracy and at the same time as the positions and accelerations of the molecules. Therefore, the kinetic and potential energy are known at each time. The velocity-Verlet algorithm requires a minimum storage memory and its numerical stability, time reversibility, and simplicity in both form and implementation make it by far the most preferred method in MD simulations to date.

A number of other algorithms for solving the equations of motion have been developed. For example, the *leap-frog algorithm* improves the accuracy by using half-steps between the time steps. Other methods are based on higher order terms of the Taylor expansions. In general, the more accurate the algorithm—the larger the time step can be used. For a complete survey of various algorithms the reader is referred to the literature [4, 5].

8.2.2 System Size

Imagine that we would like to simulate a macroscopic system containing 1 mol of gas. The number of molecules in this system, given by the Avogadro number, is of the order of 10^{23}. Molecular simulation of such an immensely large number of particles is impossible to accomplish because of the limitations of computer power. Although computer power rises continuously it will not be possible to simulate a macroscopic system for a very long time, if ever. Therefore it is necessary to reduce the size of the simulation system. Usually molecular simulations are performed with $N \sim 10^3 - 10^4$ particles (atoms or molecules). The required computational effort depends on the kind of interaction potential between the atoms and molecules. For example, short range van der Waals forces can be evaluated rapidly because not all interactions have to be taken into account but only those between molecules separated by a short distance. On the other hand, electrostatic interactions are long range and require long range corrections leading to considerable computational efforts.

In order to simulate a molecular system one has to cut out a small piece of the complete macroscopic system. The question is whether such a small piece actually represents the total system properly. An obvious problem is that a small system has a large surface-to-volume ratio. If we take a system of 1,000 molecules in a cubic box, we can easily estimate that roughly half of the molecules will be in the surface region of the box. As a result, simulation of such a system will give results strongly influenced by the surface effects rather than describing the bulk properties. These difficulties can be overcome by implementing *periodic boundary conditions*. The idea of periodic boundary conditions is that the simulated cubic box is replicated indefinitely in each direction, thus forming and infinite lattice. Any molecule moving in the simulation volume has an infinite number of copies in the neighboring boxes, which move in exactly the same way. If a molecule leaves the system, a copy of it enters the simulation volume from the opposite side. The box itself appears without any walls and the number of molecules in it is conserved. This construction is illustrated in Fig. 8.1. Note, that it is not necessary (and indeed impossible) to store the coordinates and velocities of the particles and all of their copies in neighboring systems. For calculation of the force the *closest* copy of a molecule is considered, which is called the *minimum image convention*.

This approach has been proven to yield accurate results for many properties of the system. However, one has to keep in mind that periodic boundary conditions generate a system with periodicity. Whenever a property or an effect is short range, i.e. its correlation length is smaller than half the box length, it can be calculated using periodic boundary conditions. On the other hand there are long range phenomena that may not fit into a given simulation box. A famous example is the density fluctuations in a substance close to the critical point. In the critical region the density fluctuations rise beyond the box dimensions. If periodic boundary conditions are applied, these long range fluctuations are cut off leading to the mean-field behavior. The latter yields critical exponents that deviate from the correct ones. Other simulation techniques, such as finite size scaling [6], allow the treatment of such a problem.

Fig. 8.1 Periodic boundary
conditions. The central box is
the simulation system while
the other boxes and all their
molecules are replicated in all
spatial directions

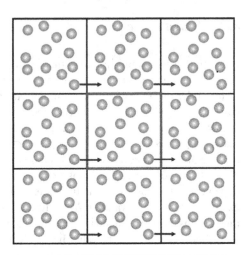

Fig. 8.2 Kinetic and potential
energy during the collision
of two Lennard-Jones argon
atoms in a constant energy
ensemble (NVE)

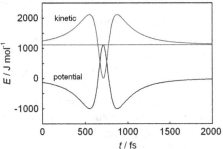

Finite size effects, i.e. the influence of the system size on the simulated properties,
are also important for calculation of the surface tension [7]. Surface fluctuations of
a vapor–liquid interface lead to the so-called capillary waves with the wavelength
related to the surface tension. Besides the wavelength, also the amplitude of the
capillary waves varies with the system size leading to a widening of the interface
with increasing system size.

8.2.3 Thermostating Techniques

Nucleation is in principle a non-isothermal process. When vapor condenses its potential energy is lowered because intermolecular separations become smaller approach-
ing the minimum of the interaction potential energy (corresponding to maximum
attraction). In a closed adiabatic system the loss in potential energy has to be com-
pensated by a gain in kinetic energy. In Fig. 8.2 this exchange of energy is depicted for
a two-atom collision. If a cluster grows the exchange of energy is similar, although,
due to many-body processes, the curves look more chaotic. Thus, in adiabatic systems

the process of condensation is accompanied by heating of the system. Therefore, to stabilize the embryo of the new phase this latent heat has to be removed. There are several approaches to perform MD simulations at constant temperature. In the simplest case one can make use of the relation between the velocities of the molecules and the instantaneous temperature T of the system. The mean kinetic energy *per molecule* of the system reads

$$\langle E_{\text{at,kin}} \rangle = \left\langle \frac{1}{2N} \sum_{i=1}^{N} m_i v_i^2 \right\rangle$$

The equipartition theorem of statistical mechanics states that each degree of freedom contributes $k_B T/2$ to the total kinetic energy of the system (see e.g. [8]) yielding

$$\langle E_{\text{at,kin}} \rangle = N_f k_B T/2 \tag{8.17}$$

where N_f is the number of degrees of freedom of a particle; for a particle with translational motion in 3D $N_f = 3$. In general, after one (or several) steps of the dynamics, the instantaneous temperature T will be different from the desired temperature T_{set}. By rescaling the velocities of the molecules one can set the system to the temperature T_{set}:

$$v_{i,\text{new}} = \sqrt{\frac{T_{\text{set}}}{T}} \, v_i$$

While such *velocity scaling* approach is suitable in some cases and for some properties, especially in equilibrium, in the context of nucleation this method of thermostating will give wrong results. Rescaling the velocities is actually a method to keep the kinetic energy of a particle constant while all temperature fluctuations are completely eliminated. At the same time if one wants to simulate the system in the canonical *NVT* ensemble, one has to realize that this ensemble exhibits thermal fluctuations. These fluctuations are the origin of the heat capacity. Thus, the velocity scaling method is by construction unable to predict the heat capacity. Furthermore, in nucleating systems the velocity scaling can lead to an artificial cooling down of the remaining monomers in the vapor phase [9].

Several thermostating techniques were proposed which are able to correctly realize the canonical ensemble. One example is the stochastic *Andersen thermostat* [10]. Instead of rescaling the velocities of *all* molecules in *every time step*, one or a few molecules are randomly chosen from the vapor and their new velocities are calculated from the Maxwell distribution function at the desired temperature. In this way, the Andersen thermostat mimics a collision of a molecule with a carrier gas particle. The frequency at which a molecule is picked from the vapor, determines the effectiveness of this method. In the limit of very high frequencies one recovers a procedure similar to velocity scaling but with a certain temperature distribution. On the other hand, a low frequency may not be sufficient to keep the temperature constant.

Another widely used method is the *Nosé-Hoover thermostat* [11, 12]. It includes the thermostat in the equations of motion. Adding an energy term, which describes the thermostat, to the Lagrange (or Hamilton) function and applying the Lagrangian (or Hamiltonian) dynamics leads to equations of motion including a friction term, that affects the acceleration of the molecules. The Nosé-Hoover thermostat correctly realizes the canonical ensemble allowing for fluctuations in the system temperature. If the thermostat itself is coupled to another thermostat, one obtains a so-called *Nosé-Hoover chain thermostat* [13], which represents an improved thermostat realizing the canonical *NVT* ensemble.

In physical experiments the exchange of energy between the system and the environment is usually provided by a carrier gas. Such carrier gas usually does not affect the phase transition beyond its function as a latent heat transfer agent. It is also possible to include a carrier gas in MD simulations. To keep the computational effort low, it is useful to employ for this purpose a mono-atomic gas, such as the Lennard-Jones argon. It is added into the simulation box, typically in abundance. When a cluster is formed by nucleation, the latent heat is removed from the cluster by collisions of the cluster particles with the carrier gas atoms. In real physical experiment the heat is removed from the carrier gas either by expanding the system or by collisions with the container walls. In MD simulations it is possible to simulate the expansion [14] directly or model the heat removal from the carrier gas by coupling it to one of the regular MD thermostats as described above. This is possible because the carrier gas remains in the gas phase and does not condense. Special care should be taken in the cases in which the carrier gas is present in the interior of the cluster or is adsorbed at the cluster surface. It should be checked whether the application of a MD-thermostat to the carrier affects the simulation results.

Westergren et al. [15] analyzed several effects on the heat exchange between a cluster and the carrier gas in MD simulation. They found an increasing energy transfer with rising the atomic mass of the noble gas (acting as a carrier gas). By using different forms of cluster-gas interaction potentials they also found that soft interactions are more efficient in the heat transfer from the cluster to the gas. While Westergren et al. employed "real" noble gas atoms, each having a distinct set of parameters of the interaction potential and the atomic mass, one can also use a pseudo-noble gas, having the same interaction parameters but different atomic masses [16]. This allows one to separate the effect of the atomic mass from the effect caused by different interaction parameters of the different noble gases. It is useful for the fundamental analysis of the heat exchange to optimize the heat transfer in a simulation, however one should be careful to directly draw conclusions for experimental systems. Figure 8.3 shows the effect of the atomic mass, given in the diagrams in atomic units, on the cooling of the largest cluster in the simulation system during a nucleation simulation. Temperature

Fig. 8.3 Heat exchange of the zinc cluster with carrier gases having different atomic masses but the same interaction parameters. (Reprinted with permission from Ref. [16], copyright (2009), American Institute of Physics.)

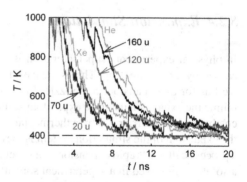

Fig. 8.4 Temperature development of the zinc subsystem as a function of time for different amount of carrier gas argon. The numbers above the plots are the ratio Zn:Ar atoms. The horizontal line gives the temperature of the argon subsystem

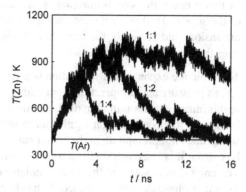

jumps in these graphs are related to cluster-cluster collisions. In principle one can identify two effects:

- The heavy atoms are slower than the light ones which means that the heavy atoms move a smaller distance and, hence, less likely collide with a cluster than the light ones. Hence, the heat exchange should become less efficient.
- On the other hand a collision with a heavy atoms itself is more efficient, i.e. more heat is transferred.

As Fig. 8.3 shows, the effect of the mass on the velocity dominates, because the lighter the carrier gas the more heat is removed from the clusters. In order to steer the heat transfer during particle formation processes, one can also vary the amount of the carrier gas. The more carrier gas is present, the more heat can be removed from the forming clusters. Figure 8.4 shows the effect of the amount of carrier gas (Ar) on the temperature of the zinc subsystem, which includes all zinc clusters in the simulation box [17]. The horizontal line gives the temperature of the argon subsystem. The labels are the ratio of Zn to Ar atoms in the system. The higher this ratio, the faster is the convergence of the zinc temperature to the carrier gas temperature.

8.2.4 Expansion Simulation

In physical experiments supersaturation is often initiated by the temperature drop caused by a pressure drop. This is the case for an expansion chamber experiment as well as for an expansion in a nozzle. In simulations it is more difficult to actually mimic the expansion using an expanding simulation. It is computationally much less costly to set the carrier gas by a thermostat to a low temperature. Still in some cases the expansion of the simulation system is inevitable. An example is the simulation of a process called a "rapid expansion of a supercritical solution" (RESS) [18]. In RESS a solute is dissolved in a supercritical solvent which is then expanded in the nozzle. In this process the supersaturation is achieved by two effects: first the solubility of the supercritical solvent decreases drastically with decreasing density during the expansion, and secondly the expansion leads to a temperature drop which further rises the supersaturation.

To mimic the expansion, a simulation box filled with the supercritical solution is set up at a pressure and temperature significantly above the critical values of the solvent. In this initial state a simulation run is performed until the system reaches equilibrium. In case of RESS simulation equilibration runs are necessary to make sure that the solute does not precipitate at the initial supercritical conditions. In the next step the box size is increased in a small step followed by a constant energy simulation (NVE). The enlargement step and the NVE simulation steps are repeated several times until the vapor phase density is reached. In has been found that an optimum with respect to computational effort and the accuracy of the simulation can be achieved by around 100 expansion steps [14].

In Fig. 8.5 the expansion path of such simulation for carbon dioxide is plotted in all three coordinates: temperature, pressure and density. Comparison with a highly accurate Span-Wagner equation of state [19] shows that the expansion path of the

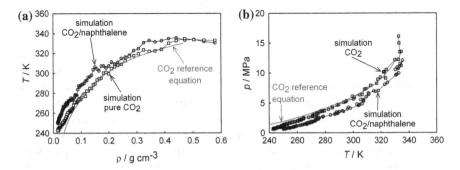

Fig. 8.5 Comparison of the path during an expansion simulation for pure CO_2 with the Span-Wagner EoS (*reference equation*)[19]. In addition the expansion path of the CO_2/naphthalene solution is plotted. (Reprinted with permission from Ref. [14], copyright (2009) American Chemical Society.)

simulation system agrees very well with the adiabatic expansion calculated with the Span-Wagner EoS in all three coordinates. In both figures also the expansion path of a dilute naphthalene solution in carbon dioxide is plotted. There is no such accurate reference equation for this mixture but the expansion path is very close to that of pure carbon dioxide. The small shift is related to the small amount of naphthalene in the solution that affects the properties through its molecular interactions.

8.3 Molecular Monte Carlo Simulation

While in MD the actual equations of motion are numerically integrated, Monte Carlo simulations sample the *configurational space* of the system. To be more specific, let us consider a canonical (*NVT*) ensemble of interacting particles (molecules). The potential energy of a given configuration $\mathbf{r}^N \equiv (\mathbf{r}_1, \ldots, \mathbf{r}_N)$ is $U(\mathbf{r}^N)$. The average value of an arbitrary function of coordinates $X(\mathbf{r}^N)$ is given by the integral

$$\langle X(\mathbf{r}^N) \rangle = \int X(\mathbf{r}^N) \, w(\mathbf{r}^N) \, d\mathbf{r}^N \tag{8.18}$$

where $w(\mathbf{r}^N)$ is the Boltzmann probability density function of a given state $\mathbf{r}^N$:

$$w(\mathbf{r}^N) = \frac{1}{Q_N} e^{-\beta U(\mathbf{r}^N)} \tag{8.19}$$

and

$$Q_N = \int e^{-\beta U(\mathbf{r}^N)} \, d\mathbf{r}^N \tag{8.20}$$

is the configuration integral of the system. Obviously, the function $w(\mathbf{r}^N)$ is positive and normalized to unity: $\int w(\mathbf{r}^N) \, d\mathbf{r}^N = 1$. From the standpoint of probability theory Eq. (8.18) defines the *mathematical expectation* of $X(\mathbf{r}^N)$. Imagine that we have a digital camera that can instantaneously take photos of the system so that we can use this camera to scan and memorize the 3D coordinates of all N molecules in the volume V. We can repeat these actions M times per second. Then, the computer memory will contain the set of coordinates $(\mathbf{r}^N)_1, (\mathbf{r}^N)_2, \ldots, (\mathbf{r}^N)_M$, where M is a number of configurations. The average *observed* value of X

$$\text{AVRG}(X) = (1/M) \sum_{k=1}^{M} X[(\mathbf{r}^N)_k] \tag{8.21}$$

gives an estimate of the *true* (exact) value $\langle X(\mathbf{r}^N) \rangle$ given by (8.18), which we are actually not able to calculate. The mean-square deviation

$$\sigma^2 = (1/M) \sum_{k=1}^{M} X^2[(\mathbf{r}^N)_k] - [\text{AVRG}(X)]^2 \qquad (8.22)$$

characterizes the accuracy of our statistical averaging. In MC the integral (8.18) is approximated by [20]

$$\langle X(\mathbf{r}^N) \rangle \approx \text{AVRG}(X) \left[1 \pm \frac{\sigma}{\sqrt{M}} \right] \qquad (8.23)$$

Note, that σ becomes independent of the number of observations for large M, implying that the error of approximation (8.23) is inversely proportional to the square root of the number of observations, which is typical for mathematical statistics.

MC simulation are based on the random generation of atom coordinates in a simulation box. In view of the large number $3N$ of space coordinates of the N-particle system it is clear, that calculations employing solely random choice of these coordinates is a hopeless task. The introduction of the so-called *importance sampling* by Metropolis et al. [21] allowed the sampling of the system with sufficient accuracy within reasonable simulation time. As indicated by the term "importance sampling", not all states of the system are sampled with equal probability, but predominantly those that bring significant contribution to the configuration integral. The criterion indicating the importance (statistical weight) of a state is its Boltzmann factor $e^{-\beta U(\mathbf{r}^N)}$. Clearly, a very large positive U results is a very low statistical weight of the state. Generating such term is hence a waist of time and should be avoided. In molecular simulation the energy of the system can become very high if two atoms overlap sufficiently. The repulsive part of the interaction potential is usually very steep yielding high energies resulting in low probabilities of such states. Based on these observations, instead of generating all coordinates of all atoms each time at random, one can modify only those configurations that are already very likely. To do so, one picks an atom and moves it at random throughout the simulation box. Then the change of the configurational energy ΔU due to this (virtual) move is computed. If the move leads to overlapping of atoms, it is clear that the energy would become very high and the move should be rejected with high probability. On the other hand, if the move lowers the energy of the system, it should be accepted. The resulting Metropolis algorithm [21] is hence given by the following set of instructions:

```
if ΔU < 0
    accept move
else
    q = exp(−ΔU/k_B T)
    x = Random[0,1]
    if x < q
        accept move
    else
        reject move
```

      endif
    endif

Here Random [0,1] means a random number at the interval [0,1].

This is the core of the Metropolis scheme but, of course, a lot of additional numerical procedures are needed to set up such an MC simulation. For example, the parameters of the random atom move should be optimized during the simulation in order to increase the efficiency. Periodic boundary conditions are required as in MD simulations. A possible cutoff of the potential has to be corrected for. A difference between MD and MC is that MC does not require calculation of the forces. Furthermore, the "natural" ensemble for MC is the canonical NVT ensemble, (while in MD it is the micro-canonical NVE ensemble). In order to switch to another ensemble in MC, one has to modify the simulation procedure in connection with a modified Boltzmann factor. For example, simulation of the NpT ensemble requires modification of the simulation box at random and an additional term pV in the Boltzmann factor.

An advantage of MC compared to MD simulation is that in MC it is not necessary to follow a real trajectory of an atom. The movement of an atom is in principle random and hence it is possible to overcome high energy barriers that would be impossible to overcome in MD simulations. For example, in MD an atom would not be able to pass through two atoms which are close to each other, because that would require a significant overlap. In MC a jump on the other side of the two atoms is possible. However, even Metropolis sampling is in a number of cases not sufficient to overcome high energy barriers most efficiently. A further improvement of MC for such cases is the so-called *umbrella sampling* introduced by Torrie and Valleau [22]. In this method the Boltzmann factor in the acceptance criterion is extended by a weighting function $w_1(\mathbf{r}^N)$. The resulting transition probability is then

$$\pi(\mathbf{r}^N) = \frac{w_1(\mathbf{r}^N)e^{-\beta U(\mathbf{r}^N)}}{\int w_1(\mathbf{s}^N)e^{-\beta U(\mathbf{s}^N)}\mathrm{d}\mathbf{s}^N} \tag{8.24}$$

The weighting function $w_1(\mathbf{r}^N) > 0$, normalized to unity, is chosen in a way providing the reduction of the energy barrier which the original system has to overcome. In order to calculate the desired properties in the canonical ensemble, it is necessary to eliminate afterwards the effect of the weighting function. The thermodynamic average $\langle X \rangle$ of a quantity X is then found from the relationship

$$\langle X \rangle = \frac{\langle \frac{X}{w_1} \rangle_{w_1}}{\langle \frac{1}{w_1} \rangle_{w_1}}$$

where $\langle \ \rangle_{w_1}$ denotes averaging over the probability distribution w_1.

8.4 Cluster Definitions and Detection Methods

One of the fundamental questions in nucleation is the definition of a cluster. While it intuitively may appear as a simple question, its physical definition turns out to be a difficult task. In general a cluster is defined as a condensed phase, i.e. liquid or solid but due to the large surface to volume ratio one cannot simply apply the criteria for a condensed bulk phase. First of all, the relatively large contribution of the surface compared to the bulk phase affects the properties of the cluster. This is partly related to a difference in the atomic structure of the cluster and the bulk phase. For the definition of a cluster and the development of numerical methods for its detection it is especially important to decide which atoms belong to a cluster and which do not. Here various criteria are possible.

Within *Stillinger's definition* [23] a molecule belongs to the cluster if a separation between one of its atoms and at least one of the atoms of the cluster is smaller than a certain bonding distance r_b. The latter is typically in the range between one and two atomic diameters σ. A very common value is $r_b = 1.5\,\sigma$. This definition is appealing, because the corresponding cluster detection algorithm is relatively simple to implement. On the other hand, there may exist certain drawbacks, depending on the system under study. For example, in the case of a liquid cluster in a metastable equilibrium with a relatively dense vapor phase, there are many vapor-phase atoms surrounding the cluster and forming a corona. The use of the Stillinger criterion can lead to inclusion of many of these atoms into the cluster even though they are not physically bonded. Also atoms passing by the cluster are counted to the cluster for a short period of time.

It is possible to exclude the atoms that pass by the cluster or those, which remain a part of the cluster during a very short time. This can be done by supplementing the Stillinger criterion with a *live-time* criterion [14]. The idea of this method is the following: (i) for each connection between two atoms the duration of this contact is determined; (ii) only those atoms are considered to belong to the cluster for which the separation remains smaller than r_b *during a certain time*, called the *live-time*. To suggest a characteristic value of the live-time, one should calculate the time required for the atom

- to enter the Stillinger sphere of radius r_b,
- collide with other atoms of the cluster and
- leave the Stillinger sphere again.

The knowledge of the average atomic velocity for the given temperature makes it possible to calculate the time required for a regular collision. A typical value for the collision time defined in this way is in the range of 1–2 ps.

Besides solely geometric criteria one can also formulate an energy-based criterion of a cluster. The two competing energies here are the kinetic energy of molecules, related to the motion and separation of the molecules, and the potential energy keeping the atoms together. If the relative kinetic energy is smaller than the potential energy, the

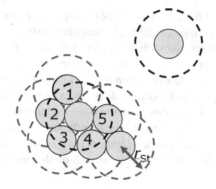

Fig. 8.6 Schematic representation of the tWF cluster detection method

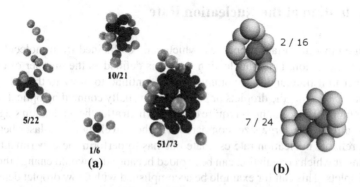

Fig. 8.7 Detection of clusters using the tWF- and Stillinger definitions. **a** *Argon*: there are 4 Stillinger clusters, within each of them dark spheres comprise a tWF-cluster; for example, the 5/22 cluster contains 22 particles according to Stillinger criterion from which only 5 comprise a tWF-cluster. (Reprinted with permission from Ref. [26], copyright (2007), American Institute of Physics). **b** *Zinc*: there are 2 Stillinger clusters, within each of them dark spheres comprise a tWF-cluster (Reprinted with permission from Ref. [17], copyright (2007), American Institute of Physics.)

atoms are likely to be physically bonded in a cluster [24]. Further developments of this approach have been proposed by several authors. Harris and Ford, for example, [25] proposed a cluster criterion that includes the subsequent dynamics to judge whether a molecules belongs to a cluster or not.

Another approach focuses on the number of nearest neighbors of an atom, N_1, called also *the coordination number*, discussed in Chap. 7). Depending on the state conditions, N_1 in the liquid phase lies in the range $N_1 \approx 4 \sim 8$ atoms. Based on this knowledge, ten Wolde and Frenkel (tWF) [27] defined a cluster as a set of atoms having at least 5 nearest neighbors shown in Fig. 8.6. In the analysis of a simulation system an algorithm, similar to the calculation of a pair correlation function, can be used. This approach eliminates the corona of the cluster and is, thus, useful to detect clusters of loosely bonded atoms or molecules. An example is the argon system: argon atoms are not strongly bonded at the liquid state due to weak van der Waals interactions. In this case the nearest-neighbor definition of a cluster is suitable for a proper analysis of the simulation data. The snapshots referring to argon nucleation depicted in Fig. 8.7a clearly show that the tWF definition of a cluster is more suitable. However, in case of a strongly bonded system, such as a metallic system, the tWF criterion leads to an underestimation of the cluster size. In metallic systems the atoms are strongly bonded and a corona does not exist as in the case of zinc shown in Fig. 8.7b. If the nearest-neighbor criterion is applied to such a system, the surface shell is removed from the cluster simply because the surface atoms do not necessarily have the required number of nearest neighbors [14]. In summary, the proper cluster definition depends on the specific system, as well on the properties calculated from the simulation system.

8.5 Evaluation of the Nucleation Rate

One of the most important properties, which can be obtained from molecular simulation of nucleation, is the nucleation rate. It is defined as the number of clusters formed per unit time and unit volume, which continue to grow to the stable bulk phase. In *experiments* the droplets or particles are usually counted by optical or scattering methods. These methods require droplets which after the nucleation stage have grown to a relatively large size, comparable to the wavelength of a laser beam. To obtain a reliable nucleation rate estimate one has to perform the experiment under conditions at which coagulation can be avoided because that would change the number of droplets. This can for example be accomplished with a low droplet density. In *molecular simulation* the determination of the nucleation rate depends strongly on the chosen simulation method.

8.5.1 Nucleation Barrier from MC Simulations

The key quantity determining the nucleation behavior of a substance is the free energy of an n-cluster formation, $\Delta G(n)$. Various theoretical models, discussed in this book, invoke various approximations to derive this quantity; the most widely used one— is the capillarity approximation of the classical nucleation theory. Calculation of the free energy of cluster formation in Monte Carlo simulations, pioneered by Lee et al. [28], is based on the analysis of cluster statistics, emerging in simulations, without referring to a particular model for $\Delta G(n)$. Below we follow the procedure outlined by Reiss and Bowles [29]. N molecules of the NVT-system can be grouped in various clusters. We will consider the system configuration containing *exactly* N_n clusters with n particles. Each n-cluster generates and exclusion volume v_n which is unaccessible for other $N - n$ molecules of the system. Using the assumption of non-interacting clusters (which is usually a good approximation for vapor–liquid nucleation), we present the partition function of the NVT-system as

$$Z(N, V, T) = \overline{Z}(N - nN_n, V - v_nN_n, T)\, Z^{(n)} \tag{8.25}$$

where $\overline{Z}$ is the partition function of the vapor which is *forbidden* to contain n-clusters—all such clusters are contained in the partition function $Z^{(n)}$ of the *gas* of n-clusters, given by Eq. (7.4):

$$Z^{(n)} = \frac{1}{N_n!}\, Z_n^{N_n} \tag{8.26}$$

Here $Z_n(n, V, T)$ is the partition function of one n-cluster in the volume V (see Eq.(7.2)). Note, that Z_n depends on the size of the system through the translational degree of freedom of the center of mass of the cluster: $Z_n(n, V, T) \sim V/\Lambda^3$. From (8.25)–(8.26) using Stirling's formula we have

$$\ln Z = N_n \ln Z_n - N_n \ln(N_n/e) + \ln \overline{Z}(N - nN_n, V - v_nN_n, T)$$

For each cluster size n a variety of N_n is possible; we will be interested in the most probable value. The latter maximizes $\ln Z$ with respect to N_n:

$$\left(\frac{\partial \ln Z}{\partial N_n}\right)_{N,V} = \ln\left(\frac{Z_n}{N_n}\right) + \left(\frac{\partial \ln \overline{Z}}{\partial N_n}\right)_{N,V} = 0 \tag{8.27}$$

Consider the second term of this equation

$$\left(\frac{\partial \ln \overline{Z}}{\partial N_n}\right)_{N,V} = -n\left(\frac{\partial \ln \overline{Z}}{\partial (N - N_n)}\right)_{N,V,V-v_nN_n} - v_n\left(\frac{\partial \ln \overline{Z}}{\partial (V - v_nN_n)}\right)_{N,V,N-nN_n}$$

Using the standard thermodynamic relationships we write this expression as

$$\left(\frac{\partial \ln \overline{Z}}{\partial N_n}\right)_{N,V} = \frac{n\,\overline{\mu^v}}{k_B T} - \frac{\overline{p^v}\, v_n}{k_B T} \tag{8.28}$$

where $\overline{\mu^v}$ and $\overline{p^v}$ are the chemical potential and pressure of the vapor of volume $V - v_nN_n$ containing $N - nN_n$ molecules. In general, $\overline{\mu^v} \neq \mu^v$ and $\overline{p^v} \neq p^v$ (μ^v is the chemical potential of a molecule in the supersaturated vapor and p^v is the vapor pressure), however for *rare* clusters (recall the assumption of noninteracting clusters) the difference between the barred and non-barred quantities is negligible. Therefore, Eqs. (8.27) and (8.28) yield:

$$N_n = Z_n(n, V, T)\, e^{-\beta(p^v v_n - n\mu^v)} \tag{8.29}$$

The ratio $Z_n(n, V, T)/V$ does not depend on V and remains constant in thermodynamic limit. This means, that if we chose another volume of the system V', we would have

$$Z_n(n, V, T)/V = Z_n(n, V', T)/V'$$

It is convenient to choose $V' = V/N = v^v$, where v^v is the volume per molecule in the vapor. Then, Eq. (8.29) takes the form

$$N_n = N Z_n(n, v^v, T) e^{-\beta(p^v v_n - n\mu^v)} \qquad (8.30)$$

Introducing the Helmholtz free energy of the n-cluster, *confined* to the volume v^v

$$F_n(n, v^v, T) = -k_B T \ln Z_n(n, v^v, T)$$

we define the *cluster size probability*

$$P(n) \equiv \frac{N_n}{N} = \exp\left\{-\beta \left[(F_n(n, v^v, T) + p^v v_n) - n\mu^v\right]\right\} \qquad (8.31)$$

The expression in the square brackets is

$$(F_n + p^v v_n) - n\mu^v = G(n) - G(n)_{bulk}$$

where $G(n)_{bulk} = n\mu^v$ is the Gibbs free energy of n molecules in the bulk super-saturated vapor *prior* to the formation of the cluster; $G(n)$ is the same quantity *after* the n-cluster was formed. Hence, $\Delta G(n) = G(n) - G(n)_{bulk}$ is the "intensive Gibbs free energy" of n-cluster formation and

$$P(n) = e^{-\beta \Delta G(n)}$$

From these considerations we can schematically interpret the process of cluster formation as consisting of two steps [29]:

1. n molecules are picked up anywhere in volume V of the system and gathered in the volume equal to the molecular volume in the vapor phase v^v;
2. within the volume v^v the cluster is formed with the volume $v_n < v^v$.

Thus, measuring the cluster-size probability distribution $P(n)$ in MC simulation, we can determine the free energy of cluster formation $\Delta G(n)$ from the relationship:

$$\beta \Delta G(n) = -\ln P(n) \qquad (8.32)$$

Its maximum gives the anticipated nucleation barrier ΔG^*. This barrier can be compared to the corresponding quantity resulting from nucleation theory. From Eq. (8.32) one can determine the nucleation barrier, but not the kinetic prefactor, which determines the flux over this barrier and can be obtained from MD simulations.

In principle, Eq. (8.32) opens a possibility of calculating the nucleation barrier by simulating the metastable vapor and counting clusters of various sizes. The total number of particles used in modern MC simulations is of the order of $N \sim 10^5 - 10^6$.

With this number of particles one can detect clusters from reliable statistics when $\Delta G(n) < 10\, k_B T$. Those are small n-mers, having the energy of formation of several $k_B T$. Only for extremely high supersaturations, close to pseudo-spinodal (discussed in Chap. 9) will the height of the nucleation barrier be in this range (such high S and therefore high nucleation rates are realized, e.g., in the supersonic Laval nozzle, where J is in the range of $10^{16} - 10^{18}\, \mathrm{cm}^{-3}\, \mathrm{s}^{-1}$ [30]). For moderate supersaturations nucleation barriers are in the range of $\sim 40 - 60\, k_B T$ and thus the clusters formed in simulations are much smaller than the critical cluster. E.g. for $\Delta G^* = 53\, k_B T$ the chance to find a critical cluster is $P = e^{-53} \approx 10^{-23}$, which means that the simulated system should contain $> 10^{23}$ particles which is equal to the Avogadro number. The previously discussed umbrella sampling technique makes it possible to overcome this difficulty. For a given n-cluster we introduce a weighting function $w_1(\mathbf{r}^n)$ which according to Eq. (8.24) replaces the internal energy of the cluster $U(\mathbf{r}^n)$ by

$$U'(\mathbf{r}^n) = U(\mathbf{r}^n) + W_1(\mathbf{r}^n)$$

where the biasing potential W_1 is

$$W_1(\mathbf{r}^n) = -k_B T\, \ln w_1(\mathbf{r}^n)$$

The simplest form of W_1 is a harmonic function [31]:

$$W_1 = \frac{1}{2} k_n\, (n - n_0)^2, \quad k_n > 0 \tag{8.33}$$

It ensures that the formation of n-clusters with the sizes outside a certain range (characterized by k_n) around n_0 becomes highly improbable: the probability of finding the n-cluster becomes proportional to

$$w_1 = \exp\left[-\beta \frac{1}{2} k_n\, (n - n_0)^2\right]$$

This function forms a Gaussian *umbrella* in the space of cluster sizes, centered at n_0. Only those clusters, which find themselves under this umbrella, will be sampled. Thus, introduction of the biasing potential opens a "window" of the cluster sizes, located at n_0, with a width of k_n, which are sampled in simulations. By changing n_0 one "opens consecutive windows" performing simulation runs within the windows, thereby consecutively scanning the cluster size space.

8.5.2 Nucleation Rate from MD Simulations

For MD simulations the nucleation process is actually mapped on a simulation system. From the time resolved simulation of the formation of a certain amount of clusters one can draw a time dependent cluster statistics. This cluster statistics can then be analyzed yielding the nucleation rate (besides other properties). For such an analysis of the cluster statistics, several methods have been proposed in the literature. Here we focus on two approaches which are commonly used in simulation studies on nucleation.

8.5.2.1 Yasuoka-Matsumoto Method

The method of Yasuoka and Matsumoto [32], which also can be called the *threshold method*, requires MD simulation system large enough to generate a significant amount of clusters. This method can be obtained from the continuity equation in the space of cluster sizes:

$$\frac{\partial}{\partial t}N(n, t) = -\frac{\partial}{\partial n}j(n, t) \tag{8.34}$$

Here $N(n, t)$ is the number of clusters of size n in the simulation box at time t and $j(n, t)$ is the rate of the formation of clusters of size n in the box. In the steady state $\frac{\partial}{\partial t}N(n, t) = 0$ yielding

$$\frac{\partial}{\partial n}j(n, t) = 0$$

showing that $j(n, t)$ is constant. If V is the volume of the simulation box, then the steady-state nucleation rate is given by $J = j(n, t)/V$. In order to determine J from the cluster statistics let us choose a certain, *threshold*, cluster size n_{thres} and integrate (8.34) over n from n_{thres} to ∞:

$$\frac{\partial}{\partial t}\int_{n_{\text{thres}}}^{\infty} N(n', t)dn' = -\int_{n_{\text{thres}}}^{\infty} \frac{\partial}{\partial n'}j(n', t)dn' \tag{8.35}$$

The integral on the left-hand side is the total number of clusters with sizes *larger than* n_{thres}:

$$N_{\Sigma}(n_{\text{thres}}, t) = \int_{n_{\text{thres}}}^{\infty} N(n', t)\,dn'$$

while the right-hand side is

$$j(n_{\text{thres}}, t) - j(\infty, t) = j(n_{\text{thres}}, t)$$

where we took into account that the rate of formation of infinitely large clusters is zero. Thus, Eq. (8.35) becomes

$$\frac{\partial}{\partial t}N_{\Sigma}(n_{\text{thres}}, t) = j(n_{\text{thres}}, t)$$

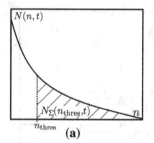
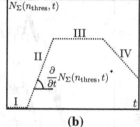

(a)   (b)

Fig. 8.8 Derivation of nucleation rate by means of the threshold method. **a** Cluster size distribution at time t: $N(n, t)$; the shaded area under the curve gives the total number of clusters $N_\Sigma(n_{\text{thres}}, t)$ larger than n_{thres} at time t. **b** Time evolution of $N_\Sigma(n_{\text{thres}}, t)$. The slope of the domain II $\frac{\partial}{\partial t} N_\Sigma(n_{\text{thres}}, t)$ is proportional to the nucleation rate

yielding for the nucleation rate:

$$J = \frac{1}{V} \frac{\partial}{\partial t} N_\Sigma(n_{\text{thres}}, t) \tag{8.36}$$

In the threshold method the number of clusters larger than a threshold value, $N_\Sigma(n_{\text{thres}}, t)$, is plotted as a function of the simulation time for different values of n_{thres}. As a result one obtains curves with four domains schematically depicted in Fig. 8.8.

The slope of the linear domain II is $\frac{\partial}{\partial t} N_\Sigma(n_{\text{thres}}, t)$. Hence, the nucleation rate is this value, divided by the simulation box volume V. The plateau-like domain III and also the following descending part IV of the curve result from the finite size of the system. The steady state situation is only possible as long as sufficient number of monomers is present in the box to deliver the clusters of size n_{thres}. At some point, due to depletion of the vapor, the monomer concentration becomes too small to provide further nucleation and at the same time clusters grow by collision. This leads to stagnation and then decreasing of the number of clusters. Therefore, from the analysis of the plateau domain of the data one can not derive the nucleation properties. In practice one finds that linear parts of the curves are not necessarily all parallel. One may use this fact for a rough estimate of the critical cluster size by calculating the curve for each single threshold value and detect the threshold value of the cluster size beyond which the slope of domain II does not change any more. Application of Yasuoka-Matsumoto method to vapor-liquid nucleation of zinc is shown in Fig. 8.9. Though the critical cluster is not known a priori, the threshold method can safely be applied. By choosing various threshold values it is possible to detect the linear domain of steady-state nucleation. If $n_{\text{thres}} > n^*$, each simulation curve exhibits a linear domain where all $N_\Sigma(t)$ lines are parallel (cf. Fig. 8.9). The average of the slopes in the linear domain can be used to calculate the nucleation rate.

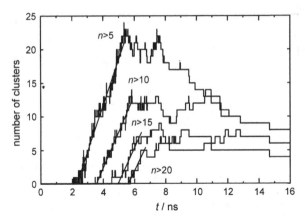

Fig. 8.9 Number of clusters larger than a certain size, indicated above each curve, for vapor-liquid nucleation of zinc (for explanation see Fig. 8.8). The plot is used to derive the nucleation rate by means of the threshold method. Zinc vapor density is $0.0315\,\text{mol/dm}^3$, temperature $T = 400\,\text{K}$. The resulting nucleation rate is $J = 25.5 \times 10^{28}\,\text{dm}^{-3}\text{s}^{-1}$. (Reprinted with permission from Ref. [17], copyright (2007), American Institute of Physics.)

8.5.2.2 Mean First-Passage Time Method

The Mean First-Passage Time method (MFPT) provides an instruction to analyze the stochastic dynamics of the nucleation process. In contrast to the threshold method, in the MFPT a relatively small simulation system is sufficient to obtain the nucleation rate. However, to get good statistics a large number of simulations is required. The stochastic dynamics of a system with an activation barrier is governed by the Fokker-Planck equation, describing the evolution of the cluster distribution function $\rho(n, t)$ caused by diffusion and drift in the space of cluster sizes, discussed in Sect. 3.7 (see Eqs. (3.79)–(3.81)):

$$\frac{\partial \rho(n, t)}{\partial t} = \frac{\partial}{\partial n}\left[B(n)\,\frac{\partial \rho}{\partial n} + \rho\,\frac{B(n)}{k_B T}\,\frac{\partial \Delta G(n)}{\partial n}\right] \tag{8.37}$$

Here $B(n)$ is a diffusion coefficient in the space of cluster sizes. In Fig. 8.10 the work of cluster formation $\Delta G(n)$ is plotted versus the cluster size. The solution of the Fokker-Planck equation requires boundary conditions. In case of nucleation the left boundary n_a is the monomer $n_a = 1$. It is called a *reflecting boundary*, since there are no clusters smaller than a monomer. The right boundary n_b is a size large enough so that the cluster $> n_b$ has a negligibly small chance to evaporate. In view of this feature, n_b is called an *absorbing boundary*.

Let us fix $n_a = 1$ and an initial value n_0 and follow the time evolution of the system for various values of the absorbing boundary n_b. For each n_b we can identify the *mean first passage time* $\tau(n_b)$ which is the average time necessary for the system, starting at n_0, to leave the domain of cluster sizes (n_a, n_b) for the first time. Clearly,

Fig. 8.10 Gibbs free energy
of a n-cluster formation.
n_0 is the initial value of n
(a starting point for MFPT
analysis) located between
the reflecting—n_a—and
absorbing—n_b—boundaries;
n^* is the critical cluster

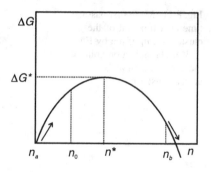

$\tau(n_b = n^*) = \tau^*$ is the average time necessary to reach the critical size. Since the
nucleation rate is the flux through the critical cluster, we may write

$$J = \frac{1}{2} \frac{1}{\tau^* V} \tag{8.38}$$

where V is the volume of the simulation box; the factor $1/2$ stands for the fact that
the critical cluster, corresponding to the maximum of $\Delta G(n)$, has a 50 % chance
of either growing to the new bulk phase or decaying, i.e. evaporating. Solving the
Fokker-Planck equation (8.37), Wedekind et al. [33] showed that for reasonably high
nucleation barriers the behavior of $\tau(n_b)$ in the vicinity of the critical size can be
approximated by the function

$$\tau(n_b) = \frac{\tau_J}{2} \left[1 + \text{erf} \left(c \left(n_b - n^*\right)\right)\right] \tag{8.39}$$

shown schematically in Fig. 8.11. Here

$$\text{erf}(x) = \frac{2}{\pi} \int_0^x e^{-x^2} dx$$

is the error function;

$$\tau_J = 1/(J\,V)$$

and c is the inverse width of the critical region given by Eq. (3.49):

$$c = 1/\Delta = \sqrt{\pi}\,\mathscr{Z}$$

The critical cluster n^* is the inflection point of $\tau(n_b)$. Note, that $\tau_J = 2\tau^*$.

In practice the MFPT method is applied to MD simulation in the following way: the
time for the system to pass for the first time a certain cluster size n_b is averaged over a
large number of simulation runs. The procedure is repeated for various values of n_b.
Then, the resulting function $\tau^{MD}(n_b)$ is fitted to the 3-parametric expression (8.39),

Fig. 8.11 Mean first passage
time as a function of the
cluster size n_b given by Eq.
(8.39). The inflection point of
the curve corresponds to the
critical cluster n^*

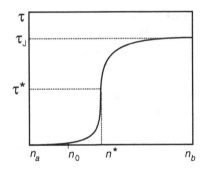

providing the 3 fit parameters: τ_J, n^* and c. The nucleation rate being the reciprocal
of τ_J reads:

$$J = \frac{1}{\tau_J V}$$

An example of the MFPT curve obtained from simulation of zinc vapor–liquid nucleation is shown in Fig. 8.12. Figures 8.10 and 8.11 demonstrate an advantage of the
MFPT method: one can recognize whether or not the nucleation process is coupled
to growth. If only the nucleation process takes place in the system, the simulation
data reach a plateau prescribed by the error function. Deviation from this shape is
related to the influence of particle growth on nucleation. In case of argon modelled
by the Lennard-Jones potential and small system sizes of approximately 300 atoms
it is possible to perform hundreds of simulations for averaging [33]. For larger systems with more complex interaction potentials the results have to be obtained from
a small number of simulation runs: e.g., for zinc nucleation (see Fig. 8.12) MFPT
results were obtained from 10 simulation runs.

While there are other methods to determine the nucleation rate from MD simulations, the two methods described in this chapter are most frequently employed. Both
methods have advantages and drawbacks. The threshold method requires a *large
simulation system* in order to provide a significant amount of clusters to obtain good
statistics. If the systems are large enough possible depletion effects can be minimized.
The MFPT method requires a *large number of simulation runs*. In order to reach such
a large number each run should be sufficiently fast. This can be accomplished by ter-
minating a simulation at the point when the chosen cluster size is passed for the first
time. Another method for optimization is to choose a relatively small system since
only clusters not larger than few times n^* are required for MFPT. On the other hand,
the smaller a simulation system is, the more important become finite sized effects.

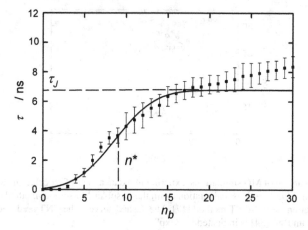

Fig. 8.12 MFPT analysis of a series of simulation runs for zinc at $T = 800\,\text{K}$ and $\log_{10} S = 2.79$. The critical cluster size, indicated by the *dashed vertical line*, is $n^* = 9$. Because of the large system size only 10 simulations were performed (Reprinted with permission from Ref. [17], copyright (2007), American Institute of Physics.)

8.6 Comparison of Simulation with Experiment

MD simulation is limited by the available computational power. This means that the system size as well as the duration of simulation is limited. In order to observe nucleation within these limits, the supersaturation has to be sufficiently high yielding high nucleation rates which for MD are typically in the range of 10^{25} to $10^{30}\,\text{cm}^{-3}\text{s}^{-1}$. Meanwhile, supersaturations which can be realized in nucleation experiments are usually lower resulting in lower nucleation rates: in an expansion chamber J ranges from 10^5 to $10^{10}\,\text{cm}^{-3}\text{s}^{-1}$; nozzle experiments make it possible to reach higher supersaturations leading to nucleation rates in the order of $10^{18} - 10^{20}\,\text{cm}^{-3}\,\text{s}^{-1}$. These values are, however, still lower than the typical MD values.[1]

That is why usually a direct comparison of MD results with experimental data can not been done in a straightforward way. Nevertheless, some important conclusions can be drawn. To illustrate this statement we show in Fig. 8.13 the results for zinc vapor-liquid nucleation obtained by MD simulation, using both the threshold and the MFPT methods [17], along with CNT predictions and available experimental data. Both MD methods are close to each other within the range of their accuracy. At the same time MD data and experiment lead to nucleation rates which are considerably higher than the CNT predictions.

---

[1] It must be noted that performing extensive simulations with up-to-date computers allows to approach the region of supersonic nozzle experiments.

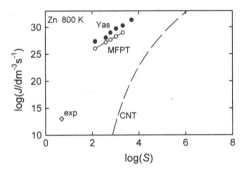

Fig. 8.13 Comparison of MD simulations, experimental data and CNT calculations for zinc vapor-liquid nucleation. *Filled symbols*: calculations with the Yasuoka-Matsumoto method ("Yas"), *open symbols*: calculations with MFPT method [17]. The dashed curve is the CNT prediction. An experimental point from Ref. [34] is indicated by "exp"

8.7 Simulation of Binary Nucleation

Most molecular simulations of nucleation have been performed for pure substances; much fewer computational studies have been devoted to *binary systems*.

MD simulation of nucleation in the binary vapor of iron and platinum performed in Ref. [35] revealed that besides the mixed clusters, in which both components are present, there are also single-component clusters containing pure iron and pure platinum. Due to the difference in attraction strength between the two substances, one observes a large amount of big platinum clusters compared to a relatively small number of small iron clusters. The iron atoms have a weaker attraction compared to platinum and either do not condense on hot platinum cluster or rapidly evaporate after condensation.

Compared to metals, the binary mixture of n-nonane and methane is characterized by much weaker van der Waals interactions. Nucleation in this mixture was studied in MD simulations of Braun [36]. The peculiar feature of this system is a very low vapor molar fraction of nonane: $y_{nonane} \approx 10^{-4}$; meanwhile it is nonane that ensures nucleation in the system. In order to tackle this problem, a simulation box containing around 10^5 methane molecules was chosen and expansion simulation method was used. From the bulk phase behavior one would expect a mole fraction of methane in the liquid-like clusters to be around 0.2–0.4 at the given nucleation conditions (high pressure and ambient temperature). At the same time simulation results show that even for weakly interacting van der Waals systems the critical clusters are very different from what one would expect from the bulk equilibrium. Clusters have a peculiar structure resulting from minimization of the surface energy. The system tends to lower its energy by phase separation and moving the more volatile component (methane) towards the shell region of the cluster.

In MC simulation of binary nucleation the free energy of formation of the (n_a, n_b)-cluster, containing n_a molecules of component a and n_b molecules of component b, in a supersaturated vapor can be found from the cluster statistics accumulated in simulation runs:

$$\beta \Delta G(n_a, n_b) = -\ln\left(\frac{N_{n_a n_b}}{N}\right)$$

where $N_{n_a n_b}$ is the number of (n_a, n_b)-clusters, N is the number of molecules in the system. This expression is a generalization of the corresponding result (8.32) for the unary case [37].

Kusaka et al. [38] performed simulations of nucleation in water-sulfuric acid system which plays an important role in atmospheric processes. Simulations employed water, hydronium ion, sulfuric acid and the bisulfate ion as species modeled by force field combining the Lennard-Jones potential with electrostatic potential by means of partial charges. From MC simulations the free energy of cluster formation was analyzed along with the cluster structures. It was found that the shape of most of the clusters differs from the spherical one. Different conformations of the clusters turn out to be very close in energy and have a fairly long lifetime. This observation leads to a conclusion that various clusters contribute to the nucleation rate.

Chen et al. [39] performed MC simulation of the nucleation in binary water/ethanol systems. They found that ethanol is enriched in the cluster surface leading to a lower surface tension than that of the ethanol/water mixture of the given mole fraction. They argue that the shortcomings of the CNT in such cases might be related to this surface enrichment.

8.8 Simulation of Heterogeneous Nucleation

Heterogeneous nucleation requires a surface at which the supersaturated vapor can nucleate. In general such surface can be implemented in molecular simulations by adding solid particles in the supersaturated vapor or by setting up a solid film in the middle of the simulation box. The heterogeneous surface lowers the activation barrier for nucleation. Therefore, the supersaturation that is necessary to observe nucleation on the time scale of MD simulations is lower than for homogeneous nucleation. The extent of this effect depends on the attraction between the nucleating substance and the substrate. This attraction strength is in turn related to the wetting behavior of a liquid on a substrate. The stronger the attraction—the smaller the contact angle of the droplet on the surface. On the other hand, for a completely repelling surface one would expect the contact angle around 180° recovering homogeneous nucleation.

Toxvaerd [40] investigated heterogenous nucleation by MD simulations using a r^{-9} potential for the wall interactions of the molecules. Depending on the interaction between the wall and the molecules, he observed either nucleation at the surface for weak attraction or prewetting, i.e. the formation of a molecular layer on the surface for strong interactions. In this investigation the wall was perfectly flat, i.e. it did not contain inhomogeneities. Kimura and Maruyama [41] instead used a surface composed by harmonically vibrating molecules coupled to a heat bath. They investigated the nucleation of argon vapor for various vapor phase temperatures and pressures. The nucleation rate was analyzed using the Yasuoka-Matsumoto threshold method. Simulation results showed good agreement with the classical heterogenous nucleation theory—the Fletcher model, discussed in Sect. 15.1.

In Ref. [42] the nucleation of argon on a polyethylene substrate was investigated. In this work the polyethylene film in the center of the simulation box was coupled to a thermostat. All latent heat of condensation was hence withdrawn from the system via the substrate. In this way the process of heterogeneous nucleation can be modelled realistically. In the transient stage of nucleation a temperature gradient develops, which after condensation is complete vanishes again. Depending on the supersaturation of the argon vapor, different types of growth take place. Comparison with the classical heterogeneous nucleation theory exhibits good agreement.

8.9 Nucleation Simulation with the Ising Model

The simplest possible model of interacting particles is the Ising model. It consists of a lattice of spins that can have two values: either $s = +1$ or $s = -1$. Each spin interacts with its nearest neighbors. Their number N_1 depends on the type of the lattice. In the simplest case of a 2D square lattice $N_1 = 4$, while for a 3D cubic lattice $N_1 = 6$. Interaction between spins is given by the coupling constant K. If K is positive the model describes ferromagnetic behavior favoring the alignment of neighboring spins parallel to each other, while a negative K mimics antiferromagnetic behavior with the preference for the anti-parallel alignment of neighboring spins. The Hamiltonian of the Ising model reads:

$$\mathcal{H} = -K \sum_{(i,j)_{nn}} s_i s_j - H \sum_k s_k \tag{8.40}$$

Here H is the magnitude of an external magnetic field. In Monte Carlo simulations of the Ising model a lattice site is chosen at random and the spin at this site is flipped while all other spins of the system remain unchanged. The energy difference ΔU resulting from this spin flip is calculated. To judge whether or not the spin flip (which in molecular MC terminology represents a MC move) can be accepted the Metropolis algorithm described above is employed.

The Ising model, which has been developed for investigation of magnetism, can be also used as a model of a fluid. A typical intermolecular interaction potential in a fluid is characterized by a strong repulsion at short distances, a potential well, and a relatively fast decaying attractive tail (cf. the Lennard-Jones potential). To a good approximation such a potential can be approximated by a square well. In turn a fluid with a square well potential can be mapped on a *lattice-gas model*.

Within this model molecules are only allowed to occupy the sites of a regular lattice instead of continuous distribution in space. This requirement mimics a short-range repulsion in a real fluid: molecules can not be closer than the lattice spacing. Attractive interactions, i.e. the attractive well, is modelled by a nearest-neighbor potential ϵ, so that the potential energy of a certain configuration takes the form

$$U = -\varepsilon \sum_{(i,j)_{nn}} \rho_i \rho_j$$

where $\rho_k = 1$ if the site k is occupied and $\rho_k = 0$ in the opposite case. Let us set

$$s_i = 2\rho_i - 1$$

Then, in the Ising model $s_i = -1$ if the site i in the lattice gas model is free and $s_i = +1$ if it is occupied. This transformation thus maps the fully occupied lattice of spins $s_i = \pm 1$ on the partially occupied lattice of fluid molecules. The two systems—lattice gas and Ising model—become thermodynamically equivalent, i.e. their partition functions are the same, if we set [43]:

$$K = \varepsilon/4, \quad H = (2\mu + N_1 \varepsilon)/4 \tag{8.41}$$

Thus, the Ising model can be used for simulation of a vapor–liquid system and hence also for vapor–liquid nucleation. A metastable state of the spin system can be achieved by varying the external field H which from (8.41) is equivalent to varying of the chemical potential of a fluid molecule resulting in a supersaturation.

Glauber [44] was the first to study the kinetics of the 1D Ising model which is solvable analytically, but does not exhibit a phase separation. Stoll et al. [45] performed Monte Carlo simulations of the 2D spin-flip Ising model and analyzed its relaxation towards equilibrium. Simulations demonstrated consistency with the dynamic scaling hypothesis. Stauffer et al. [46] analyzed nucleation in 3D Ising lattice gas by Monte Carlo simulations. They observed that the results obtained from the lattice model are roughly in agreement with CNT.

The Ising model is very useful for the investigation of fundamental concepts of nucleation. It can be employed to the analysis of the scaling behavior expressed in the form of power laws. Since the pioneering works [45, 46] the model was employed for various other systems—e.g. for heterogeneous nucleation [47], to mention just one example. Meanwhile, it must be noted that simulation of real substances goes beyond the scaling behavior and requires explicit force fields acting between the molecules.

References

1. L.D. Landau, E.M. Lifshitz, *Mechanics* (Butterworth-Heinemann, Oxford, 1976)
2. L. Verlet, Phys. Rev. **159**, 98 (1967)
3. W.C. Swope, H.C. Andersen, P.H. Berens, K.R. Wilson, J. Chem. Phys. **76**, 637 (1982)
4. M.P. Allen, D.J. Tildesley, *Computer Simulation of Liquids* (Oxford University Press, Oxford, 1989)
5. D. Frenkel, B. Smit, *Understanding Molecular Simulaton* (Academic Press, London, 1996)
6. A.Z. Panagiotopoulos, Int. J. Thermophys. **15**(15), 1057 (1994)
7. K. Binder, Phys. Rev. A **29**, 341 (1984)
8. V.I. Kalikmanov, *Statistical Physics of Fluids. Basic Concepts and Applications* (Springer, Berlin, 2001)
9. P. Erhart, K. Albe, Appl. Surf. Sci. **226**, 12 (2004)
10. H.C. Andersen, J. Chem. Phys. **72**, 2384 (1980)
11. S. Nose, Mol. Phys. **52**, 255 (1984)
12. W.G. Hoover, Phys. Rev. A **31**, 1695 (1985)
13. G.J. Martyna, M.L. Klein, M. Tuckerman, J. Chem. Phys. **97**, 3625 (1992)
14. R. Römer, T. Kraska, J. Phys. Chem. C **113**, 19028 (2009)
15. J. Westergren, H. Grönbeck, S.-G. Kim, D. Tomanek, J. Chem. Phys. **107**, 3071 (1997)
16. S. Braun, F. Römer, T. Kraska, J. Chem. Phys. **131**, 064308 (2009)
17. R. Römer, T. Kraska, J. Chem. Phys. **127**, 234509 (2007)
18. M. Türk, J. Supercrit. Fluids **15**, 79 (1999)
19. R. Span, W. Wagner, J. Phys. Chem. Ref. Data **8**, 1509 (1996)
20. G.A. Korn, T.M. Korn, *Mathematical Handbook* (McGraw-Hill, New York, 1968)
21. N. Metropolis, A.W. Rosenbluth, M.N. Rosenbluth, A.H. Teller, E. Teller, J. Chem. Phys. **21**, 1087 (1953)
22. G.M. Torrie, J.P. Valleau, Chem. Phys. Lett. **28**, 578 (1974)
23. F.H. Stillinger, J. Chem. Phys. **38**, 1486 (1963)
24. T.L. Hill, *Statistical Mechanics: Principles and Selected Applications* (McGraw-Hill, New York, 1956)
25. S.A. Harris, I.J. Ford, J. Chem. Phys. **118**, 9216 (2003)
26. J. Wedekind, J. Wölk, D. Reguera, R. Strey, J. Chem. Phys. **127**, 154516 (2007)
27. P.R. ten Wolde, D. Frenkel, J. Chem. Phys. **109**, 9901 (1998)
28. J.K. Lee, J.A. Baker, F.F. Abraham, J. Chem. Phys. **58**, 3166 (1973)
29. H. Reiss, R. Bowles, J. Chem. Phys. **111**, 7501 (1999)
30. S. Sinha, A. Bhabbe, H. Laksmono, J. Wölk, R. Strey, B. Wyslouzil, J. Chem. Phys. **132**, 064304 (2010)
31. P.R. ten Wolde, D. Oxtoby, D. Frenkel, J. Chem. Phys. **111**, 4762 (1999)
32. K. Yasuoka, M. Matsumoto, J. Chem. Phys. **109**, 8451 (1998)
33. J. Wedekind, R Strey, D. Reguera, J. Chem. Phys. **126**, 134103 (2007)
34. A.A. Onischuk, P.A. Purtov, A.M. Baklanov, V.V. Karasev, S.V. Vosel, J. Chem. Phys. **124**, 014506 (2006)
35. N. Lümmen, T. Kraska, Nanotechnology **15**, 525 (2004)
36. S. Braun, T. Kraska, J. Chem. Phys. **136**, 214506 (2012)
37. S. Yoo, K.J. Oh, X.C. Zeng, J. Chem. Phys. **115**, 8518 (2001)
38. I. Kusaka, Z.-G. Wang, J.H. Seinfeld, J. Chem. Phys. **108**, 6829 (1998)
39. B. Chen, J.I. Siepmann, M.L. Klein, J. Am. Chem. Soc. **125**, 3113 (2003)
40. S. Toxvaerd, J. Chem. Phys. **117**, 10303 (2002)
41. T. Kimura, S. Maruyama, Microscale Thermophys. Eng. **6**, 3 (2002)
42. R. Rozas, T. Kraska, J. Phys. Chem. C **111**, 15784 (2007)
43. R.J. Baxter, *Exactly Solved Models in Statistical Mechanics* (Academics Press, London, 1982)
44. R.J. Glauber, J. Math. Phys. **4**, 294 (1963)
45. E. Stoll, K. Binder, T. Schneider, Phys. Rev. B **8**, 3266 (1973)
46. D. Stauffer, A. Coniglio, D.W. Heermann, Phys. Rev. Lett. **49**, 1299 (1982)
47. D. Winter, P. Virnau, K. Binder, J. Phys. Condens. Matter **21**, 464118 (2009)

Chapter 9
Nucleation at High Supersaturations

9.1 Introduction

At high supersaturations (deep quenches) the system from being metastable becomes unstable; in the theory of phase transitions the boundary between the metastable and unstable regions is given by a thermodynamic spinodal being a locus of points corresponding to a divergent compressibility. Rigorously speaking the transition from metastable to unstable states does not reduce to a sharp line but rather represents a region of a certain width which depends on the range of interparticle interactions [1]. Within the spinodal region the fluid becomes unstable giving rise to the phenomenon of *spinodal decomposition* [2], characterized by vanishing of the free energy barrier of cluster formation at some finite value of the supersaturation. The classical theory does not signal the spinodal: the nucleation barrier decreases with S but remains finite for all values of S (see Eq. (3.28)). Therefore, nucleation in the spinodal region can not be described by CNT and a more general formalism is needed.

Such a formalism, the *field theoretical approach*, was pioneered by Cahn and Hilliard [3] and developed by Langer [4, 5], Klein and Unger [6, 7]. It is based on the mean-field Ginzburg–Landau theory of phase transitions. Cahn-Hilliard's approach (usually termed a "gradient theory of nucleation") leads to the existence of a well-defined mean-field spinodal characterized by a supersaturation S_{sp}. A mean-field theory becomes asymptotically accurate in the limit of infinite-range intermolecular interactions, hence a spinodal *line* exists in the same limit. At the spinodal the barrier vanishes which means that the capillary forces can no longer sustain the compact form of a droplet, clusters in the vicinity of a spinodal are ramified fractal objects [6, 7].

In this chapter we formulate the mean-field (Cahn-Hilliard) gradient theory considering nucleation at high supersaturations. Special attention in this domain should be paid to the role of fluctuations giving rise to the concept of *pseudospinodal*. Analysis of nucleation near the pseudospinodal results in a generalized form of the classical Kelvin equation (3.61) relating the size of the critical cluster to the supersaturation.

V. I. Kalikmanov, *Nucleation Theory*, Lecture Notes in Physics 860, 145
DOI: 10.1007/978-90-481-3643-8_9, © Springer Science+Business Media Dordrecht 2013

9.2 Mean-Field Theory

9.2.1 Landau Expansion for Metastable Equilibrium

The starting point for the mean-field analysis of nucleation in the vicinity of the thermodynamic spinodal is the Landau expansion of the free energy density in powers of the order parameter m [8]

$$g = g_0 + \frac{a}{2} m^2 + \frac{b}{4} m^4 - m h \tag{9.1}$$

where

$$a = a_0 t, \qquad t \equiv (T - T_c)/T_c, \quad a_0 > 0, \, b > 0$$

h is the external field conjugate to m. For the gas-liquid transition the order parameter can be defined as

$$m = \rho - \rho_c$$

where ρ_c is the critical density, and

$$h = \Delta\mu = \mu^v(p^v) - \mu^l(p^v) \tag{9.2}$$

is the external field. At the spinodal $h = h_{sp}$, while at the *binodal* the chemical potentials of the phases are equal yielding $h = 0$. Below the critical point $a < 0$ and the free energy density has a double-well structure shown in Fig. 9.1. In thermodynamic equilibrium one should have

$$\frac{\partial g}{\partial m} = 0, \quad \frac{\partial^2 g}{\partial m^2} > 0$$

yielding

$$a m + b m^3 - h = 0 \tag{9.3}$$
$$a + 3b m^2 > 0 \tag{9.4}$$

At $h = 0$, $g(m)$ has two equal minima corresponding to the two coexisting phases. For $h \neq 0$ the cubic equation (9.3) has a single real root if $h^2 > h_{sp}^2$ [9], where

$$h_{sp}^2 = -\frac{4}{27} \frac{a^3}{b} \tag{9.5}$$

This root refers to the single, stable, phase (liquid). If $h^2 \leq h_{sp}^2$, there are three real roots; for $h^2 = h_{sp}^2$ two of them are equal. The left *local* minimum of the free energy at $m = m_*$ corresponds to the metastable state (supersaturated vapor), while the

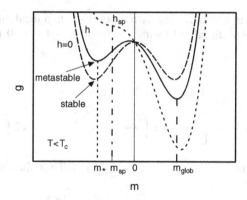

Fig. 9.1 Schematic plot of Landau free energy density g for $T < T_c$. At $h = 0$ (*long dashed line*) there are two equal minima corresponding to the coexisting states. At $0 < h^2 < h_{sp}^2$ (*solid line*) the left, local, minimum at m_* corresponds to a metastable state (supersaturated vapor), while the right, global, minimum refers to a stable state (liquid); the two states are separated by the energy barrier. At $h = h_{sp}$ (*short dashed line*) the local minimum disappears—this is the case of spinodal decomposition

right—*global*—minimum corresponds to the stable state (liquid); the two states are separated by the energy barrier (these features are illustrated in Fig. 9.1). Expression (9.5) gives the maximum supersaturation corresponding to $h = h_{sp}$, where the local minimum of g becomes an inflection point—this is the case of spinodal decomposition [2].

Nucleation takes place for $0 < h^2 < h_{sp}^2$. The discussion below refers to this case. Since $a < 0$, it is convenient to introduce $s = -a$. Then the solutions of the cubic equation (9.3) are [9, 10]:

$$m(h) = 2\sqrt{\frac{s}{3b}} \cos\left(\alpha + \frac{2\pi}{3}k\right), \quad k = 0, 1, 2 \tag{9.6}$$

where

$$\cos 3\alpha = \frac{3\sqrt{3}}{2} \frac{b^{1/2}}{s^{3/2}} h, \quad 0 \le \alpha \le \frac{\pi}{6} \tag{9.7}$$

Substituting (9.6) into the minimization condition (9.4) we obtain

$$\left| \cos\left(\alpha + \frac{2\pi}{3}k\right) \right| > \frac{1}{2}$$

which is satisfied for $k = 0$, 1 and is not satisfied for $k = 2$. The latter case corresponds to a maximum of g while the other two solutions correspond to the two minima.

In order to determine which of these two roots refers to a local minimum (a supersaturated state) we substitute (9.6) with $k = 0$ and $k = 1$ into (9.1) and after some algebra obtain:

$$g|_{k=0} = g_0 + \frac{s^2}{6b}(1 + \cos 2\alpha)(1 - 3\cos 2\alpha)$$

$$g|_{k=1} = g_0 + \frac{s^2}{6b}\left[1 + \cos 2\left(\alpha + \frac{2\pi}{3}\right)\right]\left[1 - 3\cos 2\left(\alpha + \frac{2\pi}{3}\right)\right]$$

It is easy to check that for $0 \leq \alpha \leq \frac{\pi}{6}$: $g|_{k=0} \leq g|_{k=1}$. Thus, the solution with $k = 1$,

$$m_*(h) = 2\sqrt{\frac{s}{3b}}\cos\left(\alpha + \frac{2\pi}{3}\right) \tag{9.8}$$

corresponds to a metastable state of the system in an external field h, the free energy of this state being $g_* = g|_{k=1}$. The solution with $k = 0$ gives the global minimum corresponding to the thermodynamically stable liquid state to which the system evolves.

For the states close to m_* the free energy density can be expanded in powers of $\phi = (m - m_*)/m_*$:

$$g = g_* + \frac{b_2}{2}\phi^2 - \frac{b_3}{3}\phi^3 + O(\phi)^4 \tag{9.9}$$

where

$$b_2(h) = m_*^2 \left.\frac{\partial^2 g}{\partial m^2}\right|_{m=m^*}, \qquad b_3(h) = -m_*^3 \frac{1}{2}\left.\frac{\partial^3 g}{\partial m^3}\right|_{m=m^*} \tag{9.10}$$

The term linear in ϕ vanishes since $m = m_*$ is a local minimum of g. At the spinodal

$$b_2(h = h_{sp}) = 0, \quad b_3(h = h_{sp}) > 0 \tag{9.11}$$

The quantities b_2 and b_3 can be calculated from the equation of state:

$$b_2 = \rho^2 \left.\frac{\partial\mu}{\partial\rho}\right|_{\rho=\rho^v} \tag{9.12}$$

$$b_3 = -\frac{1}{2}\rho^3 \left.\frac{\partial^2\mu}{\partial\rho^2}\right|_{\rho=\rho^v} = b_2 - \frac{1}{2}\rho\left.\frac{\partial b_2}{\partial\rho}\right|_{\rho=\rho^v} \tag{9.13}$$

Using the thermodynamic relationship

$$\frac{\partial p}{\partial\rho} = \rho\frac{\partial\mu}{\partial\rho}$$

we have

$$b_2 = \rho \left. \frac{\partial p}{\partial \rho} \right|_{\rho=\rho^v} \tag{9.14}$$

showing that b_2 is the inverse isothermal compressibility of the vapor at the given metastable state.

9.2.2 Nucleation in the Vicinity of the Thermodynamic Spinodal

Consider the behavior of the system near the mean-field spinodal. For this purpose we construct an appropriate *Ginzburg–Landau free energy functional* which should describe the state of the system undergoing a first order phase transition, characterized by a scalar order parameter [11]

$$\phi(\mathbf{r}) = [m(\mathbf{r}) - m_*]/m_*$$

Now we allow for its spatial variations. Using (9.9), this functional reads:

$$\mathscr{F}[\phi(r)] = \mathscr{F}_* + \int d\mathbf{r} \left[\frac{c_0}{2} |\nabla \phi|^2 + \frac{b_2}{2} \phi^2 - \frac{b_3}{3} \phi^3 \right] \tag{9.15}$$

where $\mathscr{F}_*$ is the free energy of the metastable state $m = m_*$ (the local minimum of the free energy), out of which nucleation starts. The square-gradient term in (9.15) is an energy cost to create an interface between the phases; $c_0 > 0$ is related to the correlation length in the system [12] and can be well approximated by [13]

$$c_0 \cong k_B T \rho_c^{1/3} \tag{9.16}$$

Following Unger [7], we associate the critical cluster with the saddle point of the functional $\mathscr{F}[\phi(r)]$. If the saddle point is found, its substitution into (9.15) yields the nucleation barrier

$$W = \mathscr{F} - \mathscr{F}_* = \int d\mathbf{r} \left[\frac{c_0}{2} |\nabla \phi|^2 + \frac{b_2}{2} \phi^2 - \frac{b_3}{3} \phi^3 \right] \tag{9.17}$$

To analyze this expression we proceed by performing a set of scaling transformation of the variables. Rescaling the order parameter

$$\phi_1 = (b_3/c_0)^{1/3} \phi$$

and denoting

$$\varepsilon = b_2 \, (b_3^2 c_0)^{-1/3} \tag{9.18}$$

we rewrite (9.17) as

$$W = c_0 \int d\mathbf{r} \left[\frac{1}{2} \left(\frac{c_0}{b_3} \right)^{2/3} |\nabla \phi_1|^2 + \frac{\varepsilon}{2} \phi_1^2 - \frac{1}{3} \phi_1^3 \right]$$

The next transformation rescales the spatial coordinates

$$\mathbf{r}_1 = (b_3/c_0)^{1/3} \mathbf{r}$$

yielding

$$W = \frac{c_0^2}{b_3} \int d\mathbf{r}_1 \left[\frac{1}{2} |\nabla_1 \phi_1|^2 + \frac{\varepsilon}{2} \phi_1^2 - \frac{1}{3} \phi_1^3 \right]$$

where $\nabla_1 = \frac{\partial}{\partial \mathbf{r}_1}$. And finally, further rescaling is useful:

$$\phi_1 = \varepsilon \widetilde{\phi}, \quad \mathbf{r}_1 = \varepsilon^{-1/2} \widetilde{\mathbf{r}} \tag{9.19}$$

with the help of which W takes the form:

$$W = \varepsilon^{3/2} \frac{c_0^2}{b_3} \int d\widetilde{\mathbf{r}} \left[\frac{1}{2} |\widetilde{\nabla} \widetilde{\phi}|^2 + \frac{1}{2} \widetilde{\phi}^2 - \frac{1}{3} \widetilde{\phi}^3 \right] \tag{9.20}$$

where $\widetilde{\nabla} = \frac{\partial}{\partial \widetilde{\mathbf{r}}}$.

The saddle point of $W[\widetilde{\nabla}(\widetilde{\mathbf{r}})]$ is given by the Euler-Lagrange equation:

$$\widetilde{\nabla}^2 \widetilde{\phi} = \widetilde{\phi} - \widetilde{\phi}^2 \tag{9.21}$$

The critical cluster is the nontrivial solution of (9.21) vanishing at infinity. The existence of such solutions was proved for sufficiently large bounded domains [14]. Without presenting its full form it is instructive to study its behavior at large r, i.e. far from the center of mass of the cluster. In this domain the amplitude of the droplet is small and we can neglect the second term in (9.21) which leads to the equation

$$\widetilde{\nabla}^2 \widetilde{\phi} = \widetilde{\phi}, \quad \text{large } \widetilde{r} \tag{9.22}$$

The spherically symmetric solution of (9.22) vanishing at infinity is the screened Coulomb function:

$$\widetilde{\phi} = C_> \frac{e^{-\widetilde{r}}}{\widetilde{r}}, \quad \text{large } \widetilde{r}$$

(where $C_>$ is a constant); in the units of (9.19):

$$\phi_1(r_1) = C_> \sqrt{\varepsilon} \, \frac{e^{-\sqrt{\varepsilon} r_1}}{r_1}, \quad \text{large } r_1$$

The spinodal corresponds to $\varepsilon = b_2 = 0$. Since $\phi_1(r_1)$ is the density fluctuation associated with a nucleus, its decay length

$$R_* \approx \varepsilon^{-1/2}, \qquad \varepsilon \to 0 \tag{9.23}$$

characterizes the size of the critical cluster. Equation (9.23) shows that *within the mean-field analysis* the critical cluster size diverges as the spinodal is approached. In the same limit the nucleation barrier (9.20) vanishes as

$$W \sim \varepsilon^{3/2}, \qquad \varepsilon \to 0 \tag{9.24}$$

Finally, we must relate ε to the physical parameters of the system. To be more precise we will determine scaling of ε near the spinodal to the leading order in $(h - h_{sp})$. Obviously

$$\varepsilon(h = h_{sp}) = 0$$

From (9.10) and (9.18) it follows that ε is proportional to the curvature of the Landau free energy at the metastable state:

$$\varepsilon \sim b_2 \sim g_2(h) \equiv \left.\frac{\partial^2 g}{\partial m^2}\right|_{m=m^*(h)} = -s + 3b\,m_*^2$$

Substituting $h = h_{sp} - u$ into (9.8), where u is a (small) deviation of h from its value at the spinodal, we obtain to the leading order in u:

$$\varepsilon \sim g_2 = 2(3bs)^{1/4}\sqrt{h_{sp} - h} \tag{9.25}$$

Expressing h in terms of the supersaturation, we have from Eq. (9.2)

$$h_{sp} - h = k_B T\,(\ln S_{sp} - \ln S) = k_B T\,\ln S_{sp}\,(1 - \eta)$$

where S_{sp} refers to the spinodal and

$$\eta = \frac{\ln S}{\ln S_{sp}}, \qquad 0 \le \eta \le 1 \tag{9.26}$$

$S_{sp}(T)$ is the upper boundary of S for nucleation at the temperature T; its value depends on the equation of state. For van der Waals fluids calculation of S_{sp} is presented in Appendix C.

From (9.25)

$$\varepsilon \sim (1 - \eta)^{1/2} \tag{9.27}$$

Substituting (9.27) into (9.24) and (9.23), we find that in the vicinity of the spinodal the nucleation barrier *vanishes* as

$$W_{sp} = c_{sp}(T) (1 - \eta)^{3/4}, \qquad c_{sp}(T) > 0, \quad \eta \to 1^- \tag{9.28}$$

while the radial extent of the critical cluster *diverges* as

$$R_* = R_0 (1 - \eta)^{-1/4}, \quad R_0(T) > 0, \quad \eta \to 1^- \tag{9.29}$$

The excess number of molecules in the critical cluster is found from (9.28) using the nucleation theorem (4.15):

$$\Delta n_c \sim (1 - \eta)^{-1/4}, \qquad \eta \to 1^- \tag{9.30}$$

Comparison of (9.29) and (9.30) gives the scaling:

$$\Delta n_c \sim R_* \tag{9.31}$$

This is distinctly different from the scaling $\Delta n_c \sim R^3$ corresponding to compact spherical droplets discussed in CNT. Equation (9.31) supports the conjecture of Klein [6] that the critical cluster near the spinodal is a ramified chain-like object.

9.3 Role of Fluctuations

Previous discussions avoided an important conceptual question: how deep the quench can be so that the concept of quasi-equilibrium (of the metastable state) can be considered valid? In other words: what is the limit of validity of the mean-field gradient theory, which completely neglects the effect of fluctuations? The answer to this question can be found using the *Ginzburg criterion* for the breakdown of Landau theory of phase transitions [12]. Very close to the spinodal fluctuations become increasingly important and the mean-field theory of Sect. 9.2.2 breaks down. The Ginzburg criterion determines the width of the domain near the spinodal, inside which the mean-field considerations are violated. Such a thermodynamic analysis was carried out by Wilemski and Li [15], who showed that for real fluids the Ginzburg criterion is violated in the entire spinodal region, where the Landau expansion is used. Having stated this, Wilemski and Li suggested that the concept of the mean-field spinodal should be replaced by the concept of a *pseudospinodal*, introduced earlier by Wang [16] in the study of polymer phase separation, which is associated with a nucleation barrier $\sim k_B T$.

The applicability of the mean-field approach can also be considered on the basis of kinetic considerations. To do this let us compare two characteristic times: (i) the time t_M necessary to form a critical cluster which is a lifetime of the metastable state, and (ii) the relaxation time t_R during which the system settles in this state. The first quantity can be related to the nucleation rate by using its definition: $t_M = 1/(JV)$. To find t_R one must study the dynamics of the metastable state. Since the order parameter $\phi(\mathbf{r})$ in the Ginzburg–Landau functional (9.15) is a conserved variable, its evolution is governed by the Cahn-Hilliard dissipative dynamics [17]:

$$\frac{\partial \phi}{\partial t} = \Gamma_0 \nabla^2 \frac{\delta \mathscr{F}}{\delta \phi} + \zeta \tag{9.32}$$

where Γ_0 is a transport coefficient and $\zeta(\mathbf{r}, t)$ is a noise source (which models thermal fluctuations) satisfying

$$\langle \zeta(\mathbf{r}, t) \zeta(\mathbf{r}', t') \rangle = -2T \, \Gamma_0 \nabla^2 \delta(\mathbf{r} - \mathbf{r}') \, \delta(t - r')$$

to ensure that the equilibrium distribution associated with (9.32) is given by the Boltzmann statistics. From the solution of (9.32) and (9.15), obtained by Patashinskii and Shumilo [18, 19] (see also [13]), it follows that

$$t_R = \frac{16 c_0}{\Gamma_0 b_2^2}$$

implying that when the system approaches the thermodynamic spinodal ($b_2 \to 0$) its relaxation time diverges. The relation between t_M and t_R established in [18, 19] is:

$$t_M = t_R \left(\frac{4\pi \chi}{\lambda_0} \right) \exp \left(\frac{\chi W}{k_B T} \right) \tag{9.33}$$

where

$$\chi = \frac{(b_2 c_0)^{3/2}}{k_B T b_3^2} \tag{9.34}$$

and $\lambda_0 \approx 8.25$. Clearly, the concept of quasi-equilibrium is meaningful for the metastable states characterized by $t_M \gg t_R$. In the opposite case this concept becomes irrelevant. The boundary between these two domains is called a *kinetic spinodal* [18–20] and can be defined by the condition $t_M \cong t_R$ yielding

$$\frac{\chi W}{k_B T} \cong 1 \tag{9.35}$$

Beyond the kinetic spinodal the phase separation proceeds not via nucleation but via the mechanism of spinodal nucleation [21], which differs both from nucleation and spinodal decomposition. As one can see, the kinetic considerations are in agreement with the thermodynamic analysis of [15]. Hence, in terms of the nucleation barrier the pseudospinodal is similar to the kinetic spinodal.

The preceding discussion shows that the spinodal limit is hard to achieve in practice: gradual quenching of the supersaturated vapor results in the barrier becoming equal to the characteristic value of natural thermal fluctuations of the free energy, which in a fluid is of the order of $k_B T$; at these conditions the time necessary to form the critical cluster becomes comparable to the relaxation time during which the system settles in the metastable state.

9.4 Generalized Kelvin Equation and Pseudospinodal

The mean-field gradient theory predicts the divergence of the critical cluster as S approaches the spinodal. Meanwhile, experiments in the supersonic Laval nozzle [22–26] with nucleation rates as high as $10^{17} - 10^{18}\,\mathrm{cm}^{-3}\mathrm{s}^{-1}$ do not support this statement: the critical cluster, determined from the experimental $J - S$ curves by means of the nucleation theorem, continuously decreases with the supersaturation, showing no signs of divergence up to the highest values of S. For those values the critical cluster is a nano-sized object containing 5–10 molecules. A possible explanation of this qualitative discrepancy was mentioned in the previous section: the mean-field considerations fail in the entire spinodal region and the physically relevant limit of the supersaturation is not the spinodal, but the pseudospinodal. Therefore, we need to study nucleation in the vicinity of the pseudospinodal.

Can the CNT be useful for this study? Recall that in the CNT the critical cluster is related to the supersaturation by the Classical Kelvin Equation (CKE) (3.63). For every finite S it predicts a certain critical cluster yielding a certain *finite* nucleation barrier (3.29). In other words the CNT does not signal either spinodal or pseudospinodal. This is not surprising since at high S the critical clusters become small and obviously do not obey the capillarity approximation. For the analysis of the system behavior in the vicinity of the pseudospinodal we need a generalization of the CKE which extends the limit of its validity down to the clusters of molecular sizes; such a generalization was proposed in Ref. [27].

The thermodynamic basis for the Kelvin equation is the metastable equilibrium between the critical cluster and the surrounding supersaturated vapor and according to (3.62) can be found from maximization of the Gibbs energy of cluster formation. This is a general statement which holds irrespective of a particular form of the Gibbs energy. Let us adopt the MKNT form for ΔG given by Eq. (7.77). Its maximum leads to

$$\ln S = \theta_{\mathrm{micro}} \left. \frac{\mathrm{d}\overline{n^s}(n)}{\mathrm{d}n} \right|_{n_c} \tag{9.36}$$

which can be termed the *Generalized Kelvin Equation (GKE)*. Taking into account (7.63) and (7.66)), it is straightforward to see that in the limit of big clusters the GKE recovers the Classical Kelvin Equation.

In order to illustrate the important features of the GKE let us consider the profiles of $\Delta G(n)$ for various values of S at a fixed temperature T. To make the illustration practically relevant, we refer to the experimental conditions for argon nucleation in the supersonic Laval nozzle studied in Ref. [26].

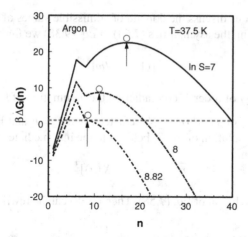

Fig. 9.2 Gibbs free energy profiles $\Delta G(n)$ for argon at $T = 37.5$ K. Labels indicate the value of $\ln S$. Arrows indicate the critical cluster for each curve (corresponding to the second maximum of $\Delta G(n)$). At $\ln S = 8.82$ the second maximum disappears; the *dashed red line* gives the nucleation barrier $\Delta G^* \approx k_B T$)

Figure 9.2 shows the free energy profiles for $T = 37.5$ K and three different values of $\ln S$. Each curve $\Delta G(n)$ has two maxima (cf. Sect. 7.7): the first (left) one is always at $n = N_1$ (coordination number in the liquid phase) and is an artifact of the MKNT, while the second (right) maximum corresponds to the critical cluster n_c: in Fig. 9.2 the critical cluster is indicated by the vertical error and the 'ball' on the top of the curve. As S increases, both n_c and the nucleation barrier $\Delta G^* = \Delta G(n_c)$ decrease. At a certain supersaturation, $\ln S = 8.82$, the second maximum disappears which means that for supersaturations higher than this value the critical cluster does not exist. In other words, there is an upper limit of S, beyond which there is no nucleation. As one can see from the dashed red line, the nucleation barrier at this value of S turns out to be $\Delta G^* \approx k_B T$ which corresponds to the pseudospinodal. Below we will show that this is not a pure coincidence.

The free energy of formation of the critical cluster in MKNT takes the form (7.82) which at the pseudospinodal gives

$$- n_c \ln S + \theta_{\text{micro}} \left[\overline{n^s}(n_c) - 1 \right] = 1 \tag{9.37}$$

Combination of (9.37) with the GKE Eq. (9.36) determines both n_c and $\ln S$ at the pseudospinodal. Excluding $\ln S$, we find that n_c satisfies the equation:

$$n_c \left. \frac{d\overline{n^s}}{dn} \right|_{n_c} - \overline{n^s}(n_c) + 1 + \upsilon = 0, \quad \text{where } \upsilon \equiv \frac{1}{\theta_{\text{micro}}} < 1 \tag{9.38}$$

Before solving it let us discuss the domain of admissible values of n_c. Suppose that n_c is *large*, then using the asymptotics (7.63) in Eq. (9.38) we find:

$$n_c = [(1 + \upsilon)/\omega]^{3/2}$$

which is the quantity of order 1, contradicting the assumption of large n.

In the opposite limit of *small* ($n \leq N_1$) clusters: $\overline{n^s}(n) = n$ and Eq. (9.38) has no solutions. Thus, the solution of (9.38) belongs to the intermediate cluster range. It is convenient to present

$$\overline{n^s}(n) = n - [X(n)]^3$$

where $X(n)$ is the solution of Eq. (7.58). Then (9.38) can be rewritten as:

$$[X(n_c)]^3 - 3\,n_c\,[X(n_c)]^2 \left.\frac{dX}{dn}\right|_{n_c} + 1 + \upsilon = 0 \tag{9.39}$$

From the previous discussion we can expect (the assumption to be verified later) that near the pseudospinodal n_c is small and close to the lower boundary of the intermediate cluster range, i.e. it lies in the vicinity of $N_1 + 1$. Correspondingly, $X(n)$ is close to unity. Presenting

$$X(n) = 1 + \delta(n)$$

and linearizing (7.58) in $\delta(n)$ we find:

$$\delta(n) = \frac{1}{3q}\,(n - N_1 - 1) \tag{9.40}$$

$$q \equiv 1 + 2\omega + \omega\lambda \tag{9.41}$$

Equation (9.39) now reads

$$3\delta(n_c) - 3n_c[1 + 2\delta(n_c)] \left.\frac{d\delta}{dn}\right|_{n_c} + (2 + \upsilon) = 0 \tag{9.42}$$

Substituting (9.40) into (9.42), we find the critical cluster at the pseudospinodal:

$$n_{c,\mathrm{psp}} = (N_1 + 1) \left(\frac{1 + \sqrt{1 + \tau}}{2}\right) \tag{9.43}$$

$$\tau = \frac{6q\,[(2 + \upsilon)q - (N_1 + 1)]}{(N_1 + 1)^2}$$

This result supports the assumption that n_c is close to $N_1 + 1$ (usually $|\tau| \ll 1$); the cluster shows the liquid-like features only when it has a core, i.e. when $n > N_1(T)$.

Setting

$$\overline{n^s}(n) = n - [1 + 3\delta(n)]$$

in Eq. (9.36) and using (9.40) we obtain the supersaturation at the pseudospinodal:

$$\ln S_{\text{psp}}(T) = \theta_{\text{micro}} \left(1 - \frac{1}{q}\right) \qquad (9.44)$$

This is the maximum value of the supersaturation at the temperature T. To a large extent it is determined by the microscopic surface tension describing the nonideality of the vapor (expressed in terms of the second virial coefficient); the temperature-dependent quantity $q(T)$ is in the range $q(T) \approx 2 \div 4$.

States with $S > S_{\text{psp}}$ are not realizable. With this in mind we rewrite the GKE in the form

$$\ln S = \begin{cases} \theta_{\text{micro}} \left. \frac{d\overline{n^s}}{dn}\right|_{n_c}, & \text{for } n_c \geq N_1 + 1 \\ \theta_{\text{micro}} \left(1 - \frac{1}{q}\right), & \text{for } n_c \leq N_1 + 1 \end{cases} \qquad (9.45)$$

showing the physical meaning of the pseudospinodal: it corresponds to the disappearance of the liquid-like structure of the critical nucleus.

The derivation of GKE assumes that the supersaturated vapor is *weakly* non-ideal implying that

$$\zeta(T, S) \equiv |B_2 p^{\text{v}}/k_B T| = S |B_2 p_{\text{sat}}/k_B T| \ll 1 \qquad (9.46)$$

This criterion determines the limit of validity of the GKE.

Setting $S = S_{\text{psp}}$, we find that Eq. (9.44) for the pseudospinodal is applicable for the temperatures satisfying

$$\zeta_{\text{psp}}(T) \equiv e^{-\theta_{\text{micro}}/q} \ll 1 \qquad (9.47)$$

Figure 9.3 shows the classical (CKE) and generalized (GKE) Kelvin equation for argon at $T = 45\,\text{K}$ and water at $T = 220\,\text{K}$. The horizontal arrow points to the pseudospinodal. For both substances the criterion (9.46) is satisfied up to the pseudospinodal:

$$\zeta_{\text{psp,argon}}(T = 45\text{ K}) \approx 0.070, \quad \zeta_{\text{psp,water}}(T = 220\text{ K}) \approx 0.034$$

CKE and GKE become indistinguishable for clusters exceeding $\sim$200 molecules. In terms of the cluster radius it corresponds to $R \approx 1.1 - 1.5\,\text{nm}$. This rather unexpected result shows that the classical Kelvin equation may be still valid down to the clusters containing $\sim$200 molecules. At large cluster sizes GKE approaches CKE: for argon— from below, whereas for water—from above. The reason for this difference is the sign of the Tolman length (7.101) which is negative for water at 220 K and positive for argon at 45 K.

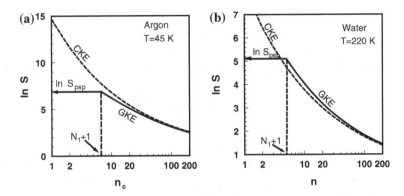

Fig. 9.3 Classical Kelvin equation (CKE) (*dashed lines*) and Generalized Kelvin Equation (GKE) (*solid lines*). **a** Argon at $T = 45$ K; **b** Water at $T = 220$ K. The horizontal arrow indicates the value of ln S at the pseudospinodal

Fig. 9.4 Nucleation barrier $\beta W^* = \beta \Delta G^*$ (*lines*) and critical cluster size (*closed and open symbols*) as a function of ln S for water at $T = 220$ K as predicted by MKNT and CNT. The *dashed horizontal line* corresponds to the barrier $W^* = k_B T$ characteristic of the pseudospinodal conditions

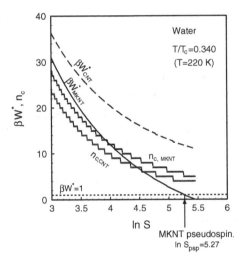

Figure 9.4 illustrates the behavior of the nucleation barrier $W^* = \Delta G^*$ and the critical cluster size of water at $T = 220$ K (predicted by CNT and MKNT) as the vapor approaches the pseudospinodal

$$\ln S_{\mathrm{psp,water}}(T = 220\,\mathrm{K}) \approx 5.27$$

Although CNT predicts smaller critical clusters than MKNT, the CNT barrier is larger than the MKNT one, which is the manifestation of the fact that the formation of small clusters is dominated by the microscopic surface tension rather than by the macroscopic one.

References

1. H. Gould, W. Klein, Physica D **66**, 61 (1993)
2. P.G. Debenedetti, *Metastable Liquids* (Princeton University Press, Princeton, Concepts and Principles, 1996)
3. J.W. Cahn, J.E. Hilliard, J. Chem. Phys. **28**, 258 (1958)
4. J.S. Langer, Ann. Phys. **41**, 108 (1967)
5. J.S. Langer, Ann. Phys. **54**, 258 (1969)
6. W. Klein, Phys. Rev. Letters **47**, 1569 (1981)
7. C. Unger, W. Klein, Phys. Rev. B **29**, 2698 (1984)
8. L.D. Landau, E.M. Lifshitz, *Statistical Physics* (Pergamon, Oxford, 1969)
9. G.A. Korn, T.M. Korn, *Mathematical Handbook* (McGraw-Hill, New York, 1968)
10. Yu.B. Rumer, M.S. Rivkin, *Thermodynamics, Statistical Physics and Kinetics* (Nauka, Moscow, 1977) (in Russian)
11. A.J. Bray, Adv. Phys. **43**, 357 (1994)
12. J.J. Binney, N.J. Dowrick, A.J. Fisher, M.E.J. Newman, *The Theory of Critical Phenomena* (Clarendon Press, Oxford, 1995)
13. S.B. Kiselev, Physica A **269**, 252 (1999)
14. M. Struwe, *Variational Methods: Application to Nonlinear Partial Differential Equations and Hamiltonian Systems* (Springer, Berlin, 2000)
15. G. Wilemski, J.-S. Li, J. Chem. Phys. **121**, 7821 (2004)
16. Z.-G. Wang, J. Chem. Phys. **117**, 481 (2002)
17. P.M. Chaikin, T.C. Lubensky, *Principles of Condensed Matter Physics* (Cambridge University Press, Cambridge, 1995)
18. A.Z. Patashinskii, B.I. Shumilo, Sov. Phys. JETP **50**, 712 (1979)
19. A.Z. Patashinskii, B.I. Shumilo, Sov. Phys. Solid State **22**, 655 (1980)
20. V.K. Schen, P.G. Debenedetti, J. Chem. Phys. **118**, 768 (2003)
21. K. Binder, Phys. Rev. A **29**, 341 (1984)
22. K.A. Streletzky, Yu. Zvinevich, B.E. Wyslouzil, J. Chem. Phys. **116**, 4058 (2002)
23. A. Khan, C.H. Heath, U.M. Dieregsweiler, B.E. Wyslouzil, R. Strey, J. Chem. Phys. **119**, 3138 (2003)
24. C.H. Heath, K.A. Streletzky, B.E. Wyslouzil, J. Wölk, R. Strey, J. Chem. Phys. **118**, 5465 (2003)
25. Y.J. Kim, B.E. Wyslouzil, G. Wilemski, J. Wölk, R. Strey, J. Phys. Chem. A **108**, 4365 (2004)
26. S. Sinha, A. Bhabbe, H. Laksmono, J. Wölk, R. Strey, B. Wyslouzil, J. Chem. Phys. **132**, 064304 (2010)
27. V.I. Kalikmanov, J. Chem. Phys. **129**, 044510 (2008)

Chapter 10
Argon Nucleation

Argon belongs to the class of so called *simple fluids* whose behavior on molecular level can be adequately described by the Lennard-Jones interaction potential

$$u_{LJ}(r) = 4\varepsilon_{LJ} \left[\left(\frac{\sigma_{LJ}}{r} \right)^{12} - \left(\frac{\sigma_{LJ}}{r} \right)^{6} \right] \tag{10.1}$$

where ε_{LJ} is the depth of the potential and σ_{LJ} is the molecular diameter; for argon $\varepsilon_{LJ}/k_B = 119.8\,K$, $\sigma_{LJ} = 3.40$ Å [1]. Since argon plays an exceptional role in various areas of soft condensed matter physics, its *equilibrium properties* have been extensively studied experimentally [2], theoretically [3]; in computer simulations— Monte Carlo and molecular dynamics—[4, 5] and by means of the density functional theory [6, 7].

Among various other issues, argon represents an important reference system for *non-equilibrium* studies. In this context the phenomenon of nucleation is of special significance. In the situation when no theoretical model can claim to be quantitatively correct in describing nucleation in all substances under various external conditions, argon can play a role of the test substance for which experimental, theoretical and simulation efforts can be combined in order to obtain a better insight into the nucleation phenomenon and abilities of various approaches to adequately describe it. This chapter is aimed at obtaining a unified picture of argon nucleation combining theory, simulation and experiment.

10.1 Temperature-Supersaturation Domain: Experiments, Simulations and Density Functional Theory

Early experimental studies of argon nucleation were carried out using various techniques: cryogenic supersonic [8] and hypersonic [9] nozzles and cryogenic shock

V. I. Kalikmanov, *Nucleation Theory*, Lecture Notes in Physics 860, DOI: 10.1007/978-90-481-3643-8_10, © Springer Science+Business Media Dordrecht 2013

tubes [10–12]. The data obtained in these experiments showed significant scatter and results of various groups turned out to be inconsistent with each other.

The experimental situation was largely improved in 2006 due to the construction of the cryogenic Nucleation Pulse Chamber (NPC) [13], and its further development [14, 15]. This chamber uses a deep adiabatic expansion of the argon–helium mixture which causes argon nucleation at temperatures below the triple point.

The onset nucleation data obtained in NPC for the temperature range 42–58 K are reproducible and refer to the estimated nucleation rates $10^{7\pm2}$ cm^{-3}s^{-1}. An important breakthrough in nucleation measurements was achieved by construction of Laval Supersonic Nozzle (SSN) [16], making it possible to accurately determine the onset conditions corresponding to significantly higher nucleation rates. Argon nucleation experiments in SSN [17], carried out in the temperature range 35–53 K which partly overlaps the NPC range, correspond to higher supersaturations yielding the estimated nucleation rates as high as $10^{17\pm1}$ cm^{-3}s^{-1}. In what follows we refer to the experimental data obtained by these two techniques [14, 15] and [17].

Nucleation is an example of a *rare-event process*, that is why molecular dynamic simulations at low temperatures are usually performed for very high supersaturations in order to get a good statistics of nucleation events. Kraska [18] carried out MD simulations in the microcanonical ensemble (MD/NVE) in the temperature range $30\,\mathrm{K} < T < 85\,\mathrm{K}$ with the nucleation rates $J_{MD/NVE} \sim 10^{25} - 10^{29}$ cm^{-3}s^{-1}. MD simulations of Wedekind et al. [19] in the canonical ensemble (MD/NVT) are performed in the temperature range $45\,\mathrm{K} < T < 70\,\mathrm{K}$ with the nucleation rates in the range of $J_{MD/NVT} = 10^{23} - 10^{25}$ cm^{-3}s^{-1}. All these simulations yield nucleation rates which are far beyond experimental values of both NPC and SSN data.

In Chap. 5 we discussed DFT of nucleation and demonstrated its predictions for Lennard-Jones fluids—see Fig. 5.5. In terms of argon properties these calculations correspond to relatively high temperatures $83\,\mathrm{K} < T < 130\,\mathrm{K}$. At this temperature range the detectable nucleation rates require relatively low supersaturations. The DFT nucleation rates lie in the range $J_{DFT} = 10^{-1} - 10^{5}$ cm^{-3}s^{-1}. Figure 10.1 illustrates experimental, simulation and DFT studies in the $T - S$ plane indicating the corresponding typical values of J.

According to Chap. 9, the experimentally achievable upper limit of supersaturation for nucleation at a temperature T is the pseudospinodal corresponding to the nucleation barrier $\Delta G^* \approx k_B T$. The MKNT pseudospinodal given by Eq. (9.44) is shown in Fig. 10.1 by the line labeled "psp". As it is seen from Fig. 10.1 the SSN experiments at low temperatures $37\,\mathrm{K} < T < 40\,\mathrm{K}$ are carried out in the pseudospinodal region which implies that one can expect critical nuclei to be nano-sized objects with the number of molecules close to the coordination number in the liquid phase. In Fig. 10.2 the pseudospinodal is compared to the estimates of the *thermodynamic spinodal*, corresponding to the limit of thermodynamic stability of the fluid. One way to estimate the spinodal is to use a suitable equation of state (EoS) [24]. Dashed line in Fig. 10.2 shows the spinodal calculated from the LJ EoS of Kolafa and Nezbeda

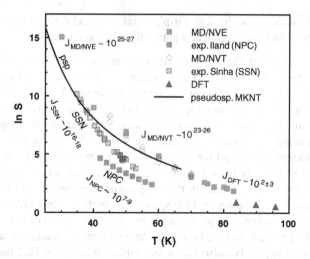

Fig. 10.1 *T-S* domain of experiments and simulations. Nucleation Pulse Chamber (NPC) experiments [14, 15] (*blue squares*), Supersonic Nozzle (SSN) experiments [16] (*green squares*), MD/*NVE* simulations [18] (*filled squares*), MD/*NVT* simulations [19] (*open rhombs*) and DFT simulations [20] (*filled triangles*). The line labelled "psp" is the MKNT pseudospinodal Eq. (9.44)

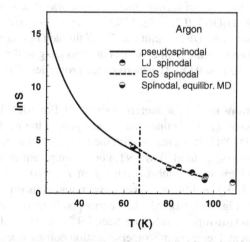

Fig. 10.2 Pseudospinodal and estimates of the thermodynamic spinodal for argon. *Solid line*: pseudospinodal Eq. (9.44); *dashed line*: the spinodal from the Lennard-Jones equation of state of Ref. [21]; *upper half-filled circles*: the spinodal from the simulations of the supersaturated Lennard-Jones vapor of Ref. [22]; *lower half-filled circles*: the spinodal from equilibrium simulations of Ref. [23]. The vertical *dashed-dotted line* corresponds to the criterion (9.47)

[21]. Extrapolations below the argon triple point $T_{tr} = 83.8$ K are limited because EoS are usually fitted to experimental data only in the stable region.

Spinodal can be also found in computer simulations of Lennard-Jones fluids. Linhart et al. [22] performed MD simulations of the *supersaturated vapor* of a LJ fluid and obtained the spinodal pressure for the LJ temperature range $0.7 \leq k_B T / \varepsilon_{LJ} \leq 1.2$. For argon this range corresponds to 84 K $< T <$ 143 K. Spinodal was estimated by the appearance of an instantaneous phase separation in the supersaturated vapor increasing the argon density in a series of simulations. Simulations were performed for a large cut-off radius $10\,\sigma_{LJ}$ and the non-shifted LJ potential. A large cut-off radius makes it feasible to apply the LJ simulation results to real argon.[1] Unfortunately, these simulations cover only partly the temperature domain of MD [18, 19] and DFT [20]. Imre et al. [23] estimated the spinodal from the extremes of the tangential component of the pressure tensor obtained from the simulations of the vapor-liquid interface. This approach is based on a single equilibrium simulation without any constraints and is applicable also below the triple point. Figure 10.2 indicates that theoretical predictions of the pseudospinodal are consistent with the calculations of thermodynamic spinodal performed by various methods: within the common temperature range the MKNT pseudospinodal lies slightly below the instability points of simulations and equation of state.

Let us discuss the predictions of various nucleation theories discussed in this book— CNT (Chap. 3), EMLD-DNT (Chap. 6), MKNT (Chap. 7),—experiment (NPC and SSN), MD simulations and DFT within the $T - S$ domain bounded from above by the pseudospinodal and for the temperatures corresponding to the range of validity of MKNT (Eq. (7.53)): $T < 92$ K. Thermodynamic properties of argon are presented in Appendix A.

The behavior of various model parameters is shown in Fig. 10.3. The bulk (macroscopic) surface tension θ_∞ determines the surface part of the nucleation barrier in the CNT and EMLD-DNT. It decreases with the temperature as well as the microscopic surface tension θ_{micro} used in MKNT. For all temperatures $\theta_\infty > \theta_{micro}$. The difference between them can be substantial: e.g. at $T = 70$ K: $\theta_\infty(T = 70\,K) = 10.68$, $\theta_{micro}(T = 70\,K) = 4.96$; at higher temperatures this difference decreases. The dashed line in Fig. 10.3 labeled "LJ" corresponds to the universal form of $\gamma_{micro}/k_B T_c$ for Lennard-Jones fluids (see Sect. 7.9.2). In view of methodological reasons the experimental temperature-supersaturation domain does not overlap with that of MD and DFT as clearly seen from Fig. 10.1. Therefore we perform separate comparisons: theory versus experiment and theory versus MD and DFT [27].

---

[1] It has been shown that the usually used cut-off radii of $5\,\sigma_{LJ}$ and $6.5\,\sigma_{LJ}$ are sufficient [25], while $2.5\,\sigma_{LJ}$ gives significant deviation in the thermophysical properties [26].

Fig. 10.3 Equilibrium
properties of argon:
θ_∞, θ_{micro}, $\gamma_{\text{micro}}/k_B T_c$.
The *dashed line* labeled
"LJ" shows $\gamma_{\text{micro}}/k_B T_c$ for
Lennard-Jones fluids accord-
ing to Eq. (7.89) with the
universal parameters given by
(7.93)–(7.93)

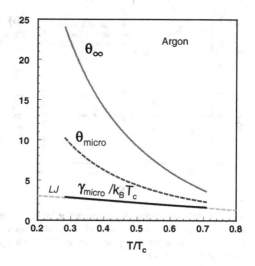

10.2 Simulations and DFT Versus Theory

Figure 10.4 shows the ratio of nucleation rate $\log_{10}(J_{\text{simul}}/J_{\text{theor}})$, where J_{simul} is the
nucleation rate found in MD simulations [18] and [19], and J_{theor} refers to one of
the theoretical models. Open symbols correspond to J_{theor} given by the CNT, and
filled symbols refer to the nonclassical models: MKNT and EMLD-DNT. The dashed
curve is the "ideal line" $J_{\text{simul}} = J_{\text{theory}}$. The agreement between simulations and
the nonclassical models in the whole temperature range is for most cases within 1–2
orders of magnitude, while the CNT rates are on average 3–5 orders of magnitude
lower than the simulation results. Figure 10.4 demonstrates that MKNT predicts a
better temperature dependence of the nucleation rates compared to EMLD-DNT.

An examination of MD results at $T = 70$ K shows that results obtained in *NVE* and
NVT simulations show a difference of one order of magnitude. There are two possible
reasons for this discrepancy. Firstly, in the *NVT* simulations the nucleation rate is
calculated from a mean first passage time analysis (MFPT) [28] while in the *NVE*
simulations the threshold method is employed. These two methods yield approximately one order of magnitude difference in the nucleation rate at given conditions
[29]. Since the nucleation rate obtained by the threshold method is larger, it is located
above the MFPT data. Secondly, in the *NVE* ensemble the latent heat heats up the
system allowing for the natural temperature fluctuations, while in the *NVT* simulations velocity scaling is applied, which forces the system to stay at a fixed temperature thereby not allowing temperature fluctuations. Figure 10.5 compares theoretical
predictions with DFT of Ref. [20]. MKNT demonstrates a perfect agreement with the
DFT while both the CNT and EMLD-DNT underestimate DFT data by 3–5 orders
of magnitude. Recalling that nucleation rate is very sensitive to the intermolecular
interaction potential the agreement between MKNT and DFT is quite remarkable

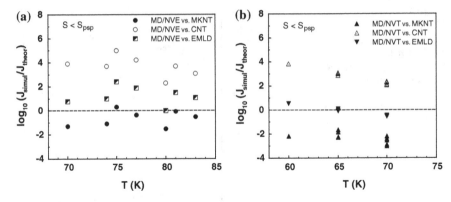

Fig. 10.4 MD simulations versus theory. **a** MD/NVE simulations of Ref. [18] versus theory. *Open circles*: CNT, *closed circles*: MKNT, *semi-filled squares*: EMLD-DNT; **b** MD/NVT simulations of Ref. [19] versus theory. *Open triangles*: CNT, *filled upward triangles*: MKNT, *filled downward triangles*: EMLD-DNT

Fig. 10.5 DFT calculations of Ref. [20] versus theory. *Open rhombs*: CNT, *filled rhombs*: MKNT, *filled stars*: EMLD-DNT

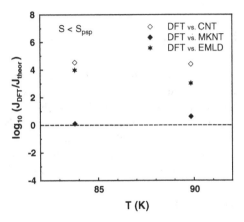

since DFT explicitly uses the interatomic interaction potential, while the MKNT is a semi-phenomenological model using as an input the macroscopic empirical EoS, the second virial coefficient, the plain layer surface tension and the coordination number in the bulk liquid. Note, however, that the amount of available DFT data is insufficient to formulate firm conclusions about the performance of different theoretical models.

10.3 Experiment Versus Theory

Consider first experiments in the nucleation pulse chamber [14, 15]. The relative nucleation rates together with the error bars of the experimental accuracy are shown in Fig. 10.6. It turns out that for argon (being a simple fluid), predictions of the

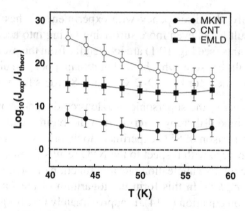

Fig. 10.6 Argon nucleation experiments in Nucleation Pulse Chamber [14, 15] versus theory: CNT (*open circles*), EMLD-DNT (*filled squares*) [19], MKNT (*filled circles*)

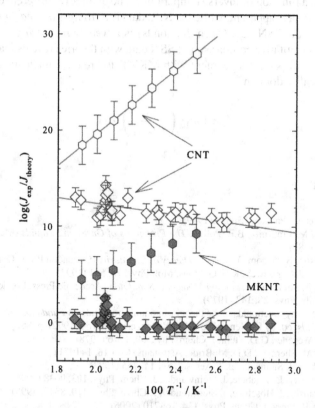

Fig. 10.7 Volmer plot for argon nucleation in nucleation pulse chamber and supersonic nozzle. *Hexagons*: NPC data; *diamonds*: SSN data. Open symbols (*hexagons and diamonds*) refer CNT, closed symbols (*hexagons and diamonds*) refer MKNT. The *solid lines* in the CNT graphs are shown to guide the eye. (Reprinted from Ref. [17] copyright (2010), American Institute of Physics.)

CNT fail dramatically: the discrepancy with experiment reaches 26–28 orders of magnitude! This result looks even more surprising taking into account that experimental nucleation points (see Fig. 10.1) are located far from the pseudospinodal. For other models the results are somewhat better but remain poor: the disagreement with experiment is 12–14 orders for EMLD-DNT and 4–8 orders for MKNT.

Nucleation experiments in the supersonic nozzle provide a possibility to reach the vicinity of pseudospinodal thereby entering into the regime with extremely small critical clusters. Comparison of SSN experiment to theories shows both quantitative and qualitative differences with respect to the NPC results. Figure 10.7, taken from Ref. [16], depicts the relative nucleation rate as a function of the inverse temperature (the so-called *Volmer plot*). In this form the logarithm of the saturation pressure, given by the Clapeyron equation (2.14), is approximately the straight line as well as the lines of constant nucleation rate (for not too high rates). The upper curve (open hexagons) reproduces the NPC data with respect to CNT (similar to the upper curve of Fig. 10.6) as a function of inverse temperature. The qualitative difference between NPC and SSN data is apparent: while the NPC data is a *strongly decreasing* function of temperature, the SSN data (open diamonds) is a *weakly increasing* function of temperature. Quantitative comparison of SSN data with theories reveals that all SSN experiments are in perfect agreement with MKNT: the relative nucleation rates lie within the "ideality domain"

$$-1 < \log_{10}\left(\frac{J_{\text{exp}}}{J_{\text{MKNT}}}\right) < 1$$

References

1. A. Michels et al., Physica **15**, 627 (1949)
2. R.C. Reid, J.M. Prausnitz, B.E. Poling, *The Properties of Gases and Liquids* (McGraw-Hill, New York, 1987)
3. J.S. Rowlinson, B. Widom, *Molecular Theory of Capillarity* (Clarendon Press, Oxford, 1982)
4. J.K. Johnson, J.A. Zollweg, K.E. Gubbins, Mol. Phys. **78**, 591 (1993)
5. D. Frenkel, B. Smit, *Understanding Molecular Simulaton* (Academic Press, London, 1996)
6. R. Evans, Adv. Phys. **28**, 143 (1979)
7. R. Evans, Density functionals in the theory of nonuniform fluids. in *Fundamentals of Inhomogeneous Fluids*, ed. by D. Henderson (Marcel Dekker, New York 1992), p. 85
8. B. Wu, P.P. Wegener, G.D. Stein, J. Chem. Phys. **69**, 1776 (1978)
9. T. Pierce, P.M. Sherman, D.D. McBride, Astronaut. Acta **16**, 1 (1971)
10. M.W. Matthew, J. Steinwandel, J. Aerosol. Sci **14**, 755 (1983)
11. R.A. Zahoransky, J. Höschele, J. Steinwandel, J. Chem. Phys. **103**, 9038 (1995)
12. R.A. Zahoransky, J. Höschele, J. Steinwandel, J. Chem. Phys. **110**, 8842 (1999)
13. A. Fladerer, R. Strey, J. Chem. Phys. **124**, 164710 (2006)
14. K. Iland, Ph.D. Thesis, University of Cologne, 2004
15. K. Iland, J. Wölk, R. Strey, D. Kashchiev, J. Chem. Phys. **127**, 154506 (2007)
16. S. Sinha, H. Laksmono, B. Wyslouzil, Rev. Sci. Instrum. **79**, 114101 (2008)
17. S. Sinha, A. Bhabbe, H. Laksmono, J. Wölk, R. Strey, B. Wyslouzil, J. Chem. Phys. **132**, 064304 (2010)

18. T. Kraska, J. Chem. Phys. **124**, 054507 (2006)
19. J. Wedekind, J. Wölk, D. Reguera, R. Strey, J. Chem. Phys. **127**, 154516 (2007)
20. X.C. Zeng, D.W. Oxtoby, J. Chem. Phys. **94**, 4472 (1991)
21. J. Kolafa, I. Nezbeda, Fluid Phase Equilib. **100**, 1 (1994)
22. A. Linhart, C.-C. Chen, J. Vrabec, H. Hasse, J. Chem. Phys. **122**, 144506 (2005)
23. A.R. Imre, G. Meyer, G. Hazi, R. Rozas, T. Kraska, J. Chem. Phys. **128**, 114708 (2008)
24. T. Kraska, Ind. Eng. Chem. Res. **43**, 6213 (2004)
25. M. Mecke, J. Winkelmann, J. Fischer, J. Chem. Phys. **107**, 9264 (1997)
26. B. Smit, J. Chem. Phys. **96**, 8639 (1992)
27. V.I. Kalikmanov, J. Wölk, T. Kraska, J. Chem. Phys. **128**, 124506 (2008)
28. J. Wedekind, R Strey, D. Reguera. J. Chem. Phys. **126**, 134103 (2007)
29. R. Römer, T. Kraska, J. Chem. Phys. **127**, 234509 (2007)

Chapter 11
Binary Nucleation: Classical Theory

11.1 Introduction

An increase of dimensionality of a problem usually brings about a new physics. Speaking about nucleation, a step from a single-component to a binary system introduces an additional thermodynamic degree of freedom: the phase equilibrium of a binary system is characterized by two thermodynamic variables—and not one as in the single-component case. Due to this feature a binary cluster of an arbitrary composition in the surrounding binary vapor at the pressure p^v and temperature T has the properties which are different from the $p^v - T$ equilibrium properties of the environment. This consideration shows a crucial role of *cluster composition* in nucleation behavior.

Even in equilibrium building of a binary cluster in the vapor, besides the creation of the gas-liquid interface, is accompanied by the free energy change associated with the difference in the chemical potential of a molecule inside and outside the cluster; the latter difference can be both positive and negative depending on cluster composition. This does not happen in a single-component case, where in equilibrium a molecule inside and outside cluster has the same chemical potential depending only on temperature.

In the binary case the free energy of cluster formation forms a surface in the space of cluster compositions. Similarly, *kinetics* of binary nucleation is characterized by an infinite number of nucleation paths. In this chapter we consider the binary classical nucleation theory (BCNT). Its history dates back to the works of Flood [1], Volmer [2], Neumann and Döring [3]. The BCNT, as it is known now, is associated with the classical work of Reiss [4]. Generalizing the CNT of Becker-Döring-Zeldovich to the binary mixtures, Reiss put forward the kinetic and thermodynamic arguments to show that the nucleation rate in the binary problem is associated with the passage over the saddle point of the free energy surface in the space of droplet compositions.

V. I. Kalikmanov, *Nucleation Theory*, Lecture Notes in Physics 860,   171
DOI: 10.1007/978-90-481-3643-8_11, © Springer Science+Business Media Dordrecht 2013

11.2 Kinetics

Kinetics of binary nucleation describes formation of binary clusters at given external conditions. A cluster with n_a particles of component a and n_b particles of component b is denoted as a point in the two-dimensional (n_a, n_b) composition space (see Fig. 11.1). As in the single-component nucleation we assume that

- the elementary process which changes the size of a nucleus is the attachment to it or loss by it of one molecule of either component a or b; thus, kinetics is governed by the following reactions:

$$(n_a, n_b) + (1, 0) \leftrightarrow (n_a + 1, n_b)$$
$$(n_a, n_b) + (0, 1) \leftrightarrow (n_a, n_b + 1)$$

- if a monomer collides a cluster it sticks to it with probability unity
- there is no correlation between successive events that change the number of particles in a cluster

The last assumption means that binary nucleation is a Markov process.

The nucleation flux at the point (n_a, n_b) is a vector

$$\mathbf{J} = (J_a(n_a, n_b), \; J_b(n_a, n_b))$$

with coordinates J_a and J_b, where $J_a(n_a, n_b)$ is a net rate at which (n_a, n_b)-clusters become $(n_a + 1, n_b)$-clusters, and $J_b(n_a, n_b)$ is a net rate at which (n_a, n_b)-clusters become $(n_a, n_b + 1)$-clusters. The vectorial nature of the flux implies the existence of a large (in fact, infinite) number of nucleation paths resulting from variety of the possible directions of $\mathbf{J}$. This makes an important difference with a single-component

Fig. 11.1 Schematic representation of binary kinetics on (n_a, n_b)-plane

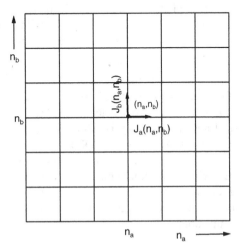

case, in which J is a scalar. From the assumptions made the kinetic equation describing the evolution of the cluster distribution function $\rho(n_a, n_b, t)$ becomes

$$\frac{\partial \rho(n_a, n_b, t)}{\partial t} = J_a(n_a - 1, n_b, t) - J_a(n_a, n_b, t) + J_b(n_a, n_b - 1, t) - J_b(n_a, n_b, t)$$

(11.1)

or in differential notations

$$\frac{\partial \rho(n_a, n_b, t)}{\partial t} = -\left[\frac{\partial J_a}{\partial n_a} + \frac{\partial J_b}{\partial n_b} \right]$$

(11.2)

The last expression manifests the conservation law for the number of particles and can be rewritten as

$$\frac{\partial \rho(\mathbf{n})}{\partial t} = -\operatorname{div} \mathbf{J}(\mathbf{n})$$

(11.3)

In the steady state

$$\operatorname{div} \mathbf{J} = 0$$

(11.4)

As usual in the rate theories we write the fluxes along n_a and n_b axis in terms of the forward (condensation) and backward (evaporation) rates:

$$J_a(n_a, n_b) = v_a A(n_a, n_b) \rho(n_a, n_b, t) - \beta_a A(n_a + 1, n_b) \rho(n_a + 1, n_b, t)$$

(11.5)

$$J_b(n_a, n_b) = v_b A(n_a, n_b) \rho(n_a, n_b, t) - \beta_b A(n_a, n_b + 1) \rho(n_a, n_b + 1, t)$$

(11.6)

Here v_i, $i = a, b$ is the *impingement rate* (per unit surface) of component i, i.e. the rate of collisions of i-monomers with a unit surface of the cluster; β_i, $i = a, b$ is the *evaporation rate* per unit surface of the component i, $A(n_a, n_b)$ is the surface area of the (n_a, n_b)-cluster. The impingement rates for gas-liquid nucleation follow the ideal gas kinetics (cf. (3.38))

$$v_i = \frac{y_i p^v}{\sqrt{2\pi m_i k_B T}}$$

(11.7)

where p^v is the total vapor pressure, y_i is the molar fraction of component i in the vapor. As in the single-component theory the evaporation rates are found from the detailed balance condition at the *constrained equilibrium* assuming that β_i is independent of the actual vapor pressure of the component i. By definition of constrained equilibrium (which throughout this chapter we denote by the subscript "eq"): $v_{i,\text{eq}} = v_i$ (cf. Chap. 3). Then from (11.5)–(11.6) the detailed balance condition $J_a = J_b = 0$ reads:

$$\beta_a = \frac{v_a \, \rho_{eq}(n_a, n_b) A(n_a, n_b)}{\rho_{eq}(n_a + 1, n_b) A(n_a + 1, n_b)} \tag{11.8}$$

$$\beta_b = \frac{v_b \, \rho_{eq}(n_a, n_b) A(n_a, n_b)}{\rho_{eq}(n_a, n_b + 1) A(n_a, n_b + 1)} \tag{11.9}$$

where $\rho_{eq}(n_a, n_b)$ is the cluster distribution function in constrained equilibrium. From the thermodynamic fluctuation theory it can be written as

$$\rho_{eq}(\mathbf{n}) = C \, \exp[-\beta \Delta G(\mathbf{n})] \tag{11.10}$$

where $\Delta G(\mathbf{n})$ is a minimal (reversible) work required to form the (n_a, n_b)-cluster and C is the normalizing factor. Given an appropriate thermodynamic model for $\Delta G(\mathbf{n})$, the most comprehensive way to evaluate the nucleation rate is by summing all fluxes J_a and J_b in Eqs. (11.5)–(11.9) that cross any arbitrary line joining the n_a and n_b axis [5, 6]. At the steady state the resulting nucleation rate must be constant.

11.3 "Direction of Principal Growth" Approximation

Carrying out the kinetic procedure outlined in Sect. 11.2 requires considerable computational effort due to the existence of a large number of nucleation paths. To proceed with an analytical approach it is then necessary to identify the domain in the (n_a, n_b) space bringing the major contribution to the nucleation rate. In its simplest form this approach requires

1. identification of the *"critical point"* in the cluster space (corresponding to the critical cluster),
2. determination of the *direction of the flow* in the critical point, and
3. making an assumption about the flow in the *vicinity of the critical point*

To cope with the problem of large number of paths contributing to the overall nucleation rate, Reiss [4] showed that the nucleation rate is primarily determined by the passage over the *saddle point* of the free energy surface $\Delta G(n_a, n_b)$. This approximation is based on the exponential dependence of ρ_{eq} on ΔG (recall that in the single-component case the main contribution to the nucleation rate comes from the vicinity of the maximum of $\Delta G(n)$). This statement addresses the first question raised above. Addressing the second one, Reiss suggested that the *direction* of the flow at the saddle point is determined by the direction of the steepest descent of the energy surface at this point, in other words this direction is determined solely by energetic factors. The latter issue was later revisited by Stauffer [7] who showed that the direction of the flow in the saddle point is also influenced by kinetics. Discussion below follows Stauffer's representation of binary nucleation kinetics [7].

Let us start with presenting the kinetic equation in vector notations. Equations (11.5)–(11.6) with the evaporation coefficients given by (11.8) and (11.9), can be written as

$$\mathbf{J} = -\rho_{eq}(\mathbf{n})\,\mathbf{F}(\mathbf{n})\,\nabla\left[\frac{\rho(\mathbf{n})}{\rho_{eq}(\mathbf{n})}\right], \qquad \mathbf{n} = (n_a, n_b) \qquad (11.11)$$

where the diagonal matrix $\mathbf{F}$ contains the forward collision rates:

$$\mathbf{F}(\mathbf{n}) = \begin{pmatrix} v_a\,A(\mathbf{n}) & 0 \\ 0 & v_b\,A(\mathbf{n}) \end{pmatrix}$$

The steady-state nucleation rate is obtained by integration of Eq. (11.11) along all possible nucleation paths subject to the boundary conditions

$$\lim_{n_a, n_b \to 0}\frac{\rho}{\rho_{eq}} = 1, \qquad \lim_{n_a, n_b \to \infty}\frac{\rho}{\rho_{eq}} = 0 \qquad (11.12)$$

which are similar to the single-component case (see discussion in Sect. 3.3). Multiplying Eq. (11.11) from the left by $(1/\rho_{eq})\mathbf{F}^{-1}$ and taking *curl* we obtain

$$\mathrm{curl}\left(\frac{1}{\rho_{eq}}\,(\mathbf{F}^{-1}\,\mathbf{J})\right) = 0 \qquad (11.13)$$

Applying the general identity

$$\mathrm{curl}\,(a\,\mathbf{x}) = a\,\mathrm{curl}\,\mathbf{x} - \mathbf{x} \times (\nabla a)$$

for

$$a \equiv \frac{1}{\rho_{eq}}, \qquad \mathbf{x} \equiv \mathbf{F}^{-1}\mathbf{J}$$

we derive using (11.10)

$$\mathrm{curl}\,(\mathbf{F}^{-1}\mathbf{J}) = (\mathbf{F}^{-1}\mathbf{J}) \times \nabla(\beta\Delta G) \qquad (11.14)$$

This equation determines the direction of the nucleation flux in *any point* (n_a, n_b) of the cluster space. It implies that the direction of the nucleation flux depends not only on the geometry of the energy surface $\Delta G(n_a, n_b)$ but also on the impingement rates of the components (through the matrix $\mathbf{F}$).

The saddle point $\mathbf{n}^* = (n_a^*, n_b^*)$ of the free energy surface satisfies

$$\left.\frac{\partial \Delta G}{\partial n_a}\right|_{\mathbf{n}^*} = \left.\frac{\partial \Delta G}{\partial n_b}\right|_{\mathbf{n}^*} = 0 \qquad (11.15)$$

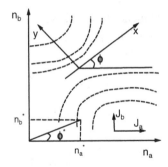

Fig. 11.2 Schematic illustration of the *direction of principal growth approximation*. *Dashed lines*: curves of constant Gibbs free energy $\Delta G(n_a, n_b)$. The origin of the $x - y$ coordinate system corresponds to the saddle point of ΔG. The angle φ gives the direction of principal growth determined by Eq. (11.25); the angle φ^* is the approximation to φ given by Eq. (11.46)

In its vicinity ΔG can be expanded as:

$$\Delta G(n_a, n_b) = \Delta G(\mathbf{n}^*) + m_a^2 \, D_{aa} + m_b^2 \, D_{bb} + 2m_a m_b \, D_{ab} \tag{11.16}$$

where

$$m_i = n_i - n_i^*, \qquad D_{ij} = \frac{1}{2} \frac{\partial^2 \Delta G}{\partial n_i \, \partial n_j} \bigg|_{\mathbf{n}^*}, \quad i, j = a, b$$

At the saddle point two eigenvalues of the symmetric Hessian matrix $\mathbf{D}$ have different signs implying that

$$\det D < 0$$

It is convenient to introduce a new, rotated, coordinate system $x(m_a, m_b)$, $y(m_a, m_b)$ with the origin at $\mathbf{n}^*$ and the x axis pointing along the direction of the flow at $\mathbf{n}^*$:

$$x = m_a \cos\varphi + m_b \sin\varphi, \qquad y = -m_a \sin\varphi + m_b \cos\varphi \tag{11.17}$$

where φ is the (yet unknown) angle between the x and n_a axis (see Fig. 11.2). The rate components in the new coordinates are:

$$J_x = J_a \cos\varphi + J_b \sin\varphi, \qquad J_y = -J_a \sin\varphi + J_b \cos\varphi \tag{11.18}$$

At the saddle point itself by definition of the rotated system

$$J_x = J_x^*(\mathbf{n}^*), \qquad J_y(\mathbf{n}^*) = 0$$

Fig. 11.3 Schematic illustration of the flux around the saddle-point in the rotated coordinate system (x, y). The flux is along the x-direction having a Gaussian form given by Eq. (11.19)

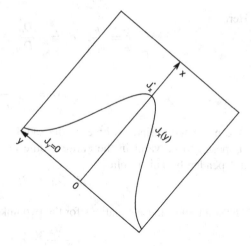

Further BCNT invokes a rather strong *Ansatz*—the *direction of principal growth approximation*: it is assumed that the *direction* of the flow (not the absolute value!) remains constant in the *entire saddle point region*:

$$J_y = 0 \quad \text{in the saddle point region}$$

Then, the continuity equation (11.4) written in the rotated system yields

$$\frac{\partial J_x(x, y)}{\partial x} = 0$$

implying that in the saddle point region the absolute value of the flux depends only on y: $J_x = J_x(y)$. Taking into account that in the same region ΔG is approximately parabolic, Eq. (11.14) suggests that J_x can be cast in the form:

$$J_x(y) = J_x^* e^{-\beta W y^2} \qquad (11.19)$$

where J_x^* is the flux at the saddle point and the dimensionless factor βW describes the width of the saddle point region (Fig. 11.3). Together with the direction of the flow, φ, it is found by substituting (11.19) into Eq. (11.14) which takes the form of a linear combination

$$Q_a\, m_a + Q_b\, m_b = 0 \qquad (11.20)$$

with

$$Q_a \equiv -w \sin^3 \varphi - wr \sin \varphi \cos^2 \varphi + r \cos \varphi + d_a \sin \varphi \qquad (11.21)$$

$$Q_b \equiv -wr \cos^3 \varphi - w \sin^2 \varphi \cos \varphi + rd_b \cos \varphi + \sin \varphi \qquad (11.22)$$

Here

$$r = \frac{v_b}{v_a}, \quad d_a = -\frac{D_{aa}}{D_{ab}}, \quad d_b = -\frac{D_{bb}}{D_{ab}} \tag{11.23}$$

and

$$w = -\frac{W}{D_{ab}} \tag{11.24}$$

(We neglected variations of $\mathbf{F}$ in the saddle point region). Expression (11.20) is supposed to be valid in the entire saddle point region with m_a and m_b varying independently. This implies

$$Q_a = Q_b = 0$$

The solution of these equations for the two unknowns φ and w is:

$$\tan \varphi = s + \sqrt{s^2 + r}, \quad \text{with } s = \frac{1}{2}(d_a - rd_b) \tag{11.25}$$

$$w = \left(\frac{1}{\sin \varphi \, \cos \varphi} \right) \left(\frac{\tan \varphi + rd_b}{\tan \varphi + \frac{r}{\tan \varphi}} \right) \tag{11.26}$$

Equation (11.25) states that the direction of the flux in the saddle point region is determined from a combination of energetic and kinetic factors. The steepest descent approximation of Ref. [4] would give $\cot(2\varphi) = (d_b - d_a)/2$. This would agree with Eq. (11.25) only when $r = 1$, i.e. when the impingement rates of the two components are equal. Let us look at the limiting cases of large and small r. From (11.25)

$$\tan \varphi = \frac{1}{d_b}, \quad \text{for } v_b \gg v_a \tag{11.27}$$

$$\tan \varphi = d_a, \quad \text{for } v_b \ll v_a \tag{11.28}$$

The remaining unknown quantity in (11.19) is the flux at the saddle point. To find it let us write the vector equation (11.11) in the rotated system. Since $J_y = 0$ we are interested only in the x-component of this equation which reads

$$(\mathbf{F}^{-1} \mathbf{J})_x \frac{1}{\rho_{eq}(x, y)} = -\frac{\partial}{\partial x} \left(\frac{\rho(x, y)}{\rho_{eq}(x, y)} \right) \tag{11.29}$$

The boundary conditions (11.12) in the rotated system are:

$$\lim_{x \to -\infty} \frac{\rho}{\rho_{eq}} = 1, \quad \lim_{x \to \infty} \frac{\rho}{\rho_{eq}} = 0$$

implying that integration of Eq. (11.29) over x gives:

$$\int_{-\infty}^{\infty} dx \, (\mathbf{F}^{-1} \mathbf{J})_x \frac{1}{\rho_{eq}(x, y)} = 1 \tag{11.30}$$

The x-coordinate of the vector $\mathbf{F}^{-1}\mathbf{J}$ reads (cf. (11.18)):

$$(\mathbf{F}^{-1}\mathbf{J})_x = (\mathbf{F}^{-1}\mathbf{J})_a \cos\varphi + (\mathbf{F}^{-1}\mathbf{J})_b \sin\varphi \qquad (11.31)$$

where

$$(\mathbf{F}^{-1}\mathbf{J})_a = \frac{J_a}{v_a A}, \quad (\mathbf{F}^{-1}\mathbf{J})_b = \frac{J_b}{v_b A} \qquad (11.32)$$

Using the standard linear algebra we express from (11.18) the "old" flux coordinates in terms of the "new" ones, taking into account that $J_y = 0$:

$$J_a = J_x \cos\varphi, \quad J_b = J_x \sin\varphi \qquad (11.33)$$

Substituting (11.33) into (11.31)–(11.32) we find

$$(\mathbf{F}^{-1}\mathbf{J})_x = J_x \frac{1}{A}\left[\frac{v_a \sin^2\varphi + v_b \cos^2\varphi}{v_a v_b}\right] \qquad (11.34)$$

Now it is convenient to introduce the average impingement rate

$$v_{av} = \frac{v_a v_b}{v_a \sin^2\varphi + v_b \cos^2\varphi} \qquad (11.35)$$

so that (11.34) takes the form of expressions (11.32)

$$(\mathbf{F}^{-1}\mathbf{J})_x = J_x(y)\frac{1}{v_{av} A}, \quad A = A(x, y) \qquad (11.36)$$

Let us substitute (11.36) into (11.30); in view of the exponential dependence of ρ_{eq} on x and y we can replace $A = A(x, y)$ by its value A^* at the saddle point and take it out from the integral:

$$J_x(y) = v_{av} A^*\left[\int_{-\infty}^{+\infty} dx \frac{1}{\rho_{eq}(x, y)}\right]^{-1} \qquad (11.37)$$

The equilibrium distribution in (x, y) coordinates is:

$$\rho_{eq}(x, y) = C e^{-\beta \Delta G(x,y)}$$

In the saddle-point region $\Delta G(x, y)$ has the parabolic form:

$$\beta \Delta G(x, y) = g^* + p_{11} x^2 + 2p_{12} xy + p_{22} y^2$$

where $g^* = \beta \Delta G(0,0)$ is the Gibbs free energy at the saddle point,

$$p_{11} = \frac{1}{2} \frac{\partial^2 \beta \Delta G}{\partial x^2}\bigg|_{(0,0)} , \quad p_{12} = \frac{1}{2} \frac{\partial^2 \beta \Delta G}{\partial x \, \partial y}\bigg|_{(0,0)} , \quad p_{22} = \frac{1}{2} \frac{\partial^2 \beta \Delta G}{\partial y^2}\bigg|_{(0,0)}$$

Here $p_{11} < 0$, $p_{22} > 0$. The integral on the right-hand side of (11.37) reads

$$\int_{-\infty}^{+\infty} dx \, \frac{1}{\rho_{eq}(x,y)} = \frac{e^{g^*}}{C} \exp(p_{22} \, y^2) \int_{-\infty}^{+\infty} dx \, \exp[p_{11} \, x^2 + 2 p_{12} \, xy]$$

Gaussian integration gives

$$\int_{-\infty}^{+\infty} dx \, \frac{1}{\rho_{eq}(x,y)} = \frac{e^{g^*}}{C} \sqrt{\frac{\pi}{(-p_{11})}} \exp\left[\frac{p_{12}^2 \, y^2}{4\,(-p_{11})} + p_{22} \, y^2 \right]$$

Then Eq. (11.37) takes the form

$$J_x(y) = C \, v_{av} \, A^* \, e^{-g^*} \sqrt{\frac{(-p_{11})}{\pi}} \exp\left\{ -\left[\frac{p_{12}^2}{4\,(-p_{11})} + p_{22} \right] y^2 \right\} \qquad (11.38)$$

Comparing it with (11.19) we identify

$$J_x^* = C \, v_{av} \, A^* \sqrt{\frac{(-p_{11})}{\pi}} \, e^{-g^*} \qquad (11.39)$$

$$\beta W = \frac{p_{12}^2}{4\,(-p_{11})} + p_{22} \qquad (11.40)$$

Finally, the total steady-state nucleation rate is given by Gaussian integration of $J_x(y)$ over y

$$J = \int_{-\infty}^{+\infty} dy \, J_x(y) = J_x^* \sqrt{\frac{\pi}{\beta W}} \qquad (11.41)$$

resulting in

$$J = K \, e^{-g^*} \qquad (11.42)$$

$$K = \mathscr{Z} \, v_{av} \, A(\mathbf{n}^*) \, C \qquad (11.43)$$

The prefactor K has the form analogous to the prefactor J_0 in the single-component case (cf. (3.54)) in which the impingement rate v is replaced by v_{av}. The Zeldovich factor $\mathscr{Z}$ determines the shape of the Gibbs free energy surface in the saddle point region:

$$\mathscr{Z} = -\frac{1}{2} \frac{(\partial^2 \Delta G / \partial x^2)_{\mathbf{n}^*}}{\sqrt{-\det \mathbf{D}}} \qquad (11.44)$$

In the original (n_a, n_b)-coordinates:

$$\mathscr{L} = -\frac{1}{2} \left[\frac{D_{aa} + 2D_{ab} \tan\varphi + D_{bb} \tan^2\varphi}{1 + \tan^2\varphi} \right] \frac{1}{\sqrt{-\det D}} \tag{11.45}$$

In simplified approaches the angle φ is approximated by the angle φ^* characterizing the critical cluster [8], as shown in Fig. 11.2:

$$\varphi^* = \arctan\left(\frac{n_b^*}{n_a^*}\right) \tag{11.46}$$

Note, that although Eqs. (11.42)–(11.43) are similar to the single-component case, it is not possible to recover the single-component nucleation rate from it by setting one of the impingement rates to zero: this would lead to $v_{av} = 0$. This result is a manifestation of the general statement concerning the reduction of the dimensionality of the physical problem. Such a reduction implies the abrupt change of symmetry which can not be derived by smooth vanishing of one of the parameters of the system.

11.4 Energetics of Binary Cluster Formation

Energetics of cluster formation determines the minimum reversible work $\Delta G(n_a, n_b)$ needed to form the (n_a, n_b)-cluster in the surrounding vapor at the constant temperature T and the vapor pressure p^v. As in the single-component case we introduce an arbitrary located Gibbs dividing surface distinguishing between the bulk (superscript "l") and excess (superscript "exc") molecules of each species in the cluster. The state of the cluster is characterized by the total numbers of molecules n_i:

$$n_i = n_i^l + n_i^{exc}, \quad i = a, b \tag{11.47}$$

Each of the quantities in the right-hand side depend on the location of the dividing surface while their sum can be assumed independent of this location to the relative accuracy of $O(\rho^v/\rho^l)$, where ρ^v and ρ^l are the number densities in the vapor and liquid phases. Therefore only $n_i \geq 0$ are observable physical properties; in this sense the *model quantities* n_i^l and n_i^{exc} can be both positive or negative.

In a unary system (see Sect. 3.2) we chose the *equimolar* dividing surface characterized by zero adsorption $n^{exc} = 0$. This choice made it possible to deal only with the bulk numbers of cluster molecules. For a mixture, however, it is impossible to choose a dividing surface in such a way that *all* excess terms n_i^{exc} vanish [9]. This is the reason for occurrence of the *surface enrichment*—preferential adsorption of one of the species relative to the other. As a result the composition inside the droplet can be different from that near its surface. For binary (and in general, multi-component) nucleation problem introduction of the Gibbs surface is a nontrivial issue.

Thermodynamic considerations [10, 11] analogous to those of Sect. 3 yield a binary mixture analogue of Eq. (3.19):

$$\Delta G = (p^v - p^l)V^l + \gamma A + \sum_{i=a,b} n_i^l \left[\mu_i^l(p^l) - \mu_i^v(p^v)\right] + \sum_{i=a,b} n_i^{exc} \left[\mu_i^{exc} - \mu_i^v(p^v)\right] \tag{11.48}$$

Here p^l is the pressure inside the cluster,

$$V^l = \sum_i n_i^l v_i^l$$

is the cluster volume, v_i^l is the partial molecular volume of component i in the liquid phase (see Appendix D), A is the surface area of the cluster calculated at the location of the dividing surface; γ is the surface tension at the dividing surface. Equation (11.48) presumes that formation of a cluster does not affect the surrounding vapor and therefore the chemical potential of a molecule in the vapor remains unchanged. Within the capillarity approximation the bulk properties of the cluster are those of the bulk liquid phase in which liquid is considered incompressible. This yields:

$$\mu_i^l(p^l) = \mu_i^l(p^v) + v_i^l(p^l - p^v) \tag{11.49}$$

Then

$$\sum_i n_i^l \left[\mu_i^l(p^l) - \mu_i^v(p^v)\right] = \sum_i n_i^l \left[\mu_i^l(p^v) - \mu_i^v(p^v)\right] - (p^v - p^l) V^l$$

The last term in this expression cancels the first term in (11.48), leading to

$$\Delta G = \gamma A - \sum_i n_i^l \Delta\mu_i + \sum_i n_i^{exc} \left[\mu_i^{exc} - \mu_i^v(p^v)\right] \tag{11.50}$$

where

$$\Delta\mu_i \equiv \mu_i^v(p^v) - \mu_i^l(p^v) \tag{11.51}$$

is the difference in the chemical potential of a molecule of component i between the vapor and liquid phases *taken at the vapor pressure* p^v. The chemical potential of a molecule of species i in the liquid phase $\mu_i^l(p^v, x_b^l)$ depends on the bulk composition of the (n_a, n_b)-cluster

$$x_b^l = \frac{n_b^l}{n_a^l + n_b^l} \tag{11.52}$$

The last term in (11.50) contains the excess quantities. The chemical potentials μ_i^{exc} refer to a *hypothetical (non-physical)* surface phase and therefore can not be measured in experiment or predicted theoretically. Therefore, one has to introduce an Ansatz for them which serves as a closure of the model [12, 13]. The diffusion coefficient in liquids is much higher than in gases (see e.g. [14]). This implies that diffusion between the surface and the interior of the cluster is much faster than diffusion between the surface and the mother vapor phase surrounding it. Hence, it is plausible to assume equilibrium between surface and the interior (liquid) phase of the cluster, resulting in the equality of the chemical potentials

$$\mu_i^{exc} = \mu_i^l(p^l, x_b^l) \tag{11.53}$$

Following [13] the Ansatz (11.53) can be termed "the equilibrium μ conjecture". Using (11.49) we may write

$$\mu_i^{exc} - \mu_i^v(p^v) = \left[\mu_i^l(p^v) - \mu_i^v(p^v)\right] + v_i^l(p^l - p^v) \equiv -\Delta\mu_i + v_i^l(p^l - p^v) \tag{11.54}$$

Substituting (11.54) into (11.50) and using Laplace equation we obtain an alternative form of the Gibbs formation energy of the binary cluster

$$\Delta G = \gamma(x_b^l) A - \sum_i \underbrace{\left[n_i^l + n_i^{exc}\right]}_{n_i} \Delta\mu_i + \frac{2\gamma(x_b^l)}{r}\left[\sum_i n_i^{exc} v_i^l\right] \tag{11.55}$$

where $\gamma(x_b^l)$ is the surface tension of the binary solution of liquid composition x_b^l. Here the second term contains the *total* numbers of molecules in the cluster, and not the bulk liquid numbers n_i^l. At the same time all the thermodynamic properties are functions of the bulk composition x_b^l and not the total composition

$$x_b^{tot} = n_b/(n_a + n_b) \tag{11.56}$$

Those two are not identical: $x_b^l \neq x_b^{tot}$. The same refers to the volume and the surface area of the cluster

$$V^l = \frac{4\pi}{3} r^3 = \sum_i n_i^l v_i^l, \qquad A = (36\pi)^{1/3}\left(\sum_i n_i^l v_i^l\right)^{2/3} \tag{11.57}$$

since by definition any dividing surface has a zero thickness. An important feature of the binary problem is the presence of the last term in the Gibbs energy (11.55).

11.5 Kelvin Equations for the Mixture

In the previous section we derived the Gibbs energy of formation for an arbitrary binary cluster. Consider now the *critical* cluster corresponding to the saddle point of ΔG:

$$\left.\frac{\partial \Delta G}{\partial n_j^l}\right|_{\mathbf{n}^*} = 0, \quad j = a, b \tag{11.58}$$

$$\left.\frac{\partial \Delta G}{\partial n_j^{\text{exc}}}\right|_{\mathbf{n}^*} = 0, \quad j = a, b \tag{11.59}$$

Since the formation of a cluster does not affect the vapor properties, we have

$$d\mu_i^{\text{v}} = dp^{\text{v}} = 0 \tag{11.60}$$

Equations (11.50) and (11.58)–(11.59) then result in:

$$\gamma \frac{\partial A}{\partial n_j^l} + A \frac{\partial \gamma}{\partial n_j^l} - \Delta\mu_j + \sum_i n_i^l \frac{\partial \mu_i^l(p^{\text{v}})}{\partial n_j^l} + \sum_i n_i^{\text{exc}} \frac{\partial \mu_i^{\text{exc}}}{\partial n_j^l} = 0, \quad j = a, b \tag{11.61}$$

$$A \frac{\partial \gamma}{\partial n_j^{\text{exc}}} - (\mu_j^{\text{v}} - \mu_j^{\text{exc}}) + \sum_i n_i^{\text{exc}} \frac{\partial \mu_i^{\text{exc}}}{\partial n_j^{\text{exc}}} + \sum_i n_i^l \frac{\partial \mu_i^l(p^{\text{v}})}{\partial n_j^{\text{exc}}} = 0, \quad j = a, b \tag{11.62}$$

Gibbs adsorption equation (2.29) for the mixture at constant T reads

$$A d\gamma + \sum_i n_i^{\text{exc}} d\mu_i^{\text{exc}} = 0 \tag{11.63}$$

We rewrite it as

$$A \frac{\partial \gamma}{\partial n_j^\alpha} + \sum_i n_i^{\text{exc}} \frac{\partial \mu_i^{\text{exc}}}{\partial n_j^\alpha} = 0, \quad \alpha = l, \text{exc} \tag{11.64}$$

Gibbs-Duhem equation (2.5) for the bulk liquid phase of the mixture at constant T is

$$-V^l dp^l + \sum_i n_i^l d\mu_i^l(p^l) = 0$$

Using (11.49) and (11.60) it can be presented as

$$\sum_i n_i^l d\mu_i^l(p^{\text{v}}) = 0 \tag{11.65}$$

resulting in

$$\sum_i n_i^l \frac{\partial \mu_i^l(p^{\text{v}})}{\partial n_j^\alpha} = 0, \quad \alpha = l, \text{exc} \tag{11.66}$$

Substituting (11.64) and (11.66) into Eqs. (11.61)–(11.62) we obtain for the saddle point:

$$\gamma \frac{\partial A}{\partial n_j^l} - \Delta \mu_j = 0 \tag{11.67}$$

$$\mu_j^{exc}(p^l) = \mu_j^v(p^v) \tag{11.68}$$

The meaning of the second equality is transparent. As we discussed, for *any* cluster the molecules belonging to the dividing surface are in equilibrium with the interior of the cluster $\mu_j^{exc}(p^l) = \mu_j^l(p^l)$. For the *critical cluster* this condition is supplemented by the condition of unstable equilibrium with surrounding vapor yielding

$$\mu_j^{exc}(p^l) = \mu_j^l(p^l) = \mu_j^v(p^v)$$

which is exactly the equation (11.68).

The first equality (Eq. (11.67)) is nontrivial. In view of (11.57) it reads

$$- \Delta \mu_a + \frac{2\gamma \, v_a^l}{r^*} = 0 \tag{11.69}$$

$$- \Delta \mu_b + \frac{2\gamma \, v_b^l}{r^*} = 0 \tag{11.70}$$

This set of equations is known as the *Kelvin equations for a mixture*; they determine the composition and the size of the critical cluster. In particular, the critical cluster composition satisfies

$$\frac{\Delta \mu_a}{v_a^l} = \frac{\Delta \mu_b}{v_b^l} \tag{11.71}$$

Once the composition x_b^{l*} is determined, the critical radius is given by

$$r^* = \frac{2\gamma \, v_j^l}{\Delta \mu_j} \tag{11.72}$$

Here the surface tension γ refers to the dividing surface of the radius r^*.

Substituting the Kelvin equations into (11.55) we find the Gibbs free energy at the saddle point (or, equivalently, the nucleation barrier):

$$\Delta G^* = \gamma A + \sum_i n_i^{l*} \left(-\frac{2\gamma v_i^l}{r^*} \right) = \gamma \left(A - \frac{2}{r} V^l \right)$$

resulting in:

$$\Delta G^* = \frac{1}{3}\gamma A \tag{11.73}$$

where both quantities on the right-hand side depend on the critical cluster composition. Within the phenomenological approach droplets are considered to be relatively large, so that one can replace γ by $\gamma_\infty(x_b^{1*})$—the surface tension of the plain layer of the binary vapor- binary liquid system when the composition of the bulk liquid is that of the critical cluster.

Note that if within the capillarity approximation we would set $n_a^{\text{exc}} = n_b^{\text{exc}} = 0$, then the terms with the excess quantities in the free energy would disappear and the Gibbs adsorption equation (11.63) can not be invoked. The resulting equations for the critical cluster would contain then the uncompensated term with the surface tension derivative:

$$- \Delta\mu_j + \frac{2\gamma_\infty v_j^1}{r^*} + A \frac{\partial\gamma_\infty}{\partial x_j^{\text{tot}}} = 0, \quad j = a, b \tag{11.74}$$

The inconsistency of this result becomes obvious if we recall that in equilibrium $\mu_i^1(p^1) = \mu_i^v(p^v)$ leading to Eqs. (11.69)–(11.70). That is why the phenomenological binary nucleation model with the critical cluster given by Eqs. (11.69)–(11.70) is called the *internally consistent* form of the BCNT ([10, 11]).

11.6 K-Surface

In the general expression for the Gibbs energy (11.50) (or its equivalent form (11.55)) the bulk and excess numbers of molecules are not specified and treated as independent variables. Their specification is related to a choice of the dividing surface for a cluster. This is not a unique procedure. One of the appropriate options is the *equimolar surface for the mixture*, termed also the *K-surface* [13, 15], defined through the requirement

$$\sum_{i=a,b} n_i^{\text{exc}} v_i^1 = 0 \tag{11.75}$$

This choice ensures that the macroscopic surface tension is independent of the curvature of the drop; however it does depend on the composition of the cluster. This can be easily seen if we present Eq. (11.55) in the form

$$\Delta G = - \sum_i \underbrace{\left[n_i^1 + n_i^{\text{exc}} \right]}_{n_i} \Delta\mu_i + \gamma(x_b^1; r) A$$

where we introduced the curvature dependent surface tension

$$\gamma(x_b^1; r) = \gamma(x_b^1) \left\{ 1 + \frac{2}{r A} \left[\sum_i n_i^{\text{exc}} v_i^1 \right] \right\}$$

For the *K*-surface the second term in the curl brackets vanishes implying that $\gamma(x_b^l; r) = \gamma(x_b^l)$. Laaksonen et al. [15] showed that the *K*-surface brings together various derivations of the free energy of cluster formation in the classical theory— due to Wilemski [10], Debenedetti [16] and Oxtoby and Kashchiev [17]. In what follows we adopt the K-surface formalism for binary clusters. Usually, the partial molecular volumes of both components in the liquid phase are positive, implying from (11.75) that the excess quantities n_i^{exc} have different signs. As we mentioned already, a negative value of one of n_i^{exc} is not unphysical as soon as the total number n_i (the quantity which does not depend on the choice of dividing surface) is nonnegative.

Within the *K*-surface formalism, the volume of the cluster and its surface area can be expressed either in terms of n_i^l or in terms of the *total* numbers of molecules $n_i^{tot} \equiv n_i$. Equation (11.57) reads:

$$V^l = \sum_i n_i^l v_i^l = \sum_i n_i^{tot} v_i^l \tag{11.76}$$

$$A = (36\pi)^{1/3} \left(\sum_i n_i^l v_i^l \right)^{2/3} = (36\pi)^{1/3} \left(\sum_i n_i^{tot} v_i^l \right)^{2/3} \tag{11.77}$$

It is important to stress that consistent evaluation of thermodynamic properties requires that μ_i^l, v_i^l, γ are functions of the *bulk* composition of the cluster x_b^l— and not the total composition x_b^{tot}) [10]—and the difference matters. Combination of (11.75) with Gibbs adsorption equation (11.63) results in the set of linear equations for the excess numbers

$$n_a^{exc} v_a^l + n_b^{exc} v_b^l = 0 \tag{11.78}$$

$$n_a^{exc} d\mu_a^l(p^l, x_b^l) + n_b^{exc} d\mu_b^l(p^l, x_b^l) + A \, d\gamma = 0 \tag{11.79}$$

From the incompressibility of the liquid phase and Laplace equation

$$\mu_i^l(p^l, x_b^l) = \mu_i^l(p^v, x_b^l) + \frac{2\gamma v_i^l}{r} \tag{11.80}$$

Excluding n_a^{exc} from (11.78), we obtain from (11.79) and (11.80)

$$n_b^{exc} \left\{ \left(-\frac{v_b^l}{v_a^l} \right) d\mu_a + d\mu_b + \frac{2\gamma}{r} v_b^l d \left(\ln \frac{v_b^l}{v_a^l} \right) \right\} + A \, d\gamma = 0 \tag{11.81}$$

where for brevity we used the notation $\mu_i \equiv \mu_i^l(p^v, x_b^l)$. Partial molecular volumes can be written as (see Appendix D):

$$v_i^1 = \frac{1}{\rho^1} \eta_i^1 \qquad (11.82)$$

with η_i^1 given by (D.7)–(D.8). Combining (11.81) and (11.82) with the Gibbs-Duhem equation

$$\sum_i x_i^1 \, d\mu_i = 0 \qquad (11.83)$$

we find

$$n_a^{\text{exc}} = -A \frac{\partial \gamma}{\partial x_b^1} \left[\frac{1}{x_b^1 \eta_b^1} \frac{\partial \mu_a}{\partial x_b^1} + \frac{2\gamma \eta_a^1}{r \rho^1} \frac{\partial \ln(\eta_a^1/\eta_b^1)}{\partial x_b^1} \right]^{-1} \qquad (11.84)$$

$$n_b^{\text{exc}} = -A \frac{\partial \gamma}{\partial x_b^1} \left[\frac{1}{x_a^1 \eta_a^1} \frac{\partial \mu_b}{\partial x_b^1} + \frac{2\gamma \eta_b^1}{r \rho^1} \frac{\partial \ln(\eta_b^1/\eta_a^1)}{\partial x_b^1} \right]^{-1} \qquad (11.85)$$

To perform calculations of n_i^{exc} according to (11.84)–(11.85) we need to specify $\partial \mu_i^1/\partial x_b^1$. This can be done using a correlation for activity coefficients [14]. To a good approximation the activity coefficients in the liquid can be set equal to unity, resulting in

$$\frac{\partial \mu_a^1}{\partial x_b^1} = -k_B T \frac{1}{x_a^1} \qquad (11.86)$$

$$\frac{\partial \mu_b^1}{\partial x_b^1} = k_B T \frac{1}{x_b^1} \qquad (11.87)$$

It is easy to see that these expressions satisfy the Gibbs-Duhem relation (11.83). A more accurate approximation can be formulated using one of the more sophisticated models for activity coefficients—e.g. van Laar model discussed in Sect. 11.9.1.

Finally, the terms $\partial \ln(\eta_i^1/\eta_j^1)/\partial x_b^1$ in (11.84)–(11.85) are found from (D.7)–(D.9):

$$\frac{\partial \ln(\eta_a^1/\eta_b^1)}{\partial x_b^1} = \frac{1}{\eta_a^1 \eta_b^1} (\tau_2 - \tau_1^2) \qquad (11.88)$$

$$\frac{\partial \ln(\eta_b^1/\eta_a^1)}{\partial x_b^1} = -\frac{\partial \ln(\eta_a^1/\eta_b^1)}{\partial x_b^1} = -\frac{1}{\eta_a^1 \eta_b^1} (\tau_2 - \tau_1^2) \qquad (11.89)$$

where

$$\tau_1 = \frac{\partial \ln \rho^1}{\partial x_b^1}, \quad \tau_2 = \frac{\partial \tau_1}{\partial x_b^1}$$

If the excess numbers, derived using this procedure, turn out to be not small compared to the bulk numbers then the classical theory, probably, falls apart [8]. This happens in the mixtures with strongly surface active components exhibiting pronounced adsorption on the K-surface. An example of such a system is the water/ethanol mixture (discussed in Sect. 11.9.3).

11.7 Gibbs Free Energy of Cluster Formation Within K-Surface Formalism

If we choose the K-dividing surface, the Gibbs energy of cluster formation (11.55) becomes

$$\Delta G = \gamma(x_b^l, T) A - \sum_i \underbrace{\left[n_i^l + n_i^{exc} \right]}_{n_i} \left[\mu_i^v(p^v, T) - \mu_i^l(p^v, x_b^l, T) \right] \quad (11.90)$$

Within the K-surface formalism for each pair of bulk cluster molecules (n_a^l, n_b^l) the excess quantities are constructed

$$n_a^{exc}(n_a^l, n_b^l), \quad n_b^{exc}(n_a^l, n_b^l)$$

implying that n_i^l and n_i^{exc} are not any more the independent quantities. Therefore, the saddle point of ΔG has to be determined in the space of independent variables—the total number of molecules:

$$\frac{\partial \Delta G}{\partial n_j} = 0, \quad j = a, b \quad (11.91)$$

leading again to the Kelvin equations (11.69)–(11.70). From the first sight it may seem that to construct $\Delta G(n_a, n_b)$ according to Eq. (11.90) one needs to know only the total numbers of molecules n_a and n_b; however this is not true, since the thermodynamic properties—the surface tension γ, partial molecular volumes v_i^l and chemical potentials μ_i^l—depend on the bulk composition x_i^l rather than on x_i^{tot}. So, in order to calculate $\Delta G(n_a, n_b)$, one has to know also the bulk numbers n_a^l and n_b^l (and therefore the bulk composition x_b^l), giving rise to these n_a and n_b.

Equation (11.90) formally coincides with the BCNT expression

$$\Delta G_{BCNT} = \gamma(x_b^{tot}, T) A - \sum_i n_i \left[\mu_i^v(p^v, T) - \mu_i^l(p^v, x_b^{tot}, T) \right] \quad (11.92)$$

except for the argument of μ_i^l and γ. This means that the standard BCNT does not discriminate between the bulk and excess molecules in the cluster and thus does not account for adsorption effects: a cluster in this model is a homogeneous object.

The Gibbs free energy contains the chemical potentials of the species which are not directly measurable quantities. Therefore, it is necessary to cast ΔG in an approximate form containing the quantities which are either measurable or can be calculated from a suitable equation of state. Let us first recall that μ_i^v is imposed by external conditions and does not depend on the composition of the cluster; at the same time μ_i^l is essentially determined by the cluster composition. For this quantity using the incompressibility of the liquid phase we may write

$$\mu_i^l(p^v, x_b^l) = \mu_i^l(p_0, x_b^l) + v_i^l (p^v - p_0) \tag{11.93}$$

where p_0 is an arbitrary chosen reference pressure. Let us choose it from the condition of bulk (x_b^l, T)-equilibrium. The latter is the equilibrium between the bulk binary liquid at temperature T having the composition x_b^l and the binary vapor. Fixing x_b^l and T, we can calculate from the EoS the corresponding coexistence pressure $p^{\text{coex}}(x_b^l, T)$ and coexistence vapor fractions of the components $y_i^{\text{coex}}(x_b^l, T)$. Now we choose p_0 as

$$p_0 = p^{\text{coex}}(x_b^l, T)$$

which transforms (11.93) into

$$\mu_i^l(p^v, x_b^l) = \mu_i^l \left(p^{\text{coex}}(x_b^l) \right) + v_i^l \left(p^v - p^{\text{coex}} \right) \tag{11.94}$$

For the gaseous phase we assume the ideal mixture behavior, implying that each component i behaves as if it were alone at the pressure $p_i^v = y_i p^v$. Then the chemical potential of component i in the binary vapor is approximately equal to its value for the pure i-vapor at the pressure p_i^v:

$$\mu_i^v(p^v, y_i) \approx \mu_{i,\text{pure}}^v(p_i^v) \tag{11.95}$$

The latter can be written as

$$\mu_{i,\text{pure}}^v(p_i^v) = \mu_{i,\text{pure}}^v(p_i') + \int_{p_i'}^{p_i^v} v_i^v(p) \, dp \tag{11.96}$$

where $v_i^v(p)$ is the molecular volume of the pure vapor i and p_i' is another arbitrary reference pressure. Let us choose it equal to the partial vapor pressure of component i at (x_b^l, T)-equilibrium:

$$p_i' = y_i^{\text{coex}}(x_b^l) \, p^{\text{coex}}(x_b^l) \equiv p_i^{\text{coex}}(x_b^l) \tag{11.97}$$

Applying the ideal gas law to (11.96), we obtain

$$\mu_i^{\rm v}(p^{\rm v}, y_i) = \mu_{i,{\rm pure}}^{\rm v}(p_i^{\rm coex}) + k_{\rm B}T\,\ln\left[\frac{y_i\,p^{\rm v}}{y_i^{\rm coex}(x_b^{\rm l})\,p^{\rm coex}(x_b^{\rm l})}\right] \qquad (11.98)$$

In $(x_b^{\rm l}, T)$-equilibrium $\mu_{i,{\rm pure}}^{\rm v}(p_i^{\rm coex}) = \mu_i^{\rm l}(p^{\rm coex})$. Subtracting (11.94) from (11.98), we find

$$\beta\Delta\mu_i = \ln\left[\frac{y_i\,p^{\rm v}}{y_i^{\rm coex}(x_b^{\rm l})\,p^{\rm coex}(x_b^{\rm l})}\right] - \left\{\frac{p^{\rm coex}v_i^{\rm l}}{k_{\rm B}T}\right\}\left(\frac{p^{\rm v}}{p^{\rm coex}} - 1\right)$$

The quantity in the curl brackets is proportional to the liquid compressibility factor, which is a small number ($\sim 10^{-6} - 10^{-2}$), implying that the second term can be neglected in favor of the first one:

$$\beta\Delta\mu_i = \ln\left[\frac{y_i\,p^{\rm v}}{y_i^{\rm coex}(x_b^{\rm l})\,p^{\rm coex}(x_b^{\rm l})}\right] \qquad (11.99)$$

Substituting (11.99) into (11.90), we deduce the desired approximation for the free energy containing now only the measurable quantities

$$\beta\Delta G(n_a, n_b) = -\sum_i n_i\,\ln\left[\frac{y_i\,p^{\rm v}}{y_i^{\rm coex}(x_b^{\rm l})\,p^{\rm coex}(x_b^{\rm l})}\right] + \beta\,\gamma(x_b^{\rm l})\,A \qquad (11.100)$$

This result coincides with the BCNT expression [4] except for the argument $x_b^{\rm l}$ of the coexistence properties:

$$\beta\Delta G_{\rm BCNT}(n_a, n_b) = -\sum_i n_i\,\ln\left[\frac{y_i\,p^{\rm v}}{y_i^{\rm coex}(x_b^{\rm tot})\,p^{\rm coex}(x_b^{\rm tot})}\right] + \beta\,\gamma(x_b^{\rm tot})\,A$$

$$(11.101)$$

Equations (11.100)–(11.101) imply that for an arbitrary (n_a, n_b)-cluster the bulk (logarithmic) terms can be both positive and negative depending on the cluster composition. Consequently, even at equilibrium conditions (say, at a fixed $p^{\rm v}$ and T) the free energy of formation of a cluster with a composition, different from the equilibrium bulk liquid composition at given $p^{\rm v}$ and T, will contain a non-zero bulk contribution. This situation is distinctly different from the single-component case, for which formation of an arbitrary cluster in equilibrium (saturated) vapor is associated only with the energy cost to build its surface and the bulk contribution to ΔG vanishes.

11.8 Normalization Factor of the Equilibrium Cluster Distribution Function

To accomplish the formulation of the theory it is necessary to determine the "normalization factor" C of the equilibrium cluster distribution function entering the prefactor K of the nucleation rate (see Eq. (11.43)). Its form is not "dictated" by the model presented in this chapter and remains a matter of controversy. In his seminal paper [4] Reiss proposed the following expression:

$$C_{\text{Reiss}} = \rho_a^{\text{v}} + \rho_b^{\text{v}} \tag{11.102}$$

where ρ_i^{v} is the number density of monomers of species i in the vapor. Such a choice, however, violates the law of mass action (recall the similar feature of the single-component CNT discussed in Sect. 3.6). Another difficulty associated with (11.102), is that the number density of pure a-clusters, $\rho_{\text{eq}}(n_a, 0)$, becomes proportional to the number density of b-monomers and vice versa. Wilemski and Wyslouzil [18] proposed an alternative form of C which is free from these inconsistencies. It was suggested that C should depend on the cluster composition:

$$C_{\text{WW}} = \left[\rho_a^{\text{v,coex}}(x_a^{\text{tot}})\right]^{x_a^{\text{tot}}} \left[\rho_b^{\text{v,coex}}(x_a^{\text{tot}})\right]^{x_b^{\text{tot}}}, \quad x_a^{\text{tot}} + x_b^{\text{tot}} = 1 \tag{11.103}$$

where $\rho_i^{\text{v, coex}}(x_i^{\text{tot}})$ is the equilibrium number density of monomers of species i in the binary vapor at coexistence with the binary liquid whose composition is x_i^{tot}. Another possibility discussed by the same authors, is the *self-consistent classical* (SCC) form of C based on the Girshick-Chiu ICCT model (3.103):

$$C_{\text{WW,SSC}} = \exp\left(x_a^{\text{tot}}\theta_{\infty,a} + x_b^{\text{tot}}\theta_{\infty,b}\right) \left[\rho_a^{\text{v, coex}}(x_a^{\text{tot}})\right]^{x_a^{\text{tot}}} \left[\rho_b^{\text{v, coex}}(x_a^{\text{tot}})\right]^{x_b^{\text{tot}}} \tag{11.104}$$

where $\theta_{\infty,i}(T)$ is the reduced macroscopic surface tension of pure component i. Mention, that Eqs. (11.103), (11.104) are just two of possible choices of the prefactor C.

11.9 Illustrative Results

11.9.1 Mixture Characterization: Gas-Phase-and Liquid-Phase Activities

Experimental results in binary nucleation are frequently expressed in terms of the gas-phase- and liquid-phase activities of the mixture components. *The gas-phase activity* of component i measures the deviation of the vapor of component i from equilibrium at a certain reference state. If we characterize the state of component i

in the vapor by its chemical potential μ_i^v, the gas-phase activity of component i is defined through

$$\mathscr{A}_i^v = \exp\left[\beta(\mu_i^v - \mu_{i,0}^v)\right] \tag{11.105}$$

where $\mu_{i,0}^v$ is the value of μ_i^v at the reference state which has yet to be specified. Let us assume that the binary vapor at the total pressure p^v is a mixture of ideal gases with the partial vapor pressures $p_i^v = y_i\, p^v$. Then, the chemical potential of component i in the binary vapor is given by Eqs. (11.95)–(11.96):

$$\mu_i^v(p_i^v) = \mu_{i,\mathrm{pure}}^v(p_i') + k_B T \ln\left(\frac{y_i\, p^v}{p_i'}\right) \tag{11.106}$$

The most frequent choice of the reference state is the vapor-liquid equilibrium of pure component i at temperature T (assuming, of course, that such a state exists!). Then

$$p_i' = p_{\mathrm{sat},i}(T), \quad \mu_{i,\mathrm{pure}}^v(p_i') = \mu_{i,0}^v = \mu_{\mathrm{sat},i} \tag{11.107}$$

where $\mu_{\mathrm{sat},i}(T)$ and $p_{\mathrm{sat},i}(T)$ are the saturation values of the chemical potential and pressure of component i, respectively. With this choice (11.105) yields

$$\mathscr{A}_i^v = \frac{p_i^v}{p_{\mathrm{sat},i}(T)} \tag{11.108}$$

This form contains the experimentally controllable parameters (in contrast to (11.105)) which makes it a convenient tool for representation of experimental results. In the single-component case the gas-phase activity coincides with the usual definition of the supersaturation $\mathscr{A}^v = S = p^v/p_{\mathrm{sat}}$.

Now, let us consider a binary mixture with a bulk liquid composition x_i at the temperature T in equilibrium with the binary vapor. The partial vapor pressure of component i over the bulk binary liquid at (x_i, T)-equilibrium can be written as

$$p_i^{\mathrm{coex}}(x_i, T) = \Gamma_i(x_i, T)\, x_i\, p_{\mathrm{sat},i}(T) \tag{11.109}$$

Here, $\Gamma_i(x_i, T)$ is called the *activity coefficient* of component i. If a mixture is ideal, which means that the compositions of the bulk liquid and bulk vapor are identical, then $\Gamma_i = 1$; this choice was employed in Eqs. (11.86)–(11.87). For *non-ideal* mixtures $\Gamma_i(x_i, T) \neq 1$. Hence, the activity coefficients describe the degree of non-ideality of the system. A microscopic origin of the non-ideality is interaction between molecules of different species.

A number of empirical correlations for activity coefficients is known in the literature [14]. One of the widely used correlations is given by *van Laar model* (see e.g. [19])

$$\ln \Gamma_a = \frac{A_L}{\left[1 + \frac{A_L x_b}{B_L x_a}\right]^2}$$

$$\ln \Gamma_b = \frac{B_L}{\left[1 + \frac{B_L x_a}{A_L x_b}\right]^2}$$

where the van Laar constants A_L and B_L are determined from the equilibrium vapor pressure measurements.

The liquid-phase activity of component i is defined as

$$\mathscr{A}_i^l = \frac{p_i^{coex}(x_i, T)}{p_{sat,i}(T)} = \Gamma_i x_i \qquad (11.110)$$

This quantity describes the influence of the bulk liquid composition on the equilibrium vapor pressure of components. For an ideal mixture $\Gamma_i = 1$, yielding $\mathscr{A}_{i,\,ideal}^l = x_i$, and for a single-component case (which is equivalent to the ideal mixture with $x_i = 1$): $\mathscr{A}_i^l = 1$. In terms of activities Eqs. (11.100)–(11.101) read

$$\beta \Delta G(n_a, n_b) = -\sum_i n_i \ln\left[\frac{\mathscr{A}_i^v}{\mathscr{A}_i^l(x_i^l)}\right] + \beta \gamma(x_b^l) A \qquad (11.111)$$

$$\beta \Delta G_{BCNT}(n_a, n_b) = -\sum_i n_i \ln\left[\frac{\mathscr{A}_i^v}{\mathscr{A}_i^l(x_i^{tot})}\right] + \beta \gamma(x_b^{tot}) A \qquad (11.112)$$

It is important to mention that if component i is supercritical at temperature T the choice of the reference state according to (11.107) becomes inappropriate and one has to resort to Eq. (11.97).

11.9.2 Ethanol/Hexanol System

Ethanol–hexanol system is a natural candidate to test predictions of the BCNT against experiment. An important feature of this system is that to a high degree of accuracy it represents the ideal liquid mixture, which makes it possible to set $\Gamma_i = 1$. Furthermore, the surface tensions of pure ethanol and hexanol are nearly identical which implies that the adsorption effects (surface enrichment) can be neglected. These observations justify the use of the BCNT with the free energy of cluster formation given by

$$\beta \Delta G_{BCNT}(n_a, n_b) = -\sum_i n_i \ln\left[\frac{\mathscr{A}_i^v}{x_i^{tot} p_{sat,i}(T)}\right] + \beta \gamma(x_b^{tot}) A \qquad (11.113)$$

Fig. 11.4 Nucleation rates for
ethanol–hexanol mixture at
$T = 260$ K as a function of the
mean vapor phase activity a
defined through Eq. (11.114).
Squares: experiment of Strey
and Viisanen [19]; *solid
lines*—BCNT with Stauffer's
expression for K. Labels are
relative activities (11.115)
(Reprinted with permission
from Ref. [18], copyright
(1995), American Institute of
Physics.)

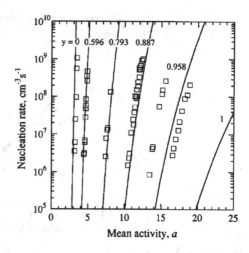

The nucleation rate is given by Eqs. (11.42)–(11.43) with the Stauffer form of the
prefactor K, and $C = C_{\text{Reiss}}$. In Fig. 11.4 we compare BCNT with the experimental
results of Strey and Viisanen [19] for nucleation of ethanol–hexanol mixture in argon
as a carrier gas at $T = 260$ K. Thermodynamic parameters used in calculations are
taken from Table I of Ref. [19]. The rates are plotted against the mean vapor phase
activity

$$a = \sqrt{(\mathscr{A}_E^V)^2 + (\mathscr{A}_H^V)^2} \tag{11.114}$$

where $\mathscr{A}_E^V$ and $\mathscr{A}_H^V$ are the ethanol and hexanol vapor phase activities, respectively.
The labels in Fig. 11.4 indicate the activity fractions

$$y = \frac{\mathscr{A}_H^V}{\mathscr{A}_E^V + \mathscr{A}_H^V} \tag{11.115}$$

The line $y = 0$ corresponds to the pure ethanol nucleation and $y = 1$—to the pure
hexanol nucleation. As one can see, BCNT is in good agreement with experiment for
$y < 0.9$ but substantially underpredicts the experimental data for $y \rightarrow 1$, i.e. at the
pure hexanol limit. It is instructive to present the same data as an activity plot. The
latter is the locus of gas phase activities of components required to produce a fixed
nucleation rate at a given temperature. Figure 11.5 is the activity plot corresponding
to the nucleation rate $J = 10^7$ cm^{-3}s^{-1}. The difference between the Stauffer from
of K and that of Reiss is quite small indicating that for this system the direction
of cluster growth at the saddle point is to a good approximation determined by the
steepest descend of the Gibbs free energy.

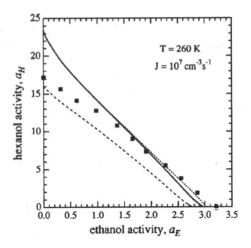

Fig. 11.5 Activity plot for ethanol–hexanol nucleation at $T = 260$ K. Activities $a_E = \mathscr{A}_E^v$ and $a_H = \mathscr{A}_H^v$ correspond to the nucleation rate $J = 10^7$ cm^{-3}s^{-1}. Squares: experiment of Strey and Viisanen [19]; *solid lines*—BCNT with Stauffer's expression for K; *short dashed line*—BCNT with Reiss expression for K; *long dashed line*—BCNT with SCC form of the prefactor C (11.104) and Stauffer's expression for K (Reprinted with permission from Ref. [18], copyright (1995), American Institute of Physics.)

11.9.3 Water/Alcohol Systems

The classical theory is quite successful in predictions of nucleation in fairly ideal mixtures. However, for mixtures showing non-ideal behavior with strong segregation effects predictions of the BCNT may lead to quantitatively or even *qualitatively wrong* results. This is in contrast to the single-component case where the CNT can be quantitatively in error, but remain qualitatively correct.

To illustrate the situation, let us consider water/alcohol mixtures. It was found that for these systems BCNT predicts the decrease of the nucleation rate when the vapor density is increased [11, 20–22]. The fact that such a behavior is unphysical can be most clearly seen by studying the activity plot. According to the nucleation theorem for a binary system (4.22)

$$\Delta n_{i,c} = \left[\frac{\partial \ln J}{\partial (\beta \mu_i^v)} \right]_T, \quad i = a, b \qquad (11.116)$$

where $n_{a,c}$ and $n_{b,c}$ are the (total) numbers of molecules of components in the critical cluster (we neglected the contribution of the prefactor K which is between 0 and 1). Recalling the definition of gas phase activities, this expression can be written as

$$\Delta n_{a,c} = \left[\frac{\partial \ln J}{\partial(\ln \mathscr{A}_a^v)}\right]_{\mathscr{A}_b^v} \tag{11.117}$$

$$\Delta n_{b,c} = \left[\frac{\partial \ln J}{\partial(\ln \mathscr{A}_b^v)}\right]_{\mathscr{A}_a^v} \tag{11.118}$$

Using the general identity

$$\left(\frac{\partial A}{\partial B}\right)_C \left(\frac{\partial B}{\partial C}\right)_A \left(\frac{\partial C}{\partial A}\right)_B = -1$$

in which $A \equiv \ln J$, $B \equiv \ln \mathscr{A}_a^v$, $C \equiv \ln \mathscr{A}_b^v$, we have

$$\left(\frac{\partial \ln J}{\partial \ln \mathscr{A}_a^v}\right)_{\mathscr{A}_b^v} \left(\frac{\partial \ln \mathscr{A}_a^v}{\partial \ln \mathscr{A}_b^v}\right)_{\ln J} \left(\frac{\partial \ln \mathscr{A}_a^v}{\partial \ln J}\right)_{\mathscr{A}_b^v} = -1 \tag{11.119}$$

The first and the third term on the left-hand side of this expression can be written using Eq. (11.116):

$$\Delta n_{a,c} \left(\frac{\partial \ln \mathscr{A}_a^v}{\partial \ln \mathscr{A}_b^v}\right)_{\ln J,T} \frac{1}{\Delta n_{b,c}} = -1 \tag{11.120}$$

resulting in

$$\left(\frac{\partial \ln \mathscr{A}_a^v}{\partial \ln \mathscr{A}_b^v}\right)_{\ln J,T} = -\frac{\Delta n_{b,c}}{\Delta n_{a,c}} \tag{11.121}$$

The left-hand side of (11.121) gives the slope of the activity plot $\ln \mathscr{A}_a^v = f(\ln \mathscr{A}_b^v)$. Since $\Delta n_{a,c}$, $\Delta n_{b,c} > 0$ this slope should be negative:

$$\left(\frac{\partial \ln \mathscr{A}_a^v}{\partial \ln \mathscr{A}_b^v}\right)_{\ln J,T} < 0 \tag{11.122}$$

Nucleation theorem is a general statement independent of the model, implying that (11.122) must be true for all binary mixtures. For the ethanol–hexanol system this requirement is satisfied—as clearly seen from Fig. 11.5.

Figure 11.6 shows the activity plot—experimental and theoretical—for nucleation in the *water/ethanol* mixture at $T = 260\,\mathrm{K}$ corresponding to the nucleation rate $J = 10^7\,\mathrm{cm}^{-3}\mathrm{s}^{-1}$. Experimental data of Viisanen et al. [21] (shown by points) are in agreement with the requirement (11.122). Meanwhile, the BCNT curve (solid line) shows the increasing part— a "hump"—(featured also by other water/alcohol systems (see [11] and reference therein)) which violates the requirement (11.122).

This unphysical "hump" corresponds to one of the $\Delta n_{i,c}$ (let it be $\Delta n_{a,c}$) being negative, while the other one is positive. Then, from (11.117) we would get

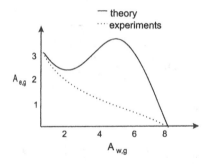

Fig. 11.6 Activity plot for the water/ethanol mixture corresponding to the nucleation rate $J = 10^7 \, \text{cm}^{-3}\text{s}^{-1}$ and $T = 260 \, \text{K}$. $A_{w,g}$ and $A_{e,g}$ are the gas-phase activities of water and ethanol, respectively. Points: experiment of Viisanen et al. [21], *full line*: BCNT predictions (Reprinted with permission from Ref. [8], copyright (2006), Springer-Verlag.)

$$\left[\frac{\partial \ln J}{\partial \ln \mathscr{A}_i^{\text{v}}} \right]_{\mathscr{A}_j^{\text{v}}, T} < 0$$

Since $\mathscr{A}_i^{\text{v}}$ is proportional to the vapor pressure (see Eq. (11.108)), the last inequality would mean the decrease of the nucleation rate when the vapor density is increased. The origin of this unphysical behavior lies in the prediction of the critical cluster composition: BCNT predicts very water-rich critical clusters. Since the surface tension of pure water is much higher than that of the pure ethanol, the resulting surface tension of the mixture becomes very high yielding low nucleation rates. These results show that BCNT fails in describing the behavior of surface enriched nuclei. For this system the adsorption effects, not taken into account by the BCNT, play an important role: surface excess numbers n_i^{exc} turn out to be large and fluctuating.

11.9.4 Nonane/Methane System

In a number of practically relevant situations one deals with gas–liquid nucleation in a mixture, in which one of the components, say, component b, is *supercritical*, i.e. its critical temperature $T_{c,b}$ is lower than the nucleation temperature T. This implies that should it be pure, it could not nucleate. In the absence of a carrier gas nucleation takes place *inside* the vapor-liquid coexistence region of the binary system and can be induced by decreasing the total pressure. This phenomenon, termed the *retrograde nucleation*, occurs in a number of applications, e.g. during production and processing of natural gas [23].

Schematically the process is depicted in Fig. 11.7. The gaseous $a - b$ mixture is initially outside the coexistence region at the state characterized by the total pressure p_0, temperature T_0 and composition $\mathbf{y} = (y_a, y_b)$. After a fast, usually adiabatic, expansion the mixture is brought inside the coexistence region to a state with the

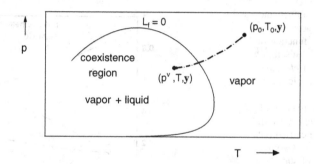

Fig. 11.7 Schematic representation of retrograde nucleation of a binary mixture; $(p_0, T_0, \mathbf{y})$ is the initial gaseous state, $(p^v, T, \mathbf{y})$ is the state after the expansion of the mixture, located inside the coexistence region; $T > T_{c,b}$. The boundary of the coexistence region corresponds to the liquid fraction $L_f = 0$

total pressure $p^v < p_0$ and temperature $T < T_0$; the latter is characterized by the *equilibrium* values of the thermodynamic parameters: the chemical potentials of the species and their vapor and liquid molar fractions. The actual vapor composition differs from the equilibrium one at the same p^v and T showing that the mixture finds itself in a nonequilibrium state. If p^v is sufficiently high, the supercritical component not only removes the latent heat (acting as a carrier gas) but also takes part in the nucleation process due to the unlike $a - b$ interactions becoming highly pronounced at high pressures. These strong real gas effects attract considerable experimental [24–28] and theoretical [29–31] attention.

As an example of a system showing retrograde nucleation behavior we consider the n-nonane/methane mixture. The choice of the system is motivated by the availability of data obtained in expansion wave tube experiments [32–36] carried out at the nucleation pressures ranging from 10 to 40 bar and temperatures—from 220 to 250 K. In this range methane is supercritical: $T_{c,b} = 190$ K [14]. In the vapor phase methane is in abundance, $y_b \approx 1$, while $y_a \sim 10^{-4} \div 10^{-3}$. Let us first study the pressure dependence of equilibrium properties influencing the nucleation behavior. For these calculations we need an EoS. The most appropriate one for mixtures of alkanes is the Redlich-Kwong-Soave equation [37].

As follows from Fig. 11.8, the miscibility of methane $x_{b,eq}$ in the bulk liquid grows with the pressure and can be as high as $\approx 50\%$ for $p^v = 100$ bar. The process of methane dissolution in liquid nonane is accompanied by the decrease of the reduced macroscopic surface tension θ_∞ of the mixture. As opposed to the previously discussed examples, θ_∞ can not be expressed as a sum of the corresponding individual properties, $\theta_{\infty,a}$ and $\theta_{\infty,b}$, since methane is supercritical. The surface tension for the mixture is found from the Parachor method [14].

Fig. 11.8 Nonane/methane
equilibrium properties at
$T = 240$ K as a function of
total pressure p^v: miscibility
of methane $x_{b,eq}$ (*left* y-axis),
reduced macroscopic surface
tension θ_∞ (*right* y-axis) and
the vapor molar fraction of
nonane $y_{a,eq}$ (*right* y-axis).
Calculations are carried out
using the Redlich-Kwong-
Soave equation of state (the
binary interaction parameter
$k_{ij} = 0.0448$ [37])

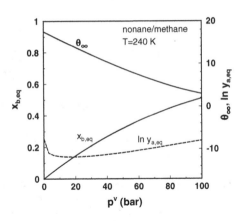

11.9.4.1 Compensation Pressure Effect

The equilibrium vapor molar fraction of nonane $y_{a,eq}$ shows nonmonotonous behavior: at low pressures it decreases with p^v since the increase of pressure results in the growth of nonane fraction in the bulk liquid at the expense of its fraction in the vapor; however, at higher pressures this process is partially blocked by penetration of the supercritical methane into the liquid phase; $y_{a,eq}$ reaches minimum at $p^v \approx 18$ bar. Qualitatively the presence and location of this minimum can be understood in terms of the "compensation pressure effect" [30]. Consider the partial molecular volume of component a in the vapor phase. By definition (D.1), v_a^v is the *change* of the total volume V^v of the binary vapor when one extra a molecule is inserted into the vapor at the fixed total pressure p^v and the number of b-molecules. One can identify two competing factors related to this process. The first factor is the tendency to *increase* the volume in order to preserve p^v. The opposite factor, manifested by the second term in (11.123), is the tendency to reduce V^v. The latter becomes pronounced for mixtures of (partially) miscible components at sufficiently high pressures when the separation between b molecules becomes of the order of the range of unlike $(a - b)$ attractions. As a result, a certain number of b molecules move in the direction of the a molecule (usually relatively big compared to the b-molecule) thereby decreasing V^v. According to Eq. (D.7):

$$v_a^v = \frac{1}{\rho^v} \left(1 - y_b \frac{\partial \ln \rho^v}{\partial y_a} \right) \tag{11.123}$$

where we took into account that $y_a + y_b = 1$. Despite the extreme smallness of y_a (usually $y_a \sim 10^{-4} \div 10^{-5}$), the term with the derivative in (11.123) can be substantial. At a certain pressure p_{comp}, satisfying

$$\left(\frac{\partial \ln \rho^v}{\partial y_a} \right)\Big|_{p_{comp}, T} = \frac{1}{y_b} \tag{11.124}$$

the two opposing trends—expansion and squeezing—compensate each other resulting in $v_a^v = 0$. At $p^v > p_{comp}$: v_a^v becomes negative implying that the "squeezing tendency" prevails and a number of b-molecules find themselves attached to an a-molecule. Since component a is supersaturated, the a-molecules tend to form a liquid-like cluster, "entraining" the attached b-molecules into it. The presence of the more volatile component decreases the specific surface free energy of a cluster. We term p_{comp} the *compensation pressure*. The simplest way to estimate $p_{comp}(T)$ is the virial expansion [14]

$$\rho^v = \beta p^v (1 - b_2), \quad b_2 \equiv \beta B_2 p^v \tag{11.125}$$

valid when $|b_2| \ll 1$. Here

$$B_2 = \sum_i \sum_j y_i y_j B_{2,ij}$$

is the second virial coefficient of the gas mixture; $B_{2,aa}(T)$, $B_{2,bb}(T)$ are the second virial coefficients of the pure substances, and the cross term $B_{2,ab}$ is constructed according to the combination rules [14]. Taking the logarithmic derivative in (11.125) and linearizing in b_2 we obtain from (11.124):

$$p_{comp}(T) = \frac{1}{2} \frac{k_B T}{[B_{2,bb}(T) - B_{2,ab}(T)]} \tag{11.126}$$

(after taking the derivative we can set $y_b \approx 1$ since methane is in abundance). This result demonstrates the leading role in the compensation pressure effect played by $a - b$ interactions giving rise to $B_{2,ab}$, which usually satisfies $|B_{2,ab}| > |B_{2,bb}|$. For the nonane/methane mixture $p_{comp}(T = 240\,\text{K}) = 17.8$ bar. This value approximately coincides with the pressure, at which $y_{a,eq}$ is at minimum (see Fig. 11.8).

11.9.4.2 Experiment Versus BCNT

As usual we characterize the state of the system by the vapor-phase activities of the components. Since methane is supercritical, the reference state should be different from the pure component vapor–liquid coexistence at temperature T discussed in Sect. 11.9.1. To this end we choose as a reference the (p^v, T)-equilibrium *of the mixture*, so that in (11.105) $\mu_{i,0}^v = \mu_{i,eq}^v(p^v, T)$. Within the ideal gas approximation Eq. (11.105) becomes

$$\mathscr{A}_i^v \approx \frac{y_i\, p^v}{y_{i,eq}(p^v, T)\, p^v} = \frac{y_i}{y_{i,eq}} \equiv S_i \tag{11.127}$$

The quantity S_i will be termed the *metastability parameter* of component i. It is a directly measurable quantity which takes into account the presence of the second

Fig. 11.9 Nonane/methane
nucleation. Nucleation rate
versus metastability parameter
of nonane S_{nonane} at various
pressures and $T = 240$ K.
Closed circles: experiments of
Luijten [33, 34]; *open circles*:
experiments of Peeters [35];
half-filled squares: experiments of Labetski [36].
Dashed lines: BCNT. *Labels*:
total pressure in bar

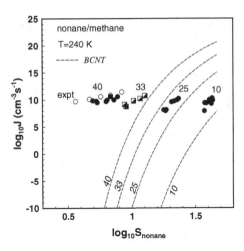

component via $y_{i,eq} = y_{i,eq}(p^v, T)$ calculated through an appropriate equation of state. The BCNT expression for the Gibbs free formation energy (11.101) reads

$$\beta \Delta G_{\text{BCNT}}(n_a, n_b) = -\sum_i n_i \ln \left[S_i \frac{y_{i,eq}\, p^v}{y_i^{coex}(x_b^{tot})\, p^{coex}(x_b^{tot})} \right] + \beta\, \gamma(x_b^{tot})\, A$$

$$(11.128)$$

Figure 11.9 shows the BCNT predictions of nucleation rate as a function of $S_a = S_{nonane}$ for temperature $T = 240$ K and pressures 10, 25, 33, 40 bar along with the experimental results of [33–35] and [36]. Theoretical predictions are fairly close to experiment for 10 and 25 bar. For higher pressures, however, BCNT largely underestimates the experimental data. In particular, the 40 bar data show extremely large deviation: more than 30 orders of magnitude. The analysis of the experimental data using nucleation theorem [34] reveals that the critical cluster is a small object, containing 10–20 molecules. With this in mind it comes as no surprise that application of a purely phenomenological BCNT approach to such clusters becomes conceptually in error. Moreover, as opposed to the ethanol/hexanol mixture studied in Sect. 11.9.2, surface enrichment in the nonane/methane mixture is expected to be highly pronounced while the standard BCNT scheme does not take it into account.

References

1. H. Flood, Z. Phys, Chem. A **170**, 286 (1934)
2. M. Volmer, *Kinetik der Phasenbildung* (Steinkopf, Dresden, 1939)
3. K. Neumann, W. Döring, Z. Phys, Chem. A **186**, 203 (1940)
4. H. Reiss, J. Chem. Phys. **18**, 840 (1950)
5. D.E. Temkin, V.V. Shevelev, J. Cryst. Growth **66**, 380 (1984)
6. B. Wyslouzil, G. Wilemski, J. Chem. Phys. **103**, 1137 (1995)

7. D. Stauffer, J. Aerosol Sci. **7**, 319 (1976)
8. H. Vehkamäki, *Classical Nucleation Theory in Multicomponent Systems* (Springer, Berlin, 2006)
9. J.S. Rowlinson, B. Widom, *Molecular Theory of Capillarity* (Clarendon Press, Oxford, 1982)
10. G. Wilemski, J. Chem. Phys. **80**, 1370 (1984)
11. G. Wilemski, J. Phys. Chem. **91**, 2492 (1987)
12. K. Nishioka, I. Kusaka, J. Chem. Phys. **96**, 5370 (1992)
13. Y.S. Djikaev, I. Napari, A. Laaksonen, J. Chem. Phys. **120**, 9752 (2004)
14. R.C. Reid, J.M. Prausnitz, B.E. Poling, *The Properties of Gases and Liquids* (McGraw-Hill, New York, 1987)
15. A. Laaksonen, R. McGraw, H. Vehkamaki, J. Chem. Phys. **111**, 2019 (1999)
16. P.G. Debenedetti, *Metastable Liquids* (Princeton University Press, Princeton, 1996)
17. D.W. Oxtoby, D. Kashchiev, J. Chem. Phys. **100**, 7665 (1994)
18. G. Wilemski, B. Wyslouzil, J. Chem. Phys. **103**, 1127 (1995)
19. R. Strey, Y. Viisanen, J. Chem. Phys. **99**, 4693 (1993)
20. J.L. Schmitt, J. Witten, G.W. Adams, R.A. Zalabsky, J. Chem. Phys. **92**, 3693 (1990)
21. Y. Viisanen, R. Strey, A. Laaksonen, M. Kulmala, J. Chem. Phys. **100**, 6062 (1994)
22. R. Strey, P.E. Wagner, Y. Viisanen, in *Nucleation and Atmospheric Aerosols*, ed. by N. Fukuta, P.E. Wagner (Deepak, Hampton, 1992), p. 111
23. M.J. Muitjens, V.I. Kalikmanov, M.E.H. van Dongen, A. Hirschberg, P. Derks, Revue de l'Institut Français du Pétrole **49**, 63 (1994)
24. R.H. Heist, H. He, J. Phys. Chem. Ref. Data **23**, 781 (1994)
25. R.H. Heist, M. Janjua, J. Ahmed, J. Phys. Chem. **98**, 4443 (1994)
26. J.L. Fisk, J.L. Katz, J. Chem. Phys. **104**, 8649 (1996)
27. D. Kane, M. El-Shall, J. Chem. Phys. **105**, 7617 (1996)
28. J.L. Katz, J.L. Fisk, V. Chakarov, in *Nucleation and Atmospheric Aerosols*, ed. by N. Fukuta, P.E. Wagner (Deepak, Hampton, 1992), p. 11
29. D.W. Oxtoby, A. Laaksonen, J. Chem. Phys. **102**, 6846 (1995)
30. V.I. Kalikmanov, D.G. Labetski, Phys. Rev. Lett. **98**, 085701 (2007)
31. J. Wedekind, R. Strey, D. Reguera, J. Chem. Phys. **126**, 134103 (2007)
32. K.N.H. Looijmans, C.C.M. Luijten, M.E.H. van Dongen, J. Chem. Phys. **103**, 1714 (1995)
33. C.C.M. Luijten, Ph.D. Thesis, Eindhoven University, 1999
34. C.C.M. Luijten, P. Peeters, M.E.H. van Dongen, J. Chem. Phys. **111**, 8535 (1999)
35. P. Peeters, Ph.D. Thesis, Eindhoven University, 2002
36. D.G. Labetski, Ph.D. Thesis, Eindhoven University, 2007
37. K.N.H. Looijmans, C.C.M. Luijten, G.C.J. Hofmans, M.E.H. van Dongen, J. Chem. Phys. **102**, 4531 (1995)

Chapter 12
Binary Nucleation: Density Functional Theory

12.1 DFT Formalism for Binary Systems. General Considerations

Classical theory of binary nucleation can be drastically in error and even lead to unphysical behavior when applied to strongly non-ideal systems with substantial surface enrichment—a vivid example is the water/alcohol system, for which BCNT predicts the decrease of nucleation rate with increasing partial pressures. An alternative to the classical treatment, based on purely phenomenological considerations, is the density functional theory based on microscopical considerations. The basic feature of DFT, discussed in Chap. 5, is the existence of the unique Helmholtz free energy functional of the nonhomogeneous one-particle density $\rho(\mathbf{r})$. The liquid-vapor equilibrium corresponds to the minimum of this functional in the space of admissible density profiles under the constraint of fixed particle number N. In a *nonequilibrium* state (like supersaturated vapor) one has to search for the *saddle point* of the free energy functional which corresponds to the critical nucleus in the surrounding supersaturated vapor.

DFT of Chap. 5 can be extended to the case of nonhomogeneous binary mixtures. Let us first discuss the two-phase equilibrium of the binary liquid and binary vapor. The Helmholtz free energy functional $\mathscr{F}$ and the grand potential functional Ω for a mixture is now written in terms of the one-body density profiles of components 1 and 2, $\rho_1(\mathbf{r})$ and $\rho_2(\mathbf{r})$, normalized as

$$\int \rho_i(\mathbf{r}_i)\, d\mathbf{r}_i = N_i, \quad i = 1, 2$$

where N_i is the total number of molecules of component i in the two-phase system. Intermolecular interactions in the mixture are described by the potentials $u_{11}(r)$, $u_{22}(r)$ and $u_{12}(r)$. The first two of them refer to interactions of the molecules of the same type, while $u_{12}(r)$ describes the *unlike* interactions between different species and is defined through an appropriate mixing rule. As in the single-

V. I. Kalikmanov, *Nucleation Theory*, Lecture Notes in Physics 860,
DOI: 10.1007/978-90-481-3643-8_12, © Springer Science+Business Media Dordrecht 2013

component DFT, we decompose each intermolecular potential $u_{ij}(r)$ ($i, j = 1, 2$) into the reference (repulsive) part, $u_{ij}^{(1)}(r)$, and perturbation, $u_{ij}^{(2)}(r)$, following the Weeks-Chandler-Andersen scheme (5.20)–(5.21)

$$u_{ij}^{(1)}(r) = \begin{cases} u_{ij}(r) + \varepsilon_{ij} & \text{for } r < r_{m,ij} \\ 0 & \text{for } r \geq r_{m,ij} \end{cases} \tag{12.1}$$

$$u_{ij}^{(2)}(r) = \begin{cases} -\varepsilon_{ij} & \text{for } r < r_{m,ij} \\ u_{ij}(r) & \text{for } r \geq r_{m,ij} \end{cases} \tag{12.2}$$

where ε_{ij} is the depth of the potential $u_{ij}(r)$ and $r_{m,ij}$ is the corresponding value of r: $u_{ij}(r_{m,ij}) = -\varepsilon_{ij}$. We assume that all interactions are pairwise additive.

The reference model is approximated by the hard-sphere mixture with appropriately chosen effective diameters d_{ij}. Within the local density approximation the reference part of the free energy is

$$\mathscr{F}_d[\rho_1, \rho_2] \approx \int d\mathbf{r} \, \psi_d \left(\rho_1(\mathbf{r}), \rho_2(\mathbf{r}) \right) \tag{12.3}$$

where $\psi_d(\rho_1(\mathbf{r}), \rho_2(\mathbf{r}))$ is the free energy density of the uniform hard-sphere mixture with the densities of components ρ_1 and ρ_2. Using the standard thermodynamic relationship (cf. (5.24)) it can be written as

$$\psi_d(r) = \sum_{i=1}^{2} \rho_i \, \mu_{d,i}(\rho_1(r), \rho_2(r)) - p_d(\rho_1(r), \rho_2(r))$$

where $\mu_{d,i}$ is the chemical potential of component i in the uniform hard-sphere fluid and p_d is the pressure of the hard-sphere mixture. The quantities $\mu_{d,i}$ and p_d are obtained by means of the binary form of the Carnahan-Starling equation due to Mansoori et al. [1] described in Appendix E.

For the pair distribution function we use the random phase approximation

$$\rho_{ij}^{(2)}(\mathbf{r}, \mathbf{r}') \approx \rho_i(\mathbf{r}) \, \rho_j(\mathbf{r}') \tag{12.4}$$

From (12.3) and (12.4) the Helmholtz free energy functional for the mixture takes the form

$$\mathscr{F}[\rho_1, \rho_2] = \int d\mathbf{r} \, \psi_d(\rho_1(\mathbf{r}), \rho_2(\mathbf{r})) + \frac{1}{2} \sum_{i,j=1}^{2} \int d\mathbf{r} \, d\mathbf{r}' \, \rho_i(\mathbf{r}) \, \rho_j(\mathbf{r}') \, u_{ij}^{(2)}(|\mathbf{r} - \mathbf{r}'|) \tag{12.5}$$

The grand potential functional for the mixture $\Omega[\rho_1, \rho_2] = \mathcal{F}[\rho_1, \rho_2] - \sum_{i=1}^{2} \mu_i N_i$ is

$$\Omega[\rho_1, \rho_2] = -\int d\mathbf{r}\, p_d(\rho_1(\mathbf{r}), \rho_2(\mathbf{r})) + \sum_{i=1}^{2} \int d\mathbf{r}\, \rho_i\, \mu_{d,i}(\rho_1(\mathbf{r}), \rho_2(\mathbf{r}))$$

$$+ \frac{1}{2} \sum_{i,j=1}^{2} \int d\mathbf{r}\, \rho_i(\mathbf{r}) \int d\mathbf{r}'\, \rho_j(\mathbf{r}')\, u_{ij}^{(2)}(|\mathbf{r} - \mathbf{r}'|)$$

$$- \sum_{i=1}^{2} \mu_i \int d\mathbf{r}\, \rho_i(\mathbf{r}) \tag{12.6}$$

where μ_i is the chemical potential of component i. Two-phase binary equilibrium corresponds to its variational minimization:

$$\frac{\delta\Omega}{\delta\rho_i(\mathbf{r})} = 0, \quad i = 1, 2 \tag{12.7}$$

or, equivalently

$$\mu_{d,i}(\rho_1(\mathbf{r}), \rho_2(\mathbf{r})) = \mu_i - \sum_{j=1}^{2} \int d\mathbf{r}'\, \rho_j(\mathbf{r}') u_{ij}^{(2)}(|\mathbf{r} - \mathbf{r}'|), \quad i = 1, 2 \tag{12.8}$$

The integral equations (12.8) are solved iteratively. First it is necessary to determine the bulk equilibrium properties of the system: μ_i, ρ_i^{v} ρ_i^{l}. Fixing two degrees of freedom (temperature and the total pressure, or temperature and bulk composition in one of the phases), the two-phase equilibrium of the mixture is given by

$$\mu_1^{\text{l}} = \mu_1^{\text{v}} \equiv \mu_1 \tag{12.9}$$
$$\mu_2^{\text{l}} = \mu_2^{\text{v}} \equiv \mu_2 \tag{12.10}$$
$$p^{\text{l}} = p^{\text{v}} \equiv p \tag{12.11}$$

Equations (12.5)–(12.8) written for a *homogeneous* bulk mixture $\rho_i(\mathbf{r}) = \rho_i$ become

$$\mathcal{F}(\rho_1, \rho_2) = \mathcal{F}_d(\rho_1, \rho_2) - V \sum_{i,j=1}^{2} \rho_i \rho_j\, a_{ij} \tag{12.12}$$

$$\mu_i = \mu_{d,i}(\rho_1, \rho_2) - 2 \sum_{j=1}^{2} \rho_j\, a_{ij} \tag{12.13}$$

where

$$a_{ij} = -\frac{1}{2} \int d\mathbf{r}\, u_{ij}^{(2)}(r)$$

is the background interaction parameter. The virial equation for a mixture reads

$$p = p_d(\rho_1, \rho_2) - \sum_{i,j=1}^{2} \rho_i \rho_j a_{ij} \tag{12.14}$$

Consider the flat geometry with the inhomogeneity along the z-axis, directed towards the bulk vapor. The bulk densities of the components in both phases provide asymptotic limits for the equilibrium density profiles in the inhomogeneous system:

$$\rho_i(z) \to \rho_i^v \quad \text{in the bulk vapor}$$
$$\rho_i(z) \to \rho_i^l \quad \text{in the bulk liquid}$$

The density profiles are calculated iteratively from Eq. (12.8) starting with an initial guess for each $\rho_i(z)$, which can be a step-function or a continuous function that varies between the bulk limits. When the equilibrium profiles are found, they can be substituted back into the thermodynamic functionals which then become the corresponding thermodynamic potentials of the two-phase system. In particular, from (12.6) and (12.8) the grand potential of the two-phase system in equilibrium reads:

$$\Omega[\rho_1, \rho_2] = -\int d\mathbf{r}\, p_d(\rho_1(\mathbf{r}), \rho_2(\mathbf{r})) - \frac{1}{2} \sum_{i,j=1}^{2} \int d\mathbf{r}\, \rho_i(\mathbf{r}) \int d\mathbf{r}'\, \rho_j(\mathbf{r}') u_{ij}^{(2)}(|\mathbf{r} - \mathbf{r}'|) \tag{12.15}$$

The plain layer surface tension of the binary system can be determined from the general thermodynamic relationship (5.35):

$$\gamma = (\Omega[\{\rho_i\}] + pV)/A \tag{12.16}$$

where A is the interfacial area. For inhomogeneity along the z direction $d\mathbf{r} = A\,dz$, and Eqs. (12.15) and (12.16) yield

$$\gamma = -\int dz \left\{ p_d(z) + \frac{1}{2} \sum_{i=1}^{2} \rho_i(z) \int d\mathbf{r}'\, \rho_j(z') u_{ij}^{(2)}(|\mathbf{r} - \mathbf{r}'|) - p \right\} \tag{12.17}$$

12.2 Non-ideal Mixtures and Surface Enrichment

As we know from Chap. 11, talking about a mixture we can not avoid the discussion of adsorption effects. On the phenomenological level it means that for a binary (or, more generally, a multi-component) mixture it is impossible to choose the Gibbs dividing surface in such a way that the excess (adsorption) terms for *all* species simultaneously vanish. This feature gives rise to the surface enrichment: a preferential adsorption of one of the species in the interfacial region between the two bulk phases.

On the microscopic level the issue of adsorption boils down to the strength and range of unlike interactions. They determine the degree of non-ideality of the system. The DFT yields the density profiles of components in the inhomogeneous system. These profiles have no rigid boundaries, their form is based on the microscopic interactions in the system, implying that adsorption (surface enrichment) is naturally built into the DFT scheme.

It is instructive to study the effects of non-ideality on the behavior of the mixture considering the simplest system: a binary mixture of Lennard-Jones fluids with the interaction potentials

$$u_{ij}(r) = 4\varepsilon_{ij} \left[\left(\frac{\sigma_{ij}}{r}\right)^{12} - \left(\frac{\sigma_{ij}}{r}\right)^{6} \right] \tag{12.18}$$

where σ_{ii} and ε_{ii} are the Lennard-Jones parameters of the individual components. For illustrative reasons (in order to have realistic numbers) we choose their values corresponding to the argon/krypton mixture:

$$\sigma_{Ar} = \sigma_{11} = 3.405 \, \text{Å}, \quad \sigma_{Kr} = \sigma_{22} = 3.632 \, \text{Å}$$

and

$$\varepsilon_{Ar}/k_B = \varepsilon_{11}/k_B = 119.8 \, \text{K}, \quad \varepsilon_{Kr}/k_B = \varepsilon_{22}/k_B = 163.1 \, \text{K}$$

The unlike interactions u_{12} are defined via the mixing rules. We assume that u_{12} has also the Lennard-Jones form. For conformal potentials it is common to present the mixing rules in the form [2]:

$$\sigma_{12} = \frac{1}{2} (\sigma_{11} + \sigma_{22}) \tag{12.19}$$

and

$$\varepsilon_{12} = \xi_{12} \sqrt{\varepsilon_{11} \varepsilon_{22}} \tag{12.20}$$

where ξ_{12} is called the *binary interaction parameter* and is found from the fit to experiment. When $\xi_{12} = 1$ one speaks about a *Lorentz-Berthelot* mixture. For real mixtures ξ_{12} is usually significantly less than unity. It is necessary to have in mind

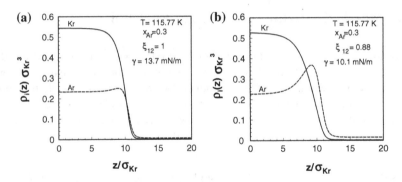

Fig. 12.1 Density profiles of argon and krypton at the flat vapor-liquid interface with the bulk liquid molar fraction of argon $x_{Ar} = 0.3$ at $T = 115.77$ K and different values of the binary interaction parameter ξ_{12}; (**a**) $\xi_{12} = 1$ (Lorentz-Berthelot mixture); (**b**) $\xi_{12} = 0.88$. Distances and densities are scaled with respect to $\sigma_{Kr} = \sigma_{22}$. The decrease of ξ_{12} leads to the increase of the surface activity of argon (surface enrichment), accompanied by the decrease of the surface tension of the mixture γ

that the laws of the ideal mixture are obtained only if all the potentials are the same; in this sense even for $\xi_{12} = 1$ the mixture should not necessarily be ideal. Note that for the equation of state of the mixture the mixing rule (12.20) leads to the corresponding form of the "energy parameter" a_{12}:

$$a_{12} = \xi_{12} \sqrt{a_{11} a_{22}}$$

Consider implications of the mixing rule (12.20). The decrease of ξ_{12} from unity will decrease the depth of unlike interactions thereby enhancing separation in the solution. Since in our example $\varepsilon_{11} < \varepsilon_{22}$, this separation leads to the increase of the surface activity of component 1 (argon) which is manifested by its pronounced surface enrichment. These features are illustrated in Fig. 12.1. Equilibrium calculations are performed for the argon/krypton mixture at $T = 115.77$ K and the bulk liquid molar fraction of argon $x_{Ar} = 0.3$ (hence, we discuss the (x, T)-equilibrium).

The left graph (a) refers to the Lorentz-Berthelot mixture: $\xi_{12} = 1$. The surface enrichment of argon in this case is very weak. The surface tension calculated from Eq. (12.17) gives $\gamma(\xi_{12} = 1) = 13.7$ mN/m. On the right graph (b) we show the DFT calculations for $\xi_{12} = 0.88$. The density profile of argon shows considerable surface enrichment, meaning that the vapor-liquid interface is argon-rich leading to the decrease of the surface tension: $\gamma(\xi_{12} = 0.88) = 10.1$ mN/m.

12.3 Nucleation Barrier and Activity Plots: DFT Versus BCNT

Formulation of the density functional theory for the study of *equilibrium* properties of inhomogeneous binary mixtures can be extended to the *nonequilibrium* case of binary nucleation. The corresponding development was carried out by Oxtoby and coworkers [3–5] and represents a generalization of the similar approach for a single-component case. In DFT of binary nucleation one studies the system *"droplet in a nonequilibrium vapor"*, where both the droplet and the vapor are binary mixtures. A droplet is associated with a density fluctuation which has no rigid boundary. The Helmholtz free energy and grand potential functionals for this system are given as before by Eqs. (12.5)–(12.6). However, the chemical potentials of the components in this expressions refer now to the actual nonequilibrium state of the system (and not to the thermodynamic equilibrium as in Sect. 12.1). This state can be characterized, e.g. by fixing the gas-phase activities of the components.

The critical nucleus is in unstable equilibrium with the environment and corresponds to the *saddle point* of $\Omega[\rho_1, \rho_2]$ (as opposed to the minimum of $\Omega[\rho_1, \rho_2]$ in the case of equilibrium conditions). The density profiles in the critical nucleus are found as before from the solution of Eq. (12.8). If the profiles are determined, then the change in the grand potential

$$\Delta\Omega^* = \Omega[\rho_1, \rho_2] - \Omega_u \qquad (12.21)$$

where Ω_u is the grand potential of the uniform nonequilibrium vapor (i.e. the binary vapor prior to the appearance of the droplet) is the free energy associated with the formation of the critical cluster. As in the single-component case it is straightforward to see that $\Delta\Omega^*$ is equal to the Gibbs energy of the critical cluster formation

$$\Delta\Omega^* = \Delta G^*$$

When we discussed Eq. (12.8) for the conditions of thermodynamic equilibrium, the iterative procedure converged rapidly (within several iterations) to the desired solution irrespective of the initial guess (as soon as it satisfies the boundary conditions)—just due to the fact that equilibrium profiles *minimize* the functional. The solution of the same equations for the critical cluster in nucleation are far from trivial. Since we deal here with the unstable equilibrium, the convergent solution of Eq. (12.8) does not exist. Meanwhile, if the initial guess for $\rho_1(\mathbf{r})$, $\rho_2(\mathbf{r})$) is close to the profiles of components in the critical cluster, the iteration process after several steps reaches a plateau with $\Delta\Omega$ remaining constant over several iterations. This plateau corresponds to the critical cluster. Continuation of the iterative process will after a certain amount of iterations lead to a deviation of $\Delta\Omega$ from the plateau (no convergency!). Clearly, the search for the critical cluster is very sensitive to the initial guess. The details of iterative procedure are described in Refs. [4, 5].

After determination of the nucleation barrier $\Delta\Omega^*$ the steady-state nucleation rate follows from:

$$J = K e^{-\beta\Delta\Omega^*} \qquad (12.22)$$

where the pre-exponential factor K can be taken from the classical binary nucleation theory (see Eq. (11.43)).

In the previous section we studied the effects of non-ideality of the mixture (expressed in the terms of the unlike interaction potential) on the equilibrium properties. Let

Fig. 12.2 Gas phase activities for the mixture of argon and krypton with $\xi_{12} = 1$. *Diamonds*: DFT, *full line*: BCNT. The results correspond to the nucleation rate of $1\,\mathrm{cm}^{-3}\mathrm{s}^{-1}$ and $T = 115.77\,\mathrm{K}$ (Reprinted with permission from Ref. [4], copyright (1995), American Institute of Physics.)

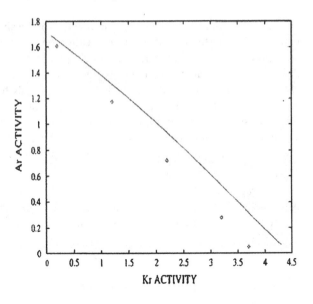

Fig. 12.3 Gas phase activities for the mixture of argon and krypton with $\xi_{12} = 0.88$. *Diamonds*: DFT; *full line*: BCNT. The results correspond to the nucleation rate of $1\,\mathrm{cm}^{-3}\mathrm{s}^{-1}$ and $T = 115.77\,\mathrm{K}$ (Reprinted with permission from Ref. [4], copyright (1995), American Institute of Physics.)

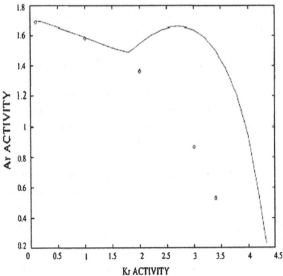

us study their impact on the nucleation behavior. As before we consider the binary mixture argon/krypton with the mixing rule (12.20). The role of non-ideality effects can be clearly demonstrated by means of activity plots. Figures 12.2 and 12.3 show the BCNT and DFT activity plots for $T = 115.77$ K corresponding to the nucleation rates

$$J_{\mathrm{BCNT}} \approx J_{\mathrm{DFT}} \approx 1\,\mathrm{cm}^{-3}\mathrm{s}^{-1}$$

Figure 12.2 refers to the Lorentz-Berthelot mixture: $\xi_{12} = 1$. As we know from the equilibrium considerations of the previous section, surface enrichment of argon in this case is very weak and the mixture is fairly ideal. It is therefore not surprising that the difference in nucleation behavior between the BCNT and DFTs is of quantitative nature; note that it increases with the krypton activity.

This situation is distinctly different from the case $\xi_{12} = 0.88$ shown in Fig. 12.3: BCNT produces a "hump"—resembling the similar predictions for water/alcohol systems (cf. Fig. 11.6). This hump, as discussed earlier, is unphysical since it violates the nucleation theorem. Its occurrence is a consequence of the neglect within the BCNT scheme of adsorption effects giving rise to surface enrichment. The DFT approach does not have this drawback because adsorption is "built into it" on the microscopic level. The DFT predictions therefore are in qualitative agreement with the nucleation theorem. Figure 12.1b demonstrated the effect of surface enrichment for the planar interface. DFT calculations reveal the same effect for the critical cluster in binary nucleation [4].

References

1. G.A. Mansoori, N.F. Carnahan, K.E. Starling, T.W. Leland, J. Chem. Phys. **54**, 1523 (1971)
2. J.S. Rowlinson, F.L. Swinton, *Liquids and Liquid Mixtures* (Butterworths, Boston, 1982)
3. X.C. Zeng, D.W. Oxtoby, J. Chem. Phys. **95**, 5940 (1991)
4. A. Laaksonen, D.W. Oxtoby, J. Chem. Phys. **102**, 5803 (1995)
5. I. Napari, A. Laaksonen, J. Chem. Phys. **111**, 5485 (1999)

Chapter 13
Coarse-Grained Theory of Binary Nucleation

13.1 Introduction

As we saw in Chap. 11, the classical theory proved to be successful for fairly ideal mixtures. Meanwhile, for non-ideal mixtures BCNT can be sufficiently in error [1–4] and even lead to unphysical results as in the case of water-alcohol systems. The reasons for this failure are the neglect of adsorption effects and the inappropriate treatment of small clusters. These issues are strongly coupled; they determine the form of the Gibbs free energy of cluster formation and subsequently the composition of the critical cluster and the nucleation barrier. One can correct the classical treatment by taken into account adsorption using the Gibbsian approximation (see Sect. 11.6).

Taking into account adsorption within the phenomenological approach does not resolve another deficiency associated with the capillarity approximation: the surface energy of a cluster is described in terms of the planar surface tension. Obviously, for small clusters the concept of macroscopic surface tension looses its meaning and this assumption fails. This difficulty is not unique for the binary problem. In the single-component mean-field kinetic nucleation theory (MKNT) of Chap. 7 this problem was tackled by formulating an interpolative model between small clusters treated using statistical mechanical considerations and big clusters described by the capillarity approximation. In the present chapter we extend these considerations to the binary case and incorporate them into a model which takes into account the adsorption effects [5].

The statistical mechanical treatment of binary clusters, which we discuss in the present chapter, originates from the analogy with the soft condensed matter theory, where the description of complex fluids can be substantially simplified if one eliminates the degrees of freedom of small solvent molecules in the solution. By performing such *coarse-graining* one is left with the pseudo-one-component system of solute particles with some effective Hamiltonian. This approach opens the possibility to study the behavior of a complex fluid using the techniques developed in the

V. I. Kalikmanov, *Nucleation Theory*, Lecture Notes in Physics 860,
DOI: 10.1007/978-90-481-3643-8_13, © Springer Science+Business Media Dordrecht 2013

theory of simple fluids [6]. Situation in nucleation theory is somewhat similar: the complexity of binary nucleation problem can be substantially reduced by tracing out the degrees of freedom of the molecules of the more volatile component in favor of the less volatile one. This pseudo-one-component system can be studied using the approach developed in Chap. 7 making it possible to adequately treat clusters of arbitrary size and composition.

13.2 Katz Kinetic Approach: Extension to Binary Mixtures

Consider a binary mixture of component a and b in the gaseous state at the temperature T and the total pressure p^v. The actual vapor mole fractions of components are y_a and y_b. In the presence of a carrier gas with the mole fraction y_c:

$$y_a + y_b = 1 - y_c$$

In the present context the term "carrier gas" refers to a passive component, which does not take part in cluster formation but serves to remove the latent heat. In some cases, a passive carrier gas is absent ($y_c = 0$), and one of the components of the mixture (a or b) plays the double role: besides taking part in the nucleation process, it removes the latent heat. In this case, this component should be in abundance in the vapor phase.

Within the general formalism of Sect. 11.9.1 the state of component i is characterized by the vapor phase activity

$$\mathscr{A}_i^v = \exp\left[\beta(\mu_i^v(p^v, T; y_i) - \mu_{i,0}^v)\right], \quad i = a, b \tag{13.1}$$

where $\mu_i^v(p^v, T; y_i)$ is the chemical potential of component i in the vapor, $\mu_{i,0}^v$ is the value of μ_i^v at some reference state. Usually one chooses as the reference the saturated state of pure components at the temperature T. This choice implicitly assumes the existence of such a state for both species. In the case when one of the components is supercritical this choice becomes inappropriate. With this in mind we choose as a reference the *true equilibrium state of the mixture* at p^v, T; y_c, so that $\mu_{i,0}^v = \mu_{i,eq}^v$. Throughout this chapter the subscript "eq" refers to the true equilibrium state of the mixture, and not a constrained equilibrium as in Chap. 11; to avoid confusion the latter will be denoted by the superscript "cons". Within the ideal gas approximation

$$\mathscr{A}_i^v \approx \frac{y_i\, p^v}{y_{i,eq}(p^v, T, y_c)\, p^v} = \frac{y_i}{y_{i,eq}} \equiv S_i \tag{13.2}$$

where S_i will be termed the *metastability parameter* of component i. Given the equation of state for the mixture, S_i is a directly measurable quantity. It is important to emphasize that its value takes into account the presence of all components in the mixture through $y_{i,eq} = y_{i,eq}(p^v, T; y_c)$.

Let us consider a two-dimensional $n_a - n_b$ space of cluster sizes. Here n_i is the *total* number of molecules of component i, which according to Gibbs thermodynamics is the sum of the bulk and excess terms

$$n_i = n_i^l + n_i^{exc}$$

Kinetics of nucleation is governed by Eq. (11.1):

$$\frac{\partial \rho(n_a, n_b, t)}{\partial t} = J_a(n_a - 1, n_b, t) - J_a(n_a, n_b, t) + J_b(n_a, n_b - 1, t) - J_b(n_a, n_b, t)$$
(13.3)

where the fluxes along the n_a and n_b directions are

$$J_a(n_a, n_b) = v_a\, A(n_a, n_b)\, \rho(n_a, n_b, t) - \beta_a\, A(n_a + 1, n_b)\, \rho(n_a + 1, n_b, t)$$
(13.4)

$$J_b(n_a, n_b) = v_b\, A(n_a, n_b)\, \rho(n_a, n_b, t) - \beta_b\, A(n_a, n_b + 1)\, \rho(n_a, n_b + 1, t)$$
(13.5)

Impingement rates (per unit surface) of component i, v_i, are given by gas kinetics:

$$v_i = \frac{y_i\, p^v}{\sqrt{2\pi m_i k_B T}}$$
(13.6)

Evaporation rates β_i are obtained from the detailed balance condition. Recall that in BCNT it is applied to the *constrained equilibrium* state which would exist for the vapor at the same temperature, pressure and vapor phase activities as the vapor in question. Instead of using this artificial state, we apply the detailed balance to the *true* (full) equilibrium of the system at (p^v, T, y_c):

$$0 = v_{a,eq}\, A(n_a, n_b)\, \rho_{eq}(n_a, n_b) - \beta_a\, A(n_a + 1, n_b)\, \rho_{eq}(n_a + 1, n_b) \quad (13.7)$$
$$0 = v_{b,eq}\, A(n_a, n_b)\, \rho_{eq}(n_a, n_b) - \beta_b\, A(n_a, n_b + 1)\, \rho_{eq}(n_a, n_b + 1) \quad (13.8)$$

This procedure is a natural extension to binary mixtures of the *Katz kinetic approach* discussed in Sect. 3.5. Assuming (following BCNT) that the evaporation rates do not depend on the surrounding vapor, $\beta_i = \beta_{i,eq}$, we find from (13.7) and (13.8)

$$\beta_a = v_{a,eq}\, \frac{A(n_a, n_b)\, \rho_{eq}(n_a, n_b)}{A(n_a + 1, n_b)\, \rho_{eq}(n_a + 1, n_b)}$$
(13.9)

$$\beta_b = v_{b,eq}\, \frac{A(n_a, n_b)\, \rho_{eq}(n_a, n_b)}{A(n_a, n_b + 1)\, \rho_{eq}(n_a, n_b + 1)}$$
(13.10)

Substituting (13.9)–(13.10) into (13.4) and (13.5) we write

$$J_a(n_a, n_b) = \rho_{eq}(n_a, n_b) \, A(n_a, n_b) \, v_a \left[\frac{\rho(n_a, n_b)}{\rho_{eq}(n_a, n_b)} - \frac{\rho(n_a + 1, n_b)}{\rho_{eq}(n_a + 1, n_b)} \frac{v_{a,eq}}{v_a} \right]$$

$$J_b(n_a, n_b) = \rho_{eq}(n_a, n_b) \, A(n_a, n_b) \, v_b \left[\frac{\rho(n_a, n_b)}{\rho_{eq}(n_a, n_b)} - \frac{\rho(n_a, n_b + 1)}{\rho_{eq}(n_a, n_b + 1)} \frac{v_{b,eq}}{v_b} \right]$$

To make the expressions in the square brackets symmetric, it is convenient introduce the function [7, 8]

$$H(n_a, n_b) = \frac{\rho(n_a, n_b)}{\rho_{eq}(n_a, n_b)} \prod_{i=a,b} \left(\frac{v_{i,eq}}{v_i} \right)^{n_i} = \frac{\rho(n_a, n_b)}{\rho_{eq}(n_a, n_b)} \prod_{i=a,b} S_i^{-n_i} \quad (13.11)$$

where the second equality results from (13.2) and (13.6). The fluxes along n_a and n_b read:

$$J_a = -\rho_{eq} \, A \, v_a \left[\prod_{i=a,b} S_i^{n_i} \right] [H(n_a + 1, n_b) - H(n_a, n_b)] \quad (13.12)$$

$$J_b = -\rho_{eq} \, A \, v_b \left[\prod_{i=a,b} S_i^{n_i} \right] [H(n_a, n_b + 1) - H(n_a, n_b)] \quad (13.13)$$

We can write down the same fluxes in terms of the *constrained* equilibrium quantities. By definition $v_{i,eq}^{cons} = v_i$. Repeating the previous steps, in which the equilibrium properties are replaced by the corresponding constrained equilibrium ones, we find

$$J_a = -\rho_{eq}^{cons} \, A \, v_a \, [H(n_a + 1, n_b) - H(n_a, n_b)] \quad (13.14)$$
$$J_b = -\rho_{eq}^{cons} \, A \, v_b \, [H(n_a, n_b + 1) - H(n_a, n_b)] \quad (13.15)$$

where the function H has now a simple form:

$$H(n_a, n_b) = \frac{\rho(n_a, n_b)}{\rho_{eq}^{cons}(n_a, n_b)} \quad (13.16)$$

Comparing (13.14)–(13.16) with (13.12)–(13.13), we find the relationship between the constrained and unconstrained distributions:

$$\rho_{eq}^{cons} = \rho_{eq} \prod_{i=a,b} S_i^{n_i} \quad (13.17)$$

Obviously, for $S_a = S_b = 1$ both equilibria become identical. A general form of $\rho_{eq}(n_a, n_b)$ resulting from the thermodynamic fluctuation theory is

$$\rho_{eq}(n_a, n_b) = C \, e^{-\beta \Delta G_{eq}(n_a, n_b)} \tag{13.18}$$

From (13.17) and (13.18)

$$\rho_{eq}^{cons}(n_a, n_b) = C \, e^{-\beta \Delta G_{eq}^{cons}(n_a, n_b)} \tag{13.19}$$

where

$$\beta \Delta G_{eq}^{cons}(n_a, n_b) = - \sum_{i=a,b} n_i \ln S_i + \beta \Delta G_{eq}(n_a, n_b) \tag{13.20}$$

It is convenient to present Eqs. (13.12)–(13.13) in the vector notations:

$$\mathbf{J} = - \left[\rho_{eq}(\mathbf{n}) \prod_{i=a,b} S_i^{n_i} \right] \mathbf{F} \nabla H, \quad \mathbf{n} = (n_a, n_b) \tag{13.21}$$

where $\nabla \equiv \left(\frac{\partial}{\partial n_a}, \frac{\partial}{\partial n_b} \right)$ and the diagonal matrix $\mathbf{F}$ contains the rate of collisions of a and b molecules with the surface of the cluster:

$$\mathbf{F} = \begin{pmatrix} v_a \, A(n_a, n_b) & 0 \\ 0 & v_b \, A(n_a, n_b) \end{pmatrix} \tag{13.22}$$

In these notations the kinetic equation (13.3) takes the form of the conservation law for the "cluster fluid" (cf. Chap. 11)

$$\frac{\partial \rho(\mathbf{n})}{\partial t} = - \operatorname{div} \mathbf{J}(\mathbf{n}) \tag{13.23}$$

with the steady state given by

$$\operatorname{div} \mathbf{J} = 0 \tag{13.24}$$

In view of (13.17) we can discuss thermodynamics of nucleation focusing on the model for the Gibbs free energy of cluster formation *in vapor-liquid equilibrium*, $\Delta G_{eq}(n_a, n_b)$. An important feature of this construction is that the prefactor C in (13.19) refers to the true equilibrium distribution $\rho_{eq}(n_a, n_b)$.

Repeating the considerations of Chap. 11, we find that the direction of the nucleation flux in any point of the cluster space satisfies (cf. Eq. (11.14)):

$$\operatorname{curl} (\mathbf{F}^{-1} \mathbf{J}) = (\mathbf{F}^{-1} \mathbf{J}) \times \nabla \left(\beta \Delta G_{eq} - \sum_{i=a,b} n_i \ln S_i \right) \tag{13.25}$$

The saddle point $\mathbf{n}^* = (n_a^*, n_b^*)$ of the constrained equilibrium free energy

$$\beta \, \Delta G_{eq}^{cons}(n_a, n_b) = -\sum_i n_i \, \ln S_i + \beta \, \Delta G_{eq} \qquad (13.26)$$

corresponds to the critical cluster. The quadratic expansion of $\beta \Delta G_{eq}^{cons}$ near the saddle point reads:

$$\beta \Delta G_{eq}^{cons}(n_a, n_b) = g^* + m_a^2 \, D_{aa} + m_b^2 \, D_{bb} + 2 \, m_a m_b \, D_{ab} \qquad (13.27)$$

Here $g^* = \beta \Delta G_{eq}^{cons}(n_a^*, n_b^*)$ is the reduced Gibbs free energy at the saddle point,

$$m_i = n_i - n_i^*, \qquad D_{ij} = \frac{1}{2} \frac{\partial^2 \beta \Delta G_{eq}}{\partial n_i \, \partial n_j}\bigg|_{\mathbf{n}^*}, \qquad i, j = a, b$$

Note, that in view of (13.20)

$$\frac{\partial^2 \beta \Delta G_{eq}^{cons}}{\partial n_i \, \partial n_j}\bigg|_{\mathbf{n}^*} = \frac{\partial^2 \beta \Delta G_{eq}}{\partial n_i \, \partial n_j}\bigg|_{\mathbf{n}^*}$$

At the saddle point the eigenvalues of the symmetric matrix $\mathbf{D}$ have different signs implying that $\det \mathbf{D} < 0$. Following the standard procedure, we introduce a rotated coordinate system (x, y) in the (n_a, n_b)-space with the origin at $\mathbf{n}^*$ and the x axis pointing along the direction of the flow at $\mathbf{n}^*$ (see Fig. 11.2):

$$x = m_a \cos \varphi + m_b \sin \varphi, \qquad y = -m_a \sin \varphi + m_b \cos \varphi$$

where φ is the yet unknown angle between the x and n_a. At the saddle point $J_b/J_a = \tan \varphi$, which from (13.12)–(13.13) is written as:

$$\frac{v_b \frac{\partial H}{\partial n_b}}{v_a \frac{\partial H}{\partial n_a}} = \tan \varphi \qquad (13.28)$$

Using the standard relationships between the derivatives in the original and the rotated systems, we present after simple algebra Eq. (13.28) as:

$$\frac{\partial H}{\partial x} = \left[\frac{v_a \tan^2 \varphi + v_b}{(v_a - v_b) \tan \varphi} \right] \frac{\partial H}{\partial y} \qquad (13.29)$$

The rate components in the new coordinates are:

$$J_x = J_a \cos \varphi + J_b \sin \varphi, \qquad J_y = -J_a \sin \varphi + J_b \cos \varphi$$

Right *at* the saddle point $J_x = J_x^*(\mathbf{n}^*)$, $J_y(\mathbf{n}^*) = 0$. As in the BCNT, we use the *direction of principal growth approximation* assuming that

$$J_y = 0 \quad \textit{in the entire saddle point region} \tag{13.30}$$

Then, from the continuity condition (13.24)

$$\frac{\partial J_x(x, y)}{\partial x} = 0$$

implying that in the saddle point region $J_x = J_x(y)$, $J_y = 0$. In order to identify the function $J_x(y)$ we express the "old" coordinates J_a and J_b in terms of the "new" ones, taking into account (13.30):

$$J_a = J_x \cos \varphi, \quad J_b = J_x \sin \varphi$$

Then, using (13.14)–(13.15) we find

$$J_x(y) = -\rho_{eq}(\mathbf{n}) \, S_a^{n_a} S_b^{n_b} \, A \, v_a \left[\frac{1}{\cos \varphi} \frac{\partial H}{\partial n_a} \right] = -\rho_{eq} \, S_a^{n_a} S_b^{n_b} \, A \, v_{av} \frac{\partial H}{\partial x} \tag{13.31}$$

where v_{av} is the average impingement rate given by Eq. (11.35). Rewriting (13.31) as

$$J_x(y) \, \frac{1}{\rho_{eq} \, S_a^{n_a} S_b^{n_b} \, A \, v_{av}} = -\frac{\partial H}{\partial x}$$

and integrating over x using the standard boundary conditions

$$\lim_{x \to -\infty} H = 1, \quad \lim_{x \to +\infty} H = 0$$

we find

$$\frac{J_x(y)}{v_{av}} \int_{-\infty}^{\infty} dx \, \frac{1}{\rho_{eq} \, S_a^{n_a} S_b^{n_b} \, A} = 1 \tag{13.32}$$

Substituting the expansion (13.27) into Eq. (13.32) and performing Gaussian integration first over x and then over y, we find for the total nucleation rate

$$J = v_{av} \, A^* \, \mathscr{L} \left(\prod_i S_i^{n_i^*} \right) \rho_{eq}(n_a^*, n_b^*) \tag{13.33}$$

where $A^* = A(n_a^*, n_b^*)$,

$$\mathscr{L} = -\frac{1}{2} \frac{(\partial^2 \beta \Delta G / \partial x^2)_{\mathbf{n}^*}}{\sqrt{-\det \mathbf{D}}} \tag{13.34}$$

is the Zeldovich factor. Equation (13.33) contains the yet undetermined direction φ of the flow in the saddle point. The latter is found by maximizing the angle-dependent part of J.

The quantities with the φ dependence are: v_{av} and $\mathscr{Z}$. It is convenient to present v_{av} as

$$v_{av} = v_b \left(\frac{1+t^2}{r+t^2} \right), \quad t \equiv \tan\varphi, \quad r = \frac{v_b}{v_a} \tag{13.35}$$

In the Zeldovich factor $\det \mathbf{D}$ is invariant to rotation, implying that the only angle-dependent part of $\mathscr{Z}$ is contained in

$$\frac{\partial^2 \beta \Delta G}{\partial x^2} = D_{aa} \cos^2\varphi + 2D_{ab} \sin\varphi \cos\varphi + D_{bb} \sin^2\varphi = -D_{ab} \left[\frac{d_a - 2t + d_b t^2}{1+t^2} \right] \tag{13.36}$$

where we denoted

$$d_a = -\frac{D_{aa}}{D_{ab}}, \quad d_b = -\frac{D_{bb}}{D_{ab}}$$

Combining (13.35) and (13.36), the φ-dependent part of J is given by the function

$$f(t) = \frac{d_a - 2t + d_b t^2}{r+t^2}$$

Its extremum $df/dt = 0$ yields

$$\tan\varphi = s + \sqrt{s^2 + r} \quad \text{with} \quad s = \frac{1}{2}(d_a - r\, d_b) \tag{13.37}$$

which coincides with Stauffer's result (11.25) for BCNT. The advantage of Eq. (13.33) is that it reduces the binary nucleation problem to the determination of the equilibrium distribution of binary clusters $\rho_{eq}(n_a, n_b)$, which we discuss in the next section.

13.3 Binary Cluster Statistics

13.3.1 Binary Vapor as a System of Noninteracting Clusters

In line with the kinetic approach we discuss the full thermodynamic equilibrium of the system at the total pressure p^v, temperature T and carrier gas composition y_c (if present). The partition function of an arbitrary (n_a, n_b)-cluster is:

$$Z_{n_a n_b} \equiv Z_{\mathbf{n}} = \frac{1}{\Lambda_a^{3n_a} \Lambda_b^{3n_b}} q_{n_a n_b} \tag{13.38}$$

where Λ_i is the thermal de Broglie wavelength of a molecule of component i; $q_{n_a n_b}$ is the configuration integral of (n_a, n_b)-cluster in a domain of volume V:

$$q_{n_a n_b}(T) = q_{\mathbf{n}}(T) = \frac{1}{n_a! \, n_b!} \oint_{cl} d\mathbf{R}^{n_a} \, d\mathbf{r}^{n_b} \, e^{-\beta U_{\mathbf{n}}} \qquad (13.39)$$

where $\mathbf{R}^{n_a}$ and $\mathbf{r}^{n_b}$ are locations of molecules a and b in the cluster, and

$$U_{\mathbf{n}} = U_{aa}(\mathbf{R}^{n_a}) + U_{bb}(\mathbf{r}^{n_b}) + U_{ab}(\mathbf{R}^{n_a}, \mathbf{r}^{n_b}) \qquad (13.40)$$

is the potential interaction energy of the cluster comprised of $a-a$, $b-b$ and (unlike) $a - b$ interactions. The prefactor $\frac{1}{n_i!}$ takes into account the indistinguishability of molecules of type i inside the cluster. The symbol $\oint_{cl}$ indicates that integration is only over those molecular configurations that belong to the cluster. The cluster as a whole can move through the entire volume V, while the molecules inside it are restricted to the configurations about cluster's center of mass that are consistent with a chosen cluster definition.

We represent the equilibrium gaseous state of the $a - b$ mixture as a system of *noninteracting* $(n_a, n_b) = \mathbf{n}$ *clusters*. Since the clusters do not interact, the partition function $Z^{(\mathbf{n})}$ of the *gas* of $N_{\mathbf{n}}$ of such $\mathbf{n}$-clusters is factorized:

$$Z^{(\mathbf{n})} = \frac{1}{N_{\mathbf{n}}!} Z_{\mathbf{n}}^{N_{\mathbf{n}}}$$

where the prefactor $1/N_{\mathbf{n}}!$ takes into account the indistinguishability of (n_a, n_b)-clusters viewed as independent entities. The Helmholtz free energy of this gas is: $\mathscr{F}^{(\mathbf{n})} = -k_B T \ln Z^{(\mathbf{n})}$ which using Stirling's formula becomes:

$$\mathscr{F}^{(\mathbf{n})} = N_{\mathbf{n}} k_B T \, \ln \left(\frac{N_{\mathbf{n}}}{Z_{\mathbf{n}} \, e} \right)$$

The chemical potential of an $\mathbf{n}$-cluster in this gas is

$$\mu_{\mathbf{n}} = \frac{\partial \mathscr{F}^{(\mathbf{n})}}{\partial N_{\mathbf{n}}} = k_B T \, \ln \left(\frac{N_{\mathbf{n}}}{Z_{\mathbf{n}}} \right)$$

which using (13.38) reads

$$\mu_{\mathbf{n}} = k_B T \, \ln \left[\rho_{eq}(n_a, n_b) \frac{V \Lambda_a^{3n_a} \Lambda_b^{3n_b}}{q_{\mathbf{n}}} \right] \qquad (13.41)$$

Here $\rho_{eq}(n_a, n_b) = N_{\mathbf{n}}/V$ is the equilibrium distribution function of binary clusters—the quantity we are aiming to determine. Equilibrium between the cluster and the surrounding vapor requires

$$\mu_\mathbf{n} = n_a \, \mu^\mathrm{v}_{a,\mathrm{eq}} + n_b \, \mu^\mathrm{v}_{b,\mathrm{eq}} \tag{13.42}$$

where $\mu^\mathrm{v}_{i,\mathrm{eq}}(p^\mathrm{v}, T)$ is the chemical potential of a molecule of the component i in the equilibrium vapor. Combining (13.41) and (13.42), we find

$$\rho_\mathrm{eq}(n_a, n_b) = \left(\frac{q_\mathbf{n}}{V}\right) [z_{a,\mathrm{eq}}]^{n_a} [z_{b,\mathrm{eq}}]^{n_b} \tag{13.43}$$

where

$$z_{i,\mathrm{eq}} = \frac{e^{\beta \mu^\mathrm{v}_i,\mathrm{eq}}}{\Lambda_i^3}, \quad i = a, b \tag{13.44}$$

is the fugacity of component i in the equilibrium vapor. Thus, we reduced the problem of finding $\rho_\mathrm{eq}(n_a, n_b)$ to the determination of the cluster configuration integral. Even though we discuss the clusters *at vapor-liquid equilibrium*, it is important to realize that the chemical potential of a molecule *inside an arbitrary binary cluster* depends on the cluster composition and therefore is not the same as in the bulk vapor surrounding it. Equation (13.43) shows that the quantity $q_\mathbf{n}/V$ plays the key role in determination of the cluster distribution function. From the definition of $q_\mathbf{n}$ it is clear that $q_\mathbf{n}/V$ involves only the degrees of freedom relative to the center of mass of the cluster. Note also, that $q_\mathbf{n}$ contains the normalization constant C of the distribution function.

13.4 Configuration Integral of a Cluster: A Coarse-Grained Description

Rewriting $q_\mathbf{n}$ in the form

$$q_\mathbf{n} = \frac{1}{n_a!} \oint_\mathrm{cl} d\mathbf{R}^{n_a} \, e^{-\beta U_{aa}} \left\{ \frac{1}{n_b!} \oint_\mathrm{cl} d\mathbf{r}^{n_b} \, e^{-\beta(U_{bb}+U_{ab})} \right\} \tag{13.45}$$

one can easily see that the expression in the curl brackets

$$q_{b/a}(\{\mathbf{R}_a^{n_a}\}) \equiv \frac{1}{n_b!} \oint_\mathrm{cl} d\mathbf{r}^{n_b} \, e^{-\beta(U_{bb}+U_{ab})} \tag{13.46}$$

is the configuration integral of b-molecules *in the external field of a-molecules* located at fixed positions $\{\mathbf{R}_a^{n_a}\}$. The configurational part of the Helmholtz free energy of this system is

$$\mathscr{F}_{b/a}(\{\mathbf{R}_a^{n_a}\}, n_a, n_b, T) = -k_\mathrm{B} T \, \ln q_{b/a} \tag{13.47}$$

Substituting (13.47) into (13.45), we present $q_\mathbf{n}$ in the *coarse-grained form*

$$q_{\mathbf{n}} = \frac{1}{n_a!} \oint_{\text{cl}} d\mathbf{R}^{n_a} \, e^{-\beta \mathscr{H}^{\text{CG}}} \tag{13.48}$$

where the positions of b-particles are integrated out. By doing so we replaced the binary cluster by the equivalent single-component one with the effective Hamiltonian

$$\mathscr{H}^{\text{CG}} = U_{aa}(\{\mathbf{R}_a^{n_a}\}) + \mathscr{F}_{b/a}(\{\mathbf{R}_a^{n_a}\}; n_a, n_b, T) \tag{13.49}$$

which is the sum of the Hamiltonian of the pure a-system, U_{aa}, and the free energy of b-molecules in the instantaneous environment of a molecules. Equation (13.48) is formally exact.

In order to derive a tractable representation of the free energy $\mathscr{F}_{b/a}$ we perform the diagrammatic expansion of $\ln q_{b/a}$ in the Mayer functions of $a - b$ and $b - b$ interactions:

$$f_{ab}(|\mathbf{R}_i - \mathbf{r}_j|) = \exp[-\beta \, u_{ab}(|\mathbf{R}_i - \mathbf{r}_j|)] - 1$$
$$f_{bb}(|\mathbf{r}_k - \mathbf{r}_l|) = \exp[-\beta \, u_{bb}(|\mathbf{r}_k - \mathbf{r}_l|)] - 1$$

As a result $\mathscr{F}_{b/a}$ is represented as the sum of m-body effective interactions between a-molecules [6, 9]:

$$\mathscr{F}_{b/a}(\{\mathbf{R}_a^{n_a}\}; n_a, n_b, T) = \mathscr{F}_0(n_a, n_b, T) + U_2(\{\mathbf{R}_a^{n_a}\}; x_b^{\text{tot}}, T) + \cdots \tag{13.50}$$

The zeroth order contribution $\mathscr{F}_0(n_a, n_b, T)$, called the *volume term*, does not depend on positions of molecules, but is important for thermodynamics since it depends on cluster composition and therefore by no means can be neglected. The first-order term U_1 in (13.50) vanishes in view of translational symmetry [6]. Combining (13.49)–(13.50) we write

$$\mathscr{H}^{\text{CG}} = \mathscr{F}_0(n_a, n_b, T) + U^{\text{CG}}(\{\mathbf{R}_a^{n_a}\}; x_b^{\text{tot}}, T) \tag{13.51}$$

with the total coarse-grained interaction energy

$$U^{\text{CG}}(\{\mathbf{R}_a^{n_a}\}; x_b, T) = U_{aa}(\{\mathbf{R}_a^{n_a}\}) + U_2(\{\mathbf{R}_a^{n_a}\}; x_b^{\text{tot}}, T) + \cdots$$

Substituting (13.51) into (13.48), we obtain

$$q_{\mathbf{n}} = e^{-\beta \mathscr{F}_0} \, q_{n_a}^{\text{CG}} \tag{13.52}$$

where

$$q_{n_a}^{\text{CG}}(x_b^{\text{tot}}, T) = \frac{1}{n_a!} \oint_{\text{cl}} d\mathbf{R}^{n_a} \, e^{-\beta U^{\text{CG}}} \tag{13.53}$$

Interpretation of Eqs. (13.52)–(13.53) is straightforward: by tracing out the degrees of freedom of b-molecules, we are left with the single-component cluster of *pseudo—a* molecules with the interaction energy U^{CG}. The latter implicitly depends on the fraction of *b*-molecules in the original binary cluster. The configuration integral of this single-component cluster is $q_{n_a}^{CG}$. Equation (13.52) is a key result of the model.

Speaking about a binary cluster, we characterized it by the total numbers of molecules n_a and n_b, not discriminating between the bulk and excess numbers of molecules of each component

$$n_i = n_i^1 + n_i^{exc}$$

Meanwhile, as we know, this distinction is important for capturing the adsorption effects resulting in nonhomogeneous distribution of molecules within the cluster. Description of adsorption requires introduction of the Gibbs dividing surface, for which we will use the K-surface of Sect. 11.6. This means, that for an arbitrary *bulk* cluster content (n_a^1, n_b^1) we find the excess numbers from Eqs. (11.84)–(11.85):

$$n_a^{exc}(n_a^1, n_b^1), \quad n_b^{exc}(n_a^1, n_b^1)$$

Thus, the point (n_a^1, n_b^1) in the space of bulk numbers yields the point (n_a, n_b) in the space of total numbers. As a result the dependence of various quantities on the total composition x_b^{tot} can be also viewed as a dependence (though a different one) on the *bulk* composition x_b^1.

Since the carrier gas is assumed to be passive, the cluster composition satisfies the normalization:

$$x_a^1 + x_b^1 = x_a^{tot} + x_b^{tot} = 1 \tag{13.54}$$

13.4.1 Volume Term

Let us discuss in more detail the volume term in the effective Hamiltonian $\mathcal{H}^{CG}$. It can be written as a sum of the free energy of ideal gas of pure *b*-molecules in the cluster, $\mathcal{F}_{b,id}$, and the excess (over ideal) contribution, $\Delta\mathcal{F}_0$, due to $b-b$ and $a-b$ interactions

$$\mathcal{F}_0 = \mathcal{F}_{b,id} + \Delta\mathcal{F}_0 \tag{13.55}$$

The ideal gas contribution reads (see e.g. [10]):

$$\beta\mathcal{F}_{b,id} = n_b \ln\left(\frac{n_b \Lambda_b^3}{V_{cl}\, e}\right) \tag{13.56}$$

where $V_{cl} = n_a^1 v_a^1 + n_b^1 v_b^1$ is the volume of the cluster. Within the K-surface formalism we can equivalently write it in terms of *total* numbers:

$$V_{cl} = n_a v_a^l + n_b v_b^l \qquad (13.57)$$

The specific feature of Eq. (13.56) is that b-molecules are contained in the cluster volume which itself depends on their number n_b. Substituting (13.57) into (13.56), we write

$$\beta \mathscr{F}_{b,\text{id}} = n_b f_{b,\text{id}} \qquad (13.58)$$

where

$$f_{b,\text{id}} = \ln \left\{ \left(\frac{x_b^{\text{tot}}}{x_a^{\text{tot}}} \right) \left[\frac{\Lambda_b^3}{\left(v_a^l + \frac{x_b^{\text{tot}}}{x_a^{\text{tot}}} v_b^l \right) e} \right] \right\} \equiv f_{b,\text{id}}(x_b^l, T)$$

depends only on intensive quantities.

Calculation of $\Delta \mathscr{F}_0$ in terms of interaction potentials is a challenging task; it has been done for a limited number of model potentials: mixtures of hard spheres [9, 11] and charged-stabilized colloidal suspensions [12, 13]. Fortunately, for our purposes we do not need to know its exact form. Instead, we make use of the general statement that $\mathscr{F}_{0,\text{exc}}$ is a homogeneous function of the first order in n_a and n_b [14]:

$$\beta \Delta \mathscr{F}_0 = \frac{n_a n_b}{V_{cl}} f_1(x_b^l, T) \qquad (13.59)$$

where f_1 is some unknown function of x_b^l and T. Using (13.57), we present Eq. (13.59) as

$$\beta \Delta \mathscr{F}_0 = n_b f_0 \qquad (13.60)$$

where

$$f_0 = \frac{f_1}{v_a^l + \frac{x_b^{\text{tot}}}{x_a^{\text{tot}}} v_b^l} \equiv f_0(x_b^l, T)$$

depends only on intensive quantities. Combining (13.55), (13.58) and (13.60), we present the volume term as

$$e^{-\beta \mathscr{F}_0} = \Phi_b^{n_b} \qquad (13.61)$$

where $\Phi_b = \exp[-(f_{b,\text{id}} + f_0)]$ is another unknown function of x_b^l and T.

13.4.2 Coarse-Grained Configuration Integral $q_{n_a}^{\text{CG}}$

The coarse-grained configuration integral $q_{n_a}^{\text{CG}}$ describes the cluster with n_a identical particles (pseudo-a molecules), characterized by unknown complex interactions. We will analyze it using the formalism of the mean-field kinetic nucleation theory (MKNT) of Chap 7. Within MKNT the cluster configuration integral is given by Eq. (7.38):

$$\frac{q_{n_a}^{CG}}{V} = C \, \Phi_a^{n_a} \, e^{-\theta_{\text{micro}} \, \overline{n_a^s}} \tag{13.62}$$

Here

$$\Phi_a = \frac{1}{z_{a,\text{sat}}} \tag{13.63}$$

$z_{a,\text{sat}}$ is the fugacity at saturation, $\overline{n_a^s}(n_a)$ is the average number of *surface particles* in the cluster, θ_{micro} is the reduced microscopic surface tension. For the coarse-grained cluster those are the functions of the cluster composition. It is important to stress, that the division of the cluster molecules into the core- and surface particles, adopted in MKNT, is different from the Gibbs construction (11.47).[1]

The properties of the pseudo-*a* fluid are *functionals* of the unknown interaction potentials. It is practically impossible to restore these potentials from the microscopic considerations. An alternative to the microscopic approach is the use of the known asymptotic features of the distribution function.

13.5 Equilibrium Distribution of Binary Clusters

Using the basic result of the coarse-graining procedure Eq. (13.52), we express the equilibrium distribution function (13.43) of binary clusters as

$$\rho_{\text{eq}}(n_a, n_b) = e^{-\beta \mathscr{F}_0} \left(\frac{q_{n_a}^{CG}}{V} \right) [z_{a,\text{eq}}]^{n_a} [z_{b,\text{eq}}]^{n_b}$$

Substitution of (13.61) and (13.62) into this expression yields

$$\rho_{\text{eq}}(n_a, n_b)_{p^v,T} = C \, [\Phi_a(x_b^1) \, z_{a,\text{eq}}]^{n_a} \, [\Phi_b(x_b^1) \, z_{b,\text{eq}}]^{n_b} \, e^{-g^{\text{surf}}(n_a; x_b^1, T)} \tag{13.64}$$

where

$$g^{\text{surf}}(n_a; x_b^1, T) = \theta_{\text{micro}}(x_b^1) \, \overline{n_a^s}(n_a; x_b^1) \tag{13.65}$$

The right-hand side of (13.64) contains the unknown intensive quantities Φ_a, Φ_b and θ_{micro} which depend on the bulk composition of the cluster and the temperature. To determine them we consider appropriate limiting cases, for which the behavior of $\rho_{\text{eq}}(n_a, n_b)$ can be deduced from thermodynamic considerations.

The (p^v, T)-equilibrium corresponds to the bulk liquid composition $x_{b,\text{eq}}(p^v, T)$. An arbitrary (n_a, n_b)-cluster at (p^v, T)-equilibrium has the bulk composition x_b^1

---

[1] Note in this respect, that $\overline{n_a^s}(n_a)$ is always positive, while the Gibbs excess numbers n_i^{exc} can be both positive and negative.

different from $x_{b,\text{eq}}$. Let us now fix x_b^1 and consider the two-phase equilibrium at the pressure $p^{\text{coex}}(x_b^1, T)$, representing the total pressure above the bulk binary solution with the composition x_b^1. Obviously, $p^{\text{coex}}(x_b^1, T) \neq p^v$ (the equality occurs only for $x_b^1 = x_{b,\text{eq}}$). At this "x_b^1-equilibrium" state the fugacities are: $z_{i,\text{coex}} = e^{\beta \mu_{i,\text{coex}}} / \Lambda_i^3$, where $\mu_{i,\text{coex}}(x_b^1)$ is the chemical potential at x_b^1-equilibrium. Thus, in the distribution function for this state $z_{i,\text{eq}}$ in (13.64) should be replaced by $z_{i,\text{coex}}$.

Now, from the entire cluster size space let us consider the clusters falling on the x_b^1-equilibrium line, i.e. those whose bulk numbers of molecules satisfy:

$$n_b^1 = n_a^1 (x_b^1 / x_a^1)$$

For them the chemical potential of the molecule inside the cluster is equal to its value in the surrounding vapor at the pressure p^{coex}. The Gibbs formation energy of such a cluster will contain only the (positive) surface term:

$$\rho(n_a, n_b)_{p^{\text{coex}}, T} = C\, e^{-g^{\text{surf}}} \tag{13.66}$$

This consideration leads to the determination of the functions $\Phi_i(x_b^1)$ in (13.64):

$$\Phi_i(x_b^1) = \frac{1}{z_{i,\text{coex}}(x_b^1)} \tag{13.67}$$

Considering the vapor to be a mixture of ideal gases, we write

$$\frac{z_{i,\text{eq}}}{z_{i,\text{coex}}(x_b^1)} = \exp\left[\beta \mu_{i,\text{eq}} - \beta \mu_{i,\text{coex}}\right] \approx \left[\frac{y_{i,\text{eq}}\, p^v}{y_i^{\text{coex}}(x_b^1)\, p^{\text{coex}}(x_b^1)}\right], \quad i = a, b$$

leading to

$$\rho_{\text{eq}}(n_a, n_b)_{p^v, T} = C \left[\frac{y_{a,\text{eq}}\, p^v}{y_a^{\text{coex}}(x_b^1)\, p^{\text{coex}}(x_b^1)}\right]^{n_a} \left[\frac{y_{b,\text{eq}}\, p^v}{y_b^{\text{coex}}(x_b^1)\, p^{\text{coex}}(x_b^1)}\right]^{n_b} e^{-g^{\text{surf}}(n_a; x_b^1, T)} \tag{13.68}$$

where $y_i^{\text{coex}}(x_b^1)$ is the equilibrium vapor fraction of component i at x_b^1-equilibrium.

Now let us identify the parameters θ_{micro} and C for the pseudo-a fluid. Within the (single-component) MKNT they are expressed in terms of the equilibrium properties of the substance:

$$\theta_{\text{micro}} = -\ln\left(-\frac{B_2 p_{\text{sat}}}{k_B T}\right), \tag{13.69}$$

$$C = \frac{p_{\text{sat}}}{k_B T}\, e^{\theta_{\text{micro}}} \tag{13.70}$$

Here $B_2(T)$ is the second virial coefficient, $p_{sat}(T)$ is the saturation pressure. The number of surface molecules $\overline{n_a^s}$ depends parametrically on the coordination number in the liquid phase N_1 and the reduced plain layer surface tension $\theta_\infty(T)$.

Comparing (13.67) and (13.63), we can identify the "saturation state" of the pseudo-a fluid: the latter is characterized by the chemical potential $\mu_{a,coex}(x_b^1)$ and the pressure

$$p_{sat} = y_a^{coex}(x_b^1)\, p^{coex}(x_b^1) \tag{13.71}$$

Since the intermolecular potential $u(r; x_b^1, T)$ of the *pseudo—a* fluid is not known, we have to introduce an approximation for the second virial coefficient, which will now depend on x_b^1. The simplest form satisfying the pure components limit, is given by the mixing rule:

$$B_2 = B_{2,aa}\,(x_a^1)^2 + 2B_{2,ab}\,x_a^1 x_b^1 + B_{2,bb}\,(x_b^1)^2 \tag{13.72}$$

where $B_{2,ii}(T)$ is the second virial coefficient of the pure component i; the cross virial term $B_{2,ab}(T)$ can be estimated using the standard methods [15]. Having identified p_{sat} and B_2 for the pseudo-a system, we find the reduced microscopic surface tension $\theta_{micro} \equiv \theta_{micro,a}$ and the normalization factor C from Eqs. (13.69)–(13.70):

$$\theta_{micro,a}(x_b^1) = -\ln\left(-\frac{B_2\, y_a^{coex}\, p^{coex}}{k_B T}\right), \tag{13.73}$$

$$C(x_b^1) = \left(\frac{y_a^{coex}\, p^{coex}}{k_B T}\right) e^{\theta_{micro,a}} \tag{13.74}$$

The cluster distribution (13.68) with g^{surf} given by (13.65) will be fully determined if we complete it by the model for $\overline{n_a^s}(n_a; x_b^1)$. Calculation of this quantity requires the knowledge of the reduced planar surface tension of the pseudo-a fluid, $\theta_{\infty,a}$ and the coordination number in the liquid phase, $N_{1,a}$. The latter can be estimated from (7.68) in which the packing fraction of pseudo-a molecules is approximated as

$$\eta = \frac{\pi}{6}\,\rho^1(x_b^1)\,\sigma_a^3 \tag{13.75}$$

Here $\rho^1(x_b^1)$ is the binary liquid number density at x_b^1-equilibrium and σ_a is the molecular diameter of component a.

To determine $\theta_{\infty,a}$ let us consider the distribution function $\rho_{eq}(n_a, n_b)$ for *big* (n_a, n_b)-clusters. These clusters satisfy the capillarity approximation in which the surface part of the Gibbs energy takes the form

$$g^{surf}(n_a, n_b; x_b^1) = \beta\, \gamma(x_b^1)\, A = \beta\, \gamma(x_b^1)\, (36\pi)^{1/3}\, (n_a^1 v_a^1 + n_b^1 v_b^1)^{2/3} \tag{13.76}$$

where $\gamma(x_b^1)$ is the planar surface tension of the binary system at x_b^1-equilibrium.

Speaking about the single-component cluster with pseudo-a particles, we characterize it by the *total* number n_a molecules (not n_a^1). Therefore, it is convenient to rewrite (13.76) in terms of the total numbers, which we can do using the properties of the K-surface:

$$g^{\text{surf}}(n_a, n_b; x_b^1) = \beta\,\gamma(x_b^1)\,(36\pi)^{1/3}\,(n_a\,v_a^1 + n_b\,v_b^1)^{2/3} \tag{13.77}$$

The partial molecular volume of component i can be expressed as (see Appendix D.1):

$$v_i^1 = \frac{1}{\rho^1}\,\eta_i^1 \tag{13.78}$$

where

$$\eta_i^1 = 1 + x_j\,\frac{\partial \ln \rho^1}{\partial x_j}\bigg|_{p^{\text{coex}}, T, j\neq i}$$

Substituting (13.78) into (13.77) we find

$$g^{\text{surf}}(n_a, n_b; x_b^1) = \theta_\infty\left(n_a\,\eta_a^1 + n_b\,\eta_b^1\right)^{2/3} = \left[\theta_\infty\left(\eta_a^1 + \frac{x_b^{\text{tot}}}{x_a^{\text{tot}}}\,\eta_b^1\right)^{2/3}\right]n_a^{2/3} \tag{13.79}$$

where

$$\theta_\infty(x_b^1) = \beta\,\gamma\,(36\pi)^{1/3}\,(\rho^1)^{-2/3} \tag{13.80}$$

is the reduced planar surface tension of the original binary system at x_b^1-equilibrium.

At the same time, for a big *single-component* cluster with n_a pseudo-a molecules the surface free energy is

$$g^{\text{surf}} = \theta_{\infty,a}\,n_a^{2/3} \tag{13.81}$$

We require that the binary (n_a, n_b)-cluster has the same surface energy as the unary n_a-cluster of pseudo-a molecules for all sufficiently large n_a. This implies that

$$\theta_{\infty,a} = \theta_\infty(x_b^1)\left(\eta_a^1 + \frac{x_b^{\text{tot}}}{x_a^{\text{tot}}}\,\eta_b^1\right)^{2/3} \tag{13.82}$$

The construction of the model thus ensures that for sufficiently large clusters, satisfying the capillarity approximation, the distribution function recovers the classical Reiss expression [16] (see also [17]) and is symmetric in both components. This will not be true for small clusters, for which the capillarity approximation fails and the Gibbs energy of cluster formation differs from its phenomenological counterpart.

Having determined all parameters, entering Eq. (13.64), we are in a position to write down the equilibrium distribution function:

$$\rho_{eq}(n_a, n_b)_{p^v, T} = \underbrace{\left[\beta \, y_a^{coex}(x_b^l) \, p^{coex}(x_b^l) \right]}_{C(x_b^l)} e^{-g_{eq}(n_a, n_b; x_b^l)} \tag{13.83}$$

where

$$g_{eq}(n_a, n_b; x_b^l) = g^{bulk}(n_a, n_b; x_b^l) + g^{surf}(n_a, n_b; x_b^l) \tag{13.84}$$

$$g^{bulk}(n_a, n_b; x_b^l) = -\sum_i n_i \ln \left[\frac{y_{i,eq} \, p^v}{y_i^{coex}(x_b^l) \, p^{coex}(x_b^l)} \right] \tag{13.85}$$

$$g^{surf}(n_a, n_b; x_b^l) = \theta_{micro,a}(x_b^l) \left[\overline{n_a^s}(n_a; x_b^l) - 1 \right] \tag{13.86}$$

We have included the exponential factor with the microscopic surface tension in (13.70) into the surface part of the free energy thereby redefining the prefactor C of the distribution function:

$$C(x_b^l) = \beta \, y_a^{coex}(x_b^l) \, p^{coex}(x_b^l) \tag{13.87}$$

The binary distribution function (13.83) recovers the single-component MKNT limit when $n_b \to 0$.

An important feature of CGNT is that it eliminates ambiguity in the normalization constant C in the nucleation rate inherent to BCNT and its modifications. This is the direct consequence of replacing the *constrained* equilibrium concept by the *full thermodynamic equilibrium*. Since all rapidly (exponentially) changing terms in the distribution function are included into the free energy, we can safely set the value of C in (13.90) to the one corresponding to the critical cluster:

$$C(x_b^{l*}) = \beta \, y_a^{coex}(x_b^{l*}) \, p^{coex}(x_b^{l*}) \tag{13.88}$$

Our choice to trace out the b-molecules in the cluster in favor of a-molecules could have been reversed: we could trace out a-molecules to be left with the effective Hamiltonian for the b-molecules resulting in Eq. (13.52) with the single-component cluster containing *pseudo-b* particles. This equation is formally exact. However, calculation of the coarse-grained configuration integral in (13.52) invokes *approximations* inherent to MKNT. Its domain of validity is given by Eq. (7.53) which for the system of pseudo-i particles ($i = a$ or b) reads

$$\left| \frac{B_2 \, y_i^{coex} \, p^{coex}}{k_B T} \right| \ll 1 \tag{13.89}$$

It is clear that to obtain accurate predictions within the present approach one has to *trace out the more volatile component*, i.e. the one with the largest bulk vapor fraction. Throughout this chapter we assume this to be component b: $y_b^{coex} > y_a^{coex}$; thus, the coarse-grained cluster contains pseudo-a particles.

13.6 Steady State Nucleation Rate

Combining Eqs. (13.33) and (13.83)–(13.86), we obtain the total steady-state nucleation rate for the binary mixture at the total pressure p^v, temperature T and vapor mole fractions y_i:

$$J = \underbrace{v_{av}^* A^* \mathscr{Z} C(x_b^{l*})}_{K} \, e^{-g(n_a^*, n_b^*)} \tag{13.90}$$

where star refers to the critical cluster being the saddle-point of the free energy surface

$$g(n_a, n_b; x_b^{l*}) = -\sum_i n_i \ln S_i + g_{eq}(n_a, n_b; x_b^{l*}) \tag{13.91}$$

in the space of *total* numbers n_i. Note, that search for the saddle point in the space of *bulk* numbers can lead to an erroneous critical cluster composition and unphysical results. Technical details of the saddle-point calculations are given in Appendix G. The proposed model is termed the *Coarse-Grained Nucleation Theory* (CGNT).

It is instructive to summarize the steps leading to Eq. (13.90).

- Step 0. Determine the metastability parameters of components using as a reference the (p^v, T)-equilibrium state of the mixture

$$S_i = \frac{y_i}{y_{i,eq}(p^v, T; y_c)}, \quad i = a, b$$

- Step 1. Choose an arbitrary *bulk composition* of a cluster n_a^l, n_b^l; the bulk fraction of component b is then

$$x_b^l = n_b^l/(n_a^l + n_b^l)$$

- Step 2. Calculate the two-phase (x_b^l, T)-equilibrium properties from an appropriate equation of state for the mixture:

$$y_i^{coex}(x_b^l, T), \quad p^{coex}(x_b^l, T), \quad \rho^l(x_b^l, T), \quad \rho^v(x_b^l, T)$$

Find the planar surface tension $\gamma_\infty(x_b^l, T)$ for the $x(_b^l, T)$-equilibrium, using e.g. the Parachor method, or some other available (semi-)empirical correlation.

- Step 3. Calculate the properties of the pseudo-a fluid: the second virial coefficient from Eq. (13.72) and the reduced microscopic surface tension from Eq. (13.73)
- Step 4. Determine the excess numbers of molecules n_a^{exc}, n_b^{exc} from the K-surface equations (11.84)–(11.85). The total numbers of molecules in the cluster are

$$n_i^1 = n_i^1 + n_i^{\text{exc}}, \quad i = a, b$$

- Step 5. Calculate the free energy of cluster formation $g(n_a, n_b; x_b^1)$ from Eq. (13.91).
- Step 6. Repeat Steps. 1–5 for a "reasonably chosen" domain of bulk compositions, accumulating the data:

$$n_a^1, \; n_b^1, \; n_a, \; n_b, \; g(n_a, n_b; x_b^1)$$

- Step 7. Determine the saddle-point (n_a^*, n_b^*) of g in the (n_a, n_b)-space (using e.g. the method outlined in Appendix G)
- Step 8. Calculate the prefactor for the nucleation rate

$$K = v_{\text{av}}^* A^* \mathscr{Z} C(x_b^{1*})$$

- Step 9. Calculate the steady-state nucleation rate from Eq. (13.90)

For practical purpose it is usually sufficient to approximate the flow direction at the saddle point by the angle satisfying $\tan \varphi \approx n_b^*/n_a^*$ (see Fig. 11.2). Another simplification refers to the Zeldovich factor. Recall that in CNT the Zeldovich factor takes the form of Eq. (3.51), where $1/\rho^1 = v^1$ is the molecular volume in the liquid phase. Replacing in (3.51) γ_∞ by the planar surface tension of the binary mixture taken at the saddle point concentration $\gamma_\infty(x_b^{1*})$, and v^1—by the average molecular volume of a (virtual) monomer in the liquid phase

$$v_{\text{av}}^1 = x_a^{1*} v_a^1 + x_b^{1*} v_b^1$$

we end up with the *virtual monomer approximation* for $\mathscr{Z}$ proposed by Kulmala and Viisanen [18]:

$$\mathscr{Z} = \sqrt{\frac{\gamma_\infty(x_b^{1*})}{k_{\text{B}} T} \frac{x_a^{1*} v_a^1 + x_b^{1*} v_b^1}{2\pi (r^*)^2}} \tag{13.92}$$

where r^* is the radius of the critical cluster.

13.7 Results: Nonane/Methane Nucleation

In Sect. 11.9.4 we considered nucleation in the binary mixture n-nonane/methane in the absence of a carrier gas. Methane, being in abundance in the vapor phase, is a natural candidate for component b in CGNT, whose degrees of freedom are traced out within the coarse-graining procedure.

In Fig. 11.9 we showed the BCNT predictions for this system for $T = 240$ K and pressures 10, 25, 33, 40 bar comparing them with the experimental data of Refs. [4, 19–21]. Figure 13.1 completes this plot with the CGNT data. The agreement between CGNT and experiment lies within the range of experimental accuracy for most of the conditions except for extremely low $S_a < 5$ at the highest pressure 40 bar.

An important insight into the binary nucleation process can be obtained from the analysis of the structure of the critical cluster. The latter determines the height of the nucleation barrier as well as the average impingement rate. Figure 13.2 shows the CGNT critical cluster content—bulk, excess, and total numbers of molecules of each species—as a function of the total pressure at a fixed nucleation rate $J = 10^{10}$ cm^{-3} s^{-1} and temperature $T = 240$ K. At $p^v < 18$ bar there are no methane molecules in the critical cluster: $n_b^l = n_b^{exc} = 0$. This feature indicates that at low pressures nucleation can be viewed as an effective single-component process in which the macroscopic thermodynamic properties—liquid density, surface tension— are those of the mixture at the given p^v and T. Even with this simplification one has to bear in mind that since the critical cluster at these conditions is quite small— $n_a \approx 19 \div 21$—the appropriate single-component treatment requires nonclassical considerations. Beyond $p^v \approx 18$ bar methane starts penetrating into the critical cluster, the nucleation process demonstrates binary features, becoming more and more pronounced as the pressure grows. In Sect. 11.9.4.1 we discussed the compensation pressure effect. For the nonane/methane system compensation pressure found from

Fig. 13.1 Nonane/methane nucleation. Nucleation rate versus metastability parameter of nonane S_{nonane} at various pressures and $T = 240$ K. *Solid lines*: CGNT; *dashed lines*: BCNT. *Symbols*: experiment of Luijten [4, 19] (*closed circles*), Peeters [20] (*open circles*) and Labetski [21] (*half-filled squares*)

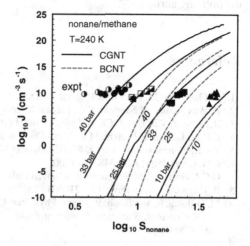

Fig. 13.2 Critical cluster
at a fixed nucleation rate
$J = 10^{10}\,\text{cm}^{-3}\,\text{s}^{-1}$ at $T =$
240 K as a function of the
total pressure; $n_i^{\text{tot}} \equiv n_i =$
$n_i^{\text{l}} + n_i^{\text{exc}}$. The *vertical arrow*
indicates the compensation
pressure

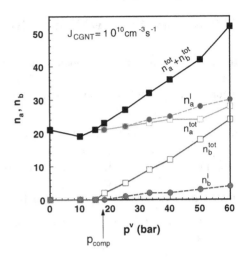

Eq. (11.126) is

$$p_{\text{comp}}(T = 240\,\text{K}) \approx 17.8\,\text{bar}$$

It is remarkable that the change in the nucleation behavior predicted by CGNT occurs exactly at p_{comp}.

As the pressure is increased the total number of nonane molecules grows very slowly while the total number of methane molecules increases rapidly, accumulating predominantly at the dividing surface; at the highest pressure of 60 bar shown in Fig. 13.2 there are only about 4 methane molecules in the interior of the cluster, while their total number is ≈ 22 being close to the total number of nonane molecules $n_a \approx 26$. Hence, at high pressures the critical cluster is a nano-sized object with a *core-shell structure*: its interior is rich in nonane while methane is predominantly adsorbed on the dividing surface.

References

1. C. Flageollet, M. Dihn Cao, P. Mirabel, J. Chem. Phys. **72**, 544 (1980)
2. B. Wyslouzil, J.H. Seinfeld, R.C. Flagan, K. Okuyama, J. Chem. Phys. **94**, 6827 (1991)
3. R. Strey, Y. Viisanen, J. Chem. Phys. **99**, 4693 (1993)
4. C.C.M. Luijten, P. Peeters, M.E.H. van Dongen, J. Chem. Phys. **111**, 8535 (1999)
5. V.I. Kalikmanov, Phys. Rev. **E 81**, 050601(R) (2010)
6. C.N. Likos, Phys. Rep. **348**, 267 (2001)
7. G.C.J. Hofmans, M.Sc. Thesis, Eindhoven University of Technology, 1993
8. R. Flagan, J. Chem. Phys. **127**, 214503 (2007)
9. M. Dijkstra, R. van Rooij, R. Evans, Phys. Rev. Lett. **81**, 2268 (1998)
10. V.I. Kalikmanov, *Statistical Physics of Fluids. Basic Concepts and Applications* (Springer, Berlin, 2001)

11. M. Dijkstra, R. van Rooij, R. Evans, Phys. Rev. E **59**, 5744 (1999)
12. R. van Rooij, J.-P. Hansen, Phys. Rev. Lett. **79**, 3082 (1997)
13. R. van Rooij, M. Dijkstra, J.-P. Hansen, Phys. Rev. E **59**, 2010 (1999)
14. V.I. Kalikmanov, Phys. Rev. E **68**, 010101 (2003)
15. R.C. Reid, J.M. Prausnitz, B.E. Poling, *The Properties of Gases and Liquids* (McGraw-Hill, New York, 1987)
16. H. Reiss, J. Chem. Phys. **18**, 840 (1950)
17. G. Wilemski, B. Wyslouzil, J. Chem. Phys. **103**, 1127 (1995)
18. M. Kulmala, Y. Viisanen, J. Aerosol Sci. **22**, S97 (1991)
19. C.C.M. Luijten, Ph.D. Thesis, Eindhoven University, 1999
20. P. Peeters, Ph.D. Thesis, Eindhoven University, 2002
21. D.G. Labetski, Ph.D. Thesis, Eindhoven University, 2007

Chapter 14
Multi-Component Nucleation

Understanding multi-component nucleation is of great importance for atmospheric and environmental sciences. Vivid examples are: polar stratospheric clouds, acid rains and air pollution. All these phenomena occur because Earth's atmosphere is a multi-component gaseous system in which nucleation leads to formation of droplets of complex composition. A rich field of applications of multi-component nucleation is associated with the natural gas industry since nucleation is the primary mechanism responsible for formation of mist during the expansion of natural gas [1]. This is the key process of the non-equilibrium gas-liquid separation technology [2].

Theoretical description of multi-component nucleation pioneered by Hirschfelder [3] and Trinkaus [4] represents the extension of the phenomenological binary nucleation theory, discussed in Chap. 11, to the N-component mixture.

14.1 Energetics of N-Component Cluster Formation

Consider an arbitrary cluster of the new phase (e.g. liquid) containing $n_1, \ldots, n_N$ molecules of components $1, \ldots, N$, respectively. We will call it the **n**-cluster, where the vector $\mathbf{n} = (n_1, \ldots, n_N)$ is represented by the point in the N-dimensional space of the cluster sizes of components. The **n**-cluster is immersed in the "mother phase" (e.g. supersaturated gas) characterized by the total pressure p^{v} and temperature T. Energetics of cluster formation determines the minimum reversible work $\Delta G(\mathbf{n})$ needed to form the **n**-cluster in the surrounding vapor.

To take into account adsorption (leading to inhomogeneous density distribution of species inside the cluster) we introduce an arbitrary located Gibbs dividing surface distinguishing between the bulk (superscript "l") and excess (superscript "exc") molecules of each species. Then, the total numbers of molecules n_i are

$$n_i = n_i^{\mathrm{l}} + n_i^{\mathrm{exc}}, \quad i = 1, \ldots, N \tag{14.1}$$

V. I. Kalikmanov, *Nucleation Theory*, Lecture Notes in Physics 860,
DOI: 10.1007/978-90-481-3643-8_14, © Springer Science+Business Media Dordrecht 2013

As discussed in Chap. 11, n_i^l and n_i^{exc} *separately* depend on the location of the dividing surface while their *sum* is independent of this location to the relative accuracy of $O(\rho^v/\rho^l)$, where ρ^v and ρ^l are the number densities in the vapor and liquid phases. Straightforward generalization of Eq. (11.48) to N-component mixture yields:

$$\Delta G = (p^v - p^l)V^l + \gamma A + \sum_{i=1}^{N} n_i^l \left[\mu_i^l(p^l) - \mu_i^v(p^v) \right] + \sum_{i=1}^{N} n_i^{exc} \left[\mu_i^{exc} - \mu_i^v(p^v) \right] \tag{14.2}$$

Here p^l is the pressure inside the cluster,

$$V^l = \sum_{i=1}^{N} n_i^l v_i^l \tag{14.3}$$

$$A = (36\pi)^{1/3} \left(\sum_{i=1}^{N} n_i^l v_i^l \right)^{2/3} \tag{14.4}$$

are, respectively, the cluster volume and surface area calculated at the location of the dividing surface; v_i^l is the partial molecular volume of component i in the liquid phase, γ is the surface tension at the dividing surface. Within the capillarity approximation

$$\mu_i^l(p^l) = \mu_i^l(p^v) + v_i^l(p^l - p^v) \tag{14.5}$$

Substituting (14.3) and (14.5) into (14.2) results in

$$\Delta G = \gamma A - \sum_{i=1}^{N} n_i^l \, \Delta\mu_i + \sum_{i=1}^{N} n_i^{exc} \left[\mu_i^{exc} - \mu_i^v(p^v) \right] \tag{14.6}$$

where

$$\Delta\mu_i \equiv \mu_i^v(p^v) - \mu_i^l(p^v) \tag{14.7}$$

is the driving force for nucleation. As in the binary case, the chemical potentials in both phases are taken at the vapor pressure p^v.

Recalling that diffusion between the surface and the interior of the cluster is much faster than diffusion between the surface and the mother vapor phase, we assume that there is always equilibrium between the cluster (dividing) surface and the interior, resulting in

$$\mu_i^{exc} = \mu_i^l(p^l) \tag{14.8}$$

Using incompressibility of the liquid this implies

$$\mu_i^{\text{exc}} - \mu_i^{\text{v}}(p^{\text{v}}) = -\Delta\mu_i + v_i^{\text{l}}(p^{\text{l}} - p^{\text{v}}) \tag{14.9}$$

Substituting (14.9) into (14.6) and using Laplace equation, we obtain a generalization of Eq. (11.55):

$$\Delta G = \gamma(\mathbf{x}^{\text{l}}) A - \sum_{i=1}^{N} \underbrace{\left[n_i^{\text{l}} + n_i^{\text{exc}} \right]}_{n_i} \Delta\mu_i + \frac{2\gamma(\mathbf{x}^{\text{l}})}{r} \left[\sum_{i=1}^{N} n_i^{\text{exc}} v_i^{\text{l}} \right] \tag{14.10}$$

where r is the radius of the cluster (assumed to be spherical), $\gamma(\mathbf{x}^{\text{l}})$ is the surface tension of the N-component liquid solution with composition

$$\mathbf{x}^{\text{l}} = (x_1^{\text{l}}, \ldots, x_N^{\text{l}}),$$

$$x_i^{\text{l}} = \frac{n_i^{\text{l}}}{\sum_{k=1}^{N} n_k^{\text{l}}}, \quad \sum_{i=1}^{N} x_i^{\text{l}} = 1$$

Note, that the second term in (14.10) contains the *total* numbers of molecules, while the thermodynamic properties depend on the *bulk* cluster composition $\mathbf{x}^{\text{l}}$.

If we choose the dividing surface according to the K-surface recipe (cf. Sect. 11.6):

$$\sum_{i=1}^{N} n_i^{\text{exc}} v_i^{\text{l}} = 0 \tag{14.11}$$

then the last term in ΔG disappears, leading to

$$\Delta G = -\sum_{i=1}^{N} n_i \Delta\mu_i + \gamma(\mathbf{x}^{\text{l}}) A \tag{14.12}$$

The volume and surface area of the cluster can now be written in terms of total numbers of molecules

$$V^{\text{l}} = \sum_{i=1}^{N} n_i v_i^{\text{l}} \tag{14.13}$$

$$A = (36\pi)^{1/3} \left(\sum_{i=1}^{N} n_i v_i^{\text{l}} \right)^{2/3} \tag{14.14}$$

The chemical potential of a molecule inside the cluster can be written as

$$\mu_i^l(\mathbf{x}^l) = \mu_i^v(\{y_j^{\text{coex}}(\mathbf{x}^l)\})$$

where $y_j^{\text{coex}}(\mathbf{x}^l)$ is the fraction of component j in the N-component vapor which coexists with the N-component liquid having the composition $\mathbf{x}^l$ of the cluster; the corresponding coexistence pressure is $p^{\text{coex}}(\mathbf{x}^l)$. The superscript "coex" emphasizes that the corresponding quantity refers to $(\mathbf{x}, T)$-equilibrium, rather than the (p^v, T)-equilibrium. The quantities $p^{\text{coex}}(\mathbf{x}^l)$ and $\mathbf{y}^{\text{coex}} = (y_1^{\text{coex}}(\mathbf{x}^l), y_2^{\text{coex}}(\mathbf{x}^l), \ldots)$ are found from the $(\mathbf{x}, T)$-equilibrium equations:

$$
\begin{aligned}
p^l(T, \rho^l, \mathbf{x}^l) &= p^{\text{coex}} \\
p^v(T, \rho^v, \mathbf{y}^{\text{coex}}) &= p^{\text{coex}} \\
\mu_i^l(T, \rho^l, \mathbf{x}^l) &= \mu_i^v(T, \rho^v, \mathbf{y}^{\text{coex}}), \quad i = 1, \ldots, N
\end{aligned}
\tag{14.15}
$$

This is the system of $N + 2$ equations for the $N + 2$ unknowns:

$$\underbrace{\rho^l, \rho^v; \; p^{\text{coex}};}_{3} \; \underbrace{y_1^{\text{coex}}, \ldots, y_{N-1}^{\text{coex}}}_{N-1}$$

Using the ideal gas approximation we write

$$\Delta G(n_1, \ldots, n_N) = -k_B T \sum_{i=1}^{N} n_i \ln\left(\frac{y_i \, p^v}{y_i^{\text{coex}}(\mathbf{x}^l) \, p^{\text{coex}}(\mathbf{x}^l)}\right) + \gamma(\mathbf{x}^l) \, A \tag{14.16}$$

If we replace the bulk fractions $\mathbf{x}^l$ by the total fractions

$$\mathbf{x}^{\text{tot}} = (x_1^{\text{tot}}, \ldots, x_N^{\text{tot}})$$

where

$$x_i^{\text{tot}} = \frac{n_i}{\sum_{k=1}^{N} n_k}$$

then Eq. (14.16) becomes a straightforward generalization of Reiss result (11.100) to the multi-component case. The *equilibrium* distribution of multi-component clusters has the general form:

$$\rho_{\text{eq}}(n_1, \ldots, n_N) = C \, e^{-\beta \Delta G(n_1, \ldots, n_N)} \tag{14.17}$$

The prefactor C in Reiss's approximation reads $C = \sum_{i=1}^{N} \rho_i$, where ρ_i is the number density of monomers of component i in the mother phase.

Let us the rotate the coordinate system $(n_1, \ldots, n_N)$ of the N-dimensional space of cluster sizes. The general form of rotation transformation is

$$w_i = \sum_{j=1}^{N} U_{ji}\, n_j \tag{14.18}$$

where the new set of coordinates is denoted as w_i, and $\mathbf{U}$ is a unitary matrix with real coefficients [5]. The latter by definition satisfies

$$\mathbf{U}^{\mathrm{T}} = \mathbf{U}^{-1} \tag{14.19}$$

where $\mathbf{U}^{\mathrm{T}}$ is the transposed matrix and $\mathbf{U}^{-1}$ is the inverse matrix. From (14.19) it follows that multiplication by $\mathbf{U}$ has no effect on inner products of the vectors, angles or lengths. In particular, lengths of the vectors $||\mathbf{n}||$ are preserved

$$(\mathbf{U}\mathbf{n})^{\mathrm{T}}\,(\mathbf{U}\mathbf{n}) = \mathbf{n}^{\mathrm{T}}\,(\mathbf{U}^{\mathrm{T}}\,\mathbf{U})\,\mathbf{n} = \mathbf{n}^{\mathrm{T}}\,\mathbf{n}$$

which stands for

$$||\mathbf{U}\mathbf{n}||^2 = ||\mathbf{n}||^2$$

The latter property actually provides rotation. Since $\mathbf{U}^{\mathrm{T}}\,\mathbf{U} = \mathscr{I}$, where $\mathscr{I}$ is the unit matrix, we have

$$\det (\mathbf{U}^{\mathrm{T}}\,\mathbf{U}) = \det \mathbf{U}^{\mathrm{T}} \det \mathbf{U} = 1$$

resulting in

$$\det \mathbf{U} = 1$$

where we used the identity $\det \mathbf{U}^{\mathrm{T}} = \det \mathbf{U}$.

For example, for the binary case, rotation at the angle ϕ results in the matrix (cf. Eq. (11.17)):

$$\mathbf{U} = \begin{pmatrix} \cos\phi & -\sin\phi \\ \sin\phi & \cos\phi \end{pmatrix} \tag{14.20}$$

It is straightforward to see that its determinant is equal to unity.

Using the direction of principal growth approximation, we define w_1 as the "reaction path" (direction of principal growth). This means that the fluxes along other directions $w_2, \ldots, w_N$ are set to zero. The rotation angle, or equivalently, the form of the rotation matrix $\mathbf{U}$ should be determined separately. The critical cluster (denoted by the subscript "c") is defined as the one corresponding to the saddle point of the free energy surface. The latter satisfies the standard relationships

$$\left(\frac{\partial \Delta G}{\partial w_i}\right)_c = 0, \quad \text{or} \quad \left(\frac{\partial \Delta G}{\partial n_i}\right)_c = 0, \quad i = 1, \ldots, N \tag{14.21}$$

Now we are in the position to impose a certain form of the rotation matrix. Since the critical cluster is associated with the saddle point of the free energy, we define $\mathbf{U}$ so that second derivatives of ΔG at the saddle point satisfy

$$\left(\frac{\partial^2 \Delta G}{\partial w_i \partial w_j}\right)_c = Q_i \delta_{ij}, \quad i, j = 1, \ldots, N \tag{14.22}$$

where δ_{ij} is the Kronecker delta; $Q_1 < 0$ (maximum of ΔG along w_1), while the rest eigenvalues are all positive: $Q_2, \ldots, Q_N > 0$ (minimum of ΔG along $w_2, \ldots, w_N$); the values of Q_i are yet to be defined. Then, the matrix $\mathbf{U}$ should satisfy

$$0 = \sum_{u,v} U_{ui} U_{vj} \left(\frac{\partial^2 \Delta G}{\partial n_u \partial n_v}\right)_c = \left(\frac{\partial^2 \Delta G}{\partial w_i \partial w_j}\right)_c, \quad i \neq j \tag{14.23}$$

The unitary conditions for $\mathbf{U}$ read:

$$\sum_j U_{ij} U_{vj} = \delta_{iv} = \sum_j U_{ji} U_{jv} \tag{14.24}$$

implying $U_{ji}^{-1} = U_{ij}$. Equation (14.23) determines the rotation angle of the original coordinate system.

Combining Eqs. (14.22) and (14.23) we find

$$\sum_{u,v} U_{iu}^{-1} \left(\frac{\partial^2 \Delta G}{\partial n_u \partial n_v}\right)_c U_{vj} = Q_i \delta_{ij} \tag{14.25}$$

We multiply both sides of (14.25) by U_{ki} and sum over k. Since $\mathbf{U}$ is the unitary matrix, we obtain using Eq. (14.24):

$$\sum_v \left[\left(\frac{\partial^2 \Delta G}{\partial n_k \partial n_v}\right)_c - Q_j \delta_{kv}\right] U_{vj} = 0 \tag{14.26}$$

This expression shows that Q_j are the eigenvalues and U_{vj} are the eigenvectors of the Hessian matrix

$$\mathbf{G_2} = \left(\frac{\partial^2 \Delta G}{\partial n_k \partial n_v}\right)_c$$

The eigenvalues are the roots of the secular equation

$$\det (\mathbf{G_2} - Q \,\mathscr{I}) = 0$$

We choose the rotational system such that the critical cluster corresponds to

$$w_1 = w_{1c}, \quad w_{2c} = \ldots = w_{Nc} = 0$$

Expansion of the Gibbs energy in the vicinity of the saddle point of ΔG reads:

$$\Delta G = \Delta G_c + \frac{1}{2} Q_1 (w_1 - w_{1c})^2 + \frac{1}{2} \sum_{i=2}^{N} Q_i w_i^2 + \ldots \qquad (14.27)$$

where ΔG_c is the value of ΔG at the saddle point. Substituting (14.27) into (14.17), we obtain the equilibrium cluster distribution in the vicinity of the saddle point:

$$\rho_{eq}(w_1, \ldots, w_N) = \rho_{eq,c} \, \exp\left\{ -\frac{1}{2} \beta \left[Q_1 (w_1 - w_{1c})^2 + \sum_{i=2}^{N} Q_i w_i^2 + \ldots \right] \right\} \qquad (14.28)$$

where

$$\rho_{eq,c} = C \, e^{-\beta \Delta G_c} \qquad (14.29)$$

is the equilibrium number density of the critical clusters.

14.2 Kinetics

We assume that formation of the multi-component cluster results from attachment or loss of a single monomer E_j of one of the species, i.e. we consider the reactions of the following type

$$E(n_1, \ldots, n_j, \ldots, n_N) + E_j \overset{k_j}{\underset{k'_j}{\rightleftharpoons}} E(n_1, \ldots, n_j + 1, \ldots, n_N) \qquad (14.30)$$

Here $k_j(n_1, \ldots, n_N)$ and $k'_j(n_1, \ldots, n_N)$ are the reaction rates. In the chemical kinetics the forward reaction rate is proportional to the impingement rate v_j of the component j and the surface area of the cluster:

$$\rho_j k_j = v_j A(n_1, \ldots, n_j, \ldots, n_N) \qquad (14.31)$$

The form of v_j depends on the physical nature of the nucleation process; for the gas-liquid transition it is given by the gas kinetics expression

$$v_j = \frac{p_j}{\sqrt{2\pi m_j k_B T}} \qquad (14.32)$$

where p_j is the partial pressure of component j in the mother phase, and m_j is the mass of the molecule of component j. Equation (14.30) describes a single act of the cluster evolution. The *net rate* at which the clusters $(n_1, \ldots, n_j + 1, \ldots, n_N)$ are created in the unit volume of the system is

$$
\begin{aligned}
I_j(n_1, \ldots, n_j, \ldots, n_N) &= \rho_j\, k_j\, \rho(n_1, \ldots, n_j, \ldots, n_N) \\
&\quad - k'_j\, \rho(n_1, \ldots, n_j + 1, \ldots, n_N)
\end{aligned}
\tag{14.33}
$$

In equilibrium all net rates vanish (here, as in the BCNT, we consider the constrained equilibrium) which yields the determination of the backward reaction rate

$$
k'_j(n_1, \ldots, n_N) = \frac{\rho_j\, k_j\, \rho_{\mathrm{eq}}(n_1, \ldots, n_j, \ldots, n_N)}{\rho_{\mathrm{eq}}(n_1, \ldots, n_j + 1, \ldots, n_N)}
\tag{14.34}
$$

The evolution of the cluster distribution function is given by the kinetic equation which includes all possible reactions with single molecules of species $1, \ldots, N$

$$
\begin{aligned}
\frac{\partial \rho(n_1, \ldots, n_j, \ldots, n_N)}{\partial t} &= \sum_{j=1}^{N} \big[I_j(n_1, \ldots, n_j - 1, \ldots, n_N) \\
&\quad - I_j(n_1, \ldots, n_j, \ldots, n_N) \big]
\end{aligned}
\tag{14.35}
$$

Let us introduce the ratios of the actual (nonequilibrium) to equilibrium distribution functions

$$
f(n_1, \ldots, n_j, \ldots, n_N) = \frac{\rho(n_1, \ldots, n_j, \ldots, n_N)}{\rho_{\mathrm{eq}}(n_1, \ldots, n_j, \ldots, n_N)}
\tag{14.36}
$$

Then using (14.34) I_j can be written as

$$
\begin{aligned}
I_j &= \rho_j\, k_j\, \rho_{\mathrm{eq}}(n_1, \ldots, n_j, \ldots, n_N) \\
&\quad \times [f(n_1, \ldots, n_j, \ldots, n_N) - f(n_1, \ldots, n_j + 1, \ldots, n_N)]
\end{aligned}
$$

As usual in the phenomenological theories, we consider sufficiently large clusters, so that derivatives can be replaced by finite differences resulting in

$$
I_j = - \rho_j\, k_j\, \rho_{\mathrm{eq}}(n_1, \ldots, n_j, \ldots, n_N)\, \frac{\partial f(n_1, \ldots, n_j, \ldots, n_N)}{\partial n_j}
\tag{14.37}
$$

Then, the kinetic equation (14.35) becomes

$$
\frac{\partial \rho(n_1, \ldots, n_j, \ldots, n_N)}{\partial t} = - \sum_{j=1}^{N} \left(\frac{\partial I_j}{\partial n_j} \right) = -\mathrm{div}\, \mathbf{I}
\tag{14.38}
$$

where the vector $\mathbf{I}$ is defined as $\mathbf{I} = (I_1, \ldots, I_N)$. The same equation in the rotated system reads

$$
\frac{\partial \rho}{\partial t} = - \sum_{i=1}^{N} \sum_{j=1}^{N} U_{ij}^{-1} \left(\frac{\partial I_j}{\partial w_i} \right) = - \sum_{i=1}^{N} \frac{\partial}{\partial w_i} \left[\sum_{j=1}^{N} U_{ij}^{-1} I_j \right] \equiv -\mathrm{div}\, \mathbf{J}
\tag{14.39}
$$

where $\mathbf{J} = (J_1, \ldots, J_N)$,

$$J_i = \sum_{j=1}^{N} U_{ij}^{-1} I_j, \quad i = 1, \ldots, N \tag{14.40}$$

is the component of the nucleation flux along the w_i-axis. Using (14.31) and (14.37) J_i can be presented as

$$J_i = -A \rho_{eq} \sum_{u=1}^{N} B_{iu} \left(\frac{\partial f}{\partial w_u} \right), \quad i = 1, \ldots, N \tag{14.41}$$

in which

$$B_{iu} = \sum_{j=1}^{N} U_{ij}^{-1} v_j U_{ju} \tag{14.42}$$

We search for the steady state solution of Eq. (14.39) ignoring the short time-lag stage. Then, Eq. (14.39) becomes

$$\text{div } \mathbf{J} = 0$$

Recall that we have chosen the rotated system in such a way that the flux $\mathbf{J}$ is directed along the w_1 axis, implying

$$J_2 = \ldots = J_N = 0 \tag{14.43}$$

Then Eq. (14.41) for $i = 2, \ldots, N$ become

$$B_{i1} \left(\frac{\partial f}{\partial w_1} \right) + \sum_{u=2}^{N} B_{iu} \left(\frac{\partial f}{\partial w_u} \right) = 0, \quad i = 2, \ldots, N$$

By separating the term with $\frac{\partial f}{\partial w_1}$ from the rest of the sum, we get a linear set of $N-1$ equations for $N-1$ unknown variables $\frac{\partial f}{\partial w_u}$, $i = 2, \ldots, N$. Its determinant is

$$D_2 = \begin{vmatrix} B_{22} & \cdots & B_{2N} \\ \vdots & & \vdots \\ B_{N2} & \cdots & B_{NN} \end{vmatrix}$$

and the solution reads:

$$\frac{\partial f}{\partial w_u} = \frac{(-1)^{u+1}}{D_2} \left(\frac{\partial f}{\partial w_1} \right) L_u \quad u = 2, \ldots, N \tag{14.44}$$

where

$$L_u = \begin{vmatrix} B_{21} & B_{22} & \ldots & B_{2,u-1} & B_{2,u+1} & \ldots & B_{2N} \\ \cdot & \cdot & \cdot & \cdot & \cdot & \cdot & \cdot \\ \cdot & \cdot & \cdot & \cdot & \cdot & \cdot & \cdot \\ \cdot & \cdot & \cdot & \cdot & \cdot & \cdot & \cdot \\ B_{N1} & B_{N2} & \ldots & B_{N,u-1} & B_{N,u+1} & \ldots & B_{NN} \end{vmatrix}$$

Having expressed $N-1$ quantities $\frac{\partial f}{\partial w_u}$ in terms of $\frac{\partial f}{\partial w_1}$, we substitute the solution (14.44) into the only remained equation from the set (14.41), namely the equation for J_1:

$$\begin{aligned} J_1 &= -A\,\rho_{eq}\left[B_{11}\left(\frac{\partial f}{\partial w_1}\right) + \sum_{u=2}^{N} B_{1u}\left(\frac{\partial f}{\partial w_u}\right) \right] \\ &= -A\,\rho_{eq}\left(\frac{\partial f}{\partial w_1}\right)\frac{1}{D_2}\left[B_{11}D_2 + \sum_{u=2}^{N}(-1)^{u+1}\,B_{1u}\,L_u \right] \end{aligned}$$

As can be easily seen, the expression in the square brackets is the determinant of the $N \times N$ matrix of all B_{ij}'s:

$$D_1 = \begin{vmatrix} B_{11} & \ldots & B_{1N} \\ \cdot & & \cdot \\ \cdot & & \cdot \\ \cdot & & \cdot \\ B_{N1} & \ldots & B_{NN} \end{vmatrix} \tag{14.45}$$

Thus, J_1 can be written in the compact form

$$J_1(w_1,\ldots,w_N) = -A\,\rho_{eq}\left(\frac{\partial f}{\partial w_1}\right)\frac{D_1}{D_2} \tag{14.46}$$

Integrating Eq. (14.46) along the reaction path w_1, we obtain

$$f(w_1) = -\int_1^{w_1}\frac{J_1\,D_2}{D_1\,\rho_{eq}\,A}\,dw_1' + f_1$$

where f_1 is an unknown integration constant. In view of the exponential form of $\rho_{eq}(w_1,\ldots,w_N)$ we may assign the slowly varying functions A, D_1, D_2, J_1 their values calculated at $w_1 = w_{1c}$ and take them outside of the integral:

$$f(w_1) = -\left(\frac{J_1\,D_2}{D_1\,A}\right)_{w_1=w_{1c}}\int_1^{w_1}\frac{1}{\rho_{eq}}\,dw_1' + f_1 \tag{14.47}$$

Now we apply the (standard) boundary conditions for the cluster distribution function (cf. (3.42)–(3.43)): the concentration of *small* clusters is nearly equal to equilibrium one; for *large* clusters ρ_{eq} diverges, while the actual distribution function ρ remains finite. These requirements yield:

$$f(w_1) \to 1 \quad \text{for } w_1 \to 1, \quad \text{and} \quad f(w_1) \to 0 \quad \text{for } w_1 \to \infty \qquad (14.48)$$

The first condition results in $f_1 = 1$, then from the second one we obtain

$$J_1(w_2, \ldots, w_N) = \left(\frac{D_1 A}{D_2}\right)_{w_1 = w_{1c}} \left[\int_1^\infty \left(\frac{1}{\rho_{eq}(w_1', w_2, \ldots, w_N)}\right) dw_1'\right]^{-1} \qquad (14.49)$$

We can simplify this result by using the expansion (14.28) and extending the integration limits in (14.49) to $\pm\infty$:

$$J_1(w_2, \ldots, w_N) = \rho_{eq,c} \left(\frac{D_1 A}{D_2}\right)_{w_1 = w_{1c}} \exp\left\{-\frac{1}{2}\beta\left[\sum_{i=2}^N Q_i w_i^2\right]\right\}$$
$$\times \left[\int_{-\infty}^\infty dw_1' \, e^{\frac{Q_1(w_1' - w_{1c})^2}{2k_B T}}\right]^{-1}$$

Taking into account that $Q_1 < 0$ and performing Gaussian integration we obtain

$$J_1(w_2, \ldots, w_N) = \rho_{eq,c} \left(\frac{D_1 A}{D_2}\right)_{w_1 = w_{1c}} \exp\left\{-\frac{1}{2}\beta\left[\sum_{i=2}^N Q_i w_i^2\right]\right\} \sqrt{\frac{(-Q_1)}{2\pi k_B T}} \qquad (14.50)$$

The total nucleation rate is found by integrating J_1 over all possible values of $w_2, \ldots, w_N$. In view of the previously presented considerations we set the limits of integration to $\pm\infty$:

$$J = \int_{-\infty}^\infty \ldots \int_{-\infty}^\infty dw_2' \ldots dw_N' \, J_1(w_2', \ldots, w_N') \qquad (14.51)$$

Assuming further that $D_1 A / D_2$ varies slowly with w_i's compared to the exponential terms $\exp[-Q_i w_i^2 / k_B T]$, we perform $N - 1$ Gaussian integrations

$$\int_{-\infty}^\infty dw_i' \, e^{-\frac{Q_i w_i'^2}{2k_B T}} = \left(\frac{2\pi k_B T}{Q_i}\right)^{1/2}$$

resulting in

$$J = C e^{-\beta \Delta G_c} \left(\frac{D_1 A}{D_2}\right)_c (2\pi k_B T)^{(N-2)/2} \sqrt{\frac{(-Q_1)}{Q_2 \ldots Q_N}} \qquad (14.52)$$

This result represents the extension of Reiss's BCNT to multicomponent systems.

14.3 Example: Binary Nucleation

Let us consider application of the general formalism to the binary nucleation. Rotation matrix $\mathbf{U}$ in this case is given by Eq. (14.20). The rotation angle ϕ is found from Eq. (14.23) which reads

$$\tan(2\phi) = \frac{2\left(\frac{\partial^2 \Delta G}{\partial n_1 \partial n_2}\right)_c}{\left[\left(\frac{\partial^2 \Delta G}{\partial n_1^2}\right)_c - \left(\frac{\partial^2 \Delta G}{\partial n_2^2}\right)_c\right]} \qquad (14.53)$$

The eigenvalues Q_j of the matrix $\mathbf{G_2}$ are

$$Q_1 = \left(\frac{\partial^2 \Delta G}{\partial n_1^2}\right)_c \cos^2 \phi + \left(\frac{\partial^2 \Delta G}{\partial n_1 \partial n_2}\right)_c \sin(2\phi) + \left(\frac{\partial^2 \Delta G}{\partial n_2^2}\right)_c \sin^2 \phi$$

$$Q_2 = \left(\frac{\partial^2 \Delta G}{\partial n_1^2}\right)_c \sin^2 \phi - \left(\frac{\partial^2 \Delta G}{\partial n_1 \partial n_2}\right)_c \sin(2\phi) + \left(\frac{\partial^2 \Delta G}{\partial n_2^2}\right)_c \cos^2 \phi$$

Equation (14.53) has two solutions for the angle ϕ. We choose the solution which gives $Q_1 < 0$ and $Q_2 > 0$ as required by construction of the model. Having determined the rotation angle, we write down the coefficients B_{ij} from Eq. (14.42):

$$B_{11} = v_1 \cos^2 \phi + v_2 \sin^2 \phi$$
$$B_{22} = v_1 \sin^2 \phi + v_2 \cos^2 \phi$$
$$B_{12} = B_{21} = (-v_1 + v_2) \sin \phi \cos \phi$$

$$(14.54)$$

The determinants D_1 and D_2 become

$$D_1 = B_{11} B_{22} - B_{12}^2 = v_1 v_2$$
$$D_2 = B_{22}$$

Then, the nucleation rate (14.52) is

$$J = C\, e^{-\beta \Delta G_c}\, v_{av}\, A_c \sqrt{\frac{(-Q_1)}{Q_2}} \qquad (14.55)$$

where

$$v_{av} = \left(\frac{v_1 v_2}{v_1 \sin^2 \phi + v_2 \cos^2 \phi}\right)_c \qquad (14.56)$$

is the average impingement rate. This result agrees with Eqs. (11.35), (11.42)–(11.44) derived in Chap. 11.

14.4 Concluding Remarks

Using similar assumptions as in the BCNT, the present model focuses on nucleation in the vicinity of the saddle point of the free energy surface and completely ignores nucleation along all paths other than the principal nucleation path (direction of principal growth). This *local* approach identifies the *predominant composition* of the observable nuclei—the one that corresponds to the saddle point. Although this approach obviously has its merits, it, however, is unable to predict the rate of formation of nuclei with arbitrary composition. The latter issue requires a *global* approach which treats all cluster compositions on an equal footing. This problem was addressed among others by Wu [6] who studied various nucleation paths (not only the saddle-point nucleation). In [6] conditions were identified under which a multi-component system behaves "as if it were simple" which means that the system can be modelled by a one-dimensional Fokker-Planck equation (3.79)–(3.80).

Direction of principal growth approximation may not always be the best choice. Depending on the form of the Gibbs free energy surface and values of the impingement rates of the components it is possible that the main nucleation flux bypasses the saddle-point. In particular, as pointed out by Trinkaus [4], if one reaction rate is essentially smaller than the other ones, the flux line can turn into the directions of the fast-reacting component and pass a ridge before the saddle-point coordinate of the slowly reacting component is reached. For binary systems these issues were studied numerically by Wyslouzil and Wilemski [7].

References

1. M.J. Muitjens, V.I. Kalikmanov, M.E.H. van Dongen, A. Hirschberg, P. Derks, Revue de l'Institut Français du Pétrole **49**, 63 (1994)
2. V. Kalikmanov, J. Bruining, M. Betting, D. Smeulders, in *2007 SPE Annual Technical Conference and Exhibition* (Anaheim, California, USA, 2007), pp. 11–14. Paper No: SPE 110736
3. O. Hirschfelder, J. Chem. Phys. **61**, 2690 (1974)
4. H. Trinkaus, Phys. Rev. B **27**, 7372 (1983)
5. K.F. Riley, M.P. Hobson, S.J. Bence, *Mathematical Methods for Physics and Engineering* (Cambridge University Press, Cambridge, 2007)
6. D. Wu, J. Chem. Phys. **99**, 1990 (1993)
7. B. Wyslouzil, G. Wilemski, J. Chem. Phys. **103**, 1137 (1995)

Chapter 15
Heterogeneous Nucleation

15.1 Introduction

Heterogeneous nucleation is a first order phase transition in which molecules of the parent phase nucleate onto surfaces forming embryos of the new phase. These preexisting foreign particles are usually called *condensation nuclei* (CN). To discriminate between the cluster and condensation nuclei, we call CN "solid", the cluster phase "liquid" and the parent phase "vapor". These notations are purely terminological: clusters as well as CN can be liquid and solid. CN can be a planar macroscopic surface, a spherical particle (liquid or solid); it can be also an ion—the latter case is termed *ion-induced nucleation* [1–4]. In this chapter we do not discuss ion-induced nucleation and consider CN to be electrically neutral and insoluble to the cluster formed on its surface, i.e. there is no mass transfer between the CN and the liquid or vapor phases. A vivid example of heterogeneous nucleation, one experiences in the everyday life, is vapor-liquid nucleation on the surface of aerosol particles in atmosphere. The process of condensational growth of aerosol particles got considerable experimental and theoretical attention due to its role in the environmental effects [5].

Classical theory of single-component heterogeneous nucleation was developed by Fletcher in 1958 [6]. Later on Lazaridis et al. [7] extended Fletcher's theory to binary systems. The general approach to kinetics of nucleation processes discussed in Sect. 3.3 remains valid for the heterogeneous case. The steady-state nucleation rate has the general form of Eq. (3.52)

$$J = K \, \exp\left(-\frac{\Delta G^*}{k_B T}\right) \qquad (15.1)$$

where K is a kinetic prefactor and ΔG^* is the free energy of formation of the critical embryo on the foreign particle (CN). As in the case of homogeneous nucleation, the heterogeneous nucleation rate is determined largely by the energy barrier ΔG^*. That is why it is sufficient to know the prefactor K (its various forms are discussed in

V. I. Kalikmanov, *Nucleation Theory*, Lecture Notes in Physics 860,
DOI: 10.1007/978-90-481-3643-8_15, © Springer Science+Business Media Dordrecht 2013

Sect. 15.5) to one or two orders of magnitude. We will show in the next section that the presence of a foreign particle reduces the energy cost to build the cluster surface.

15.2 Energetics of Embryo Formation

We consider a cluster (embryo) of phase 2 which has the form of a spherical liquid cap of volume V_2 and radius r and contains n molecules,

$$n = \rho^l V_2 \tag{15.2}$$

where ρ^l is the number density of the phase 2. The embryo is resting on the spherical foreign particle (CN) 3 of a radius R_p surrounded by the parent phase 1. The line where all three phases meet, called a three-phase contact line, is characterized by the *contact angle* θ. This configuration is schematically shown in Fig. 15.1. Within the phenomenological approach embryos are considered to be the objects characterized by macroscopic properties of phase 2. Applying Gibbs thermodynamics to this system, it is necessary to define two dividing surfaces: one for the gas-liquid (1–2) interface and one for the solid-liquid (2–3) interface. Since we discuss the single-component nucleation, we can choose for both of them the corresponding equimolar surface so that the adsorption terms in the Gibbs energy vanish. Within the capillarity approximation the Gibbs free energy of an embryo formation is

$$\Delta G(n) = -n\,\Delta\mu + \Delta G^{\mathrm{surf}} \tag{15.3}$$

Here $\Delta\mu = \mu_1 - \mu_2 > 0$ is the difference in chemical potential of the phases 1 (supersaturated vapor) and phase 2 taken at the same pressure p^v of the vapor phase. The surface contribution contains two terms:

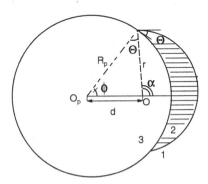

Fig. 15.1 Embryo 2 (*dashed area*) on the foreign particle (*condensation nucleus*) 3 in the parent phase 1. R_p is the radius of the spherical CN, O_p is its center; r is the radius of the sphere with the center in O corresponding to the embryo, $d = O_p\,O$, θ is the contact (*wetting*) angle

$$\Delta G^{\text{surf}} = \gamma_{12} A_{12} + (\gamma_{23} - \gamma_{13}) A_{23} \qquad (15.4)$$

Here γ_{ij} is the interfacial tension between phases i and j, A_{ij} is the corresponding surface area. The first term in (15.4) is the energy cost of the vapor-liquid interface, the second one stems from the fact that when a cluster is built on the surface of CN, the initially existed solid-vapor interface, characterized by the interfacial tension γ_{13}, is replaced by the solid-liquid interface with the interfacial tension γ_{23}. From geometrical considerations:

$$A_{12} = 2\pi r^2 (1 - \cos\alpha) \qquad (15.5)$$

$$A_{23} = 2\pi R_p^2 (1 - \cos\phi) \qquad (15.6)$$

$$V_2 = \frac{4\pi}{3} r^3 f_V \qquad (15.7)$$

where the angles α and ϕ are defined in Fig. 15.1:

$$\cos\phi = \frac{R_p - rm}{d}, \quad \cos\alpha = -\frac{r - R_p m}{d} \qquad (15.8)$$

Here

$$m = \cos\theta \qquad (15.9)$$

$$d = \sqrt{R_p^2 + r^2 - 2R_p rm} \qquad (15.10)$$

Geometrical factor f_V relates the embryo volume (shaded area in Fig. 15.1) to its homogeneous counterpart, being the full sphere of radius r. Standard algebra yields:

$$f_V = q(\cos\alpha) - a^3 q(\cos\phi), \quad a \equiv \frac{R_p}{r} \qquad (15.11)$$

where the function $q(y)$ is:

$$q(y) = \frac{1}{4}(2 - 3y + y^3) \qquad (15.12)$$

The angles ϕ and α depend not on r and R_p individually, but on their ratio a; therefore it is convenient to rewrite (15.8), (15.10) in dimensionless units:

$$\cos\phi = \frac{a - m}{w}, \quad \cos\alpha = -\left(\frac{1 - ma}{w}\right) \qquad (15.13)$$

$$w = \sqrt{1 + a^2 - 2ma} \qquad (15.14)$$

Looking at (15.3)–(15.14), one can see that our model for ΔG is incomplete: the surface areas A_{12}, A_{23} and the volume of the embryo depend on the yet undefined contact angle θ. Its value can be derived from the *interfacial force balance* at the three phase boundary known as the *Dupre-Young equation*:

$$m \equiv \cos\theta_{eq} = (\gamma_{13} - \gamma_{23})/\gamma_{12} \qquad (15.15)$$

(to emphasize the bulk equilibrium nature of the Dupre-Young equation we added the subscript "eq" to the contact angle). An alternative way to derive this result is to consider a variational problem: formation of an embryo with a *given volume* V_2 with an *arbitrary* contact angle. The equilibrium angle θ_{eq} will be the one that minimizes ΔG at fixed V_2:

$$\left.\frac{\partial \Delta G}{\partial m}\right|_{V_2} = 0 \qquad (15.16)$$

Combining (15.15) with (15.3)–(15.4) we find

$$\Delta G = -n\,\Delta\mu + \gamma_{12}\,A_{12}\left(1 - m\,\frac{A_{23}}{A_{12}}\right) \qquad (15.17)$$

Let us compare the free energies necessary to form a cluster with the same number of molecules n at the temperature T and the supersaturation S (or equivalently, $\Delta\mu$) for the homogeneous and heterogeneous nucleation. Clearly, for both cases the bulk contribution to ΔG will be the same: $-n\,\Delta\mu$. Consider now the surface term. In the *homogeneous* case:

$$\Delta G^{\mathrm{surf}}(n)_{\mathrm{hom}} = \gamma_{12}\,s_1\,n^{2/3}$$

where

$$s_1 = (36\pi)^{1/3}\left(\rho^1\right)^{-2/3} \qquad (15.18)$$

In the *heterogeneous* case according to (15.17)

$$\Delta G^{\mathrm{surf}}(n)_{\mathrm{het}} = \gamma_{12}\,A_{12}\left(1 - m\,\frac{A_{23}}{A_{12}}\right)$$

Using (15.5)–(15.12) and (15.18) we find

$$\frac{\Delta G^{\mathrm{surf}}(n)_{\mathrm{het}}}{\Delta G^{\mathrm{surf}}(n)_{\mathrm{hom}}} = \frac{2^{1/3}\left[(1 - \cos\alpha) - m\,a^2\,(1 - \cos\phi)\right]}{f_V^{2/3}}$$

Straightforward inspection of this result shows that

$$\frac{\Delta G^{\mathrm{surf}}(n)_{\mathrm{het}}}{\Delta G^{\mathrm{surf}}(n)_{\mathrm{hom}}} \leq 1 \qquad (15.19)$$

implying that it is energetically more favorable to form an n-cluster on a foreign surface than to form it directly in the mother phase 1. The equality sign in (15.19) is realized for non-wetting conditions: $\theta_{eq} = \pi \Leftrightarrow m = -1$. Mention several limiting cases:

- $R_p \to 0$ corresponds to the absence of foreign particles: $f_V(m, 0) = 1$, thus leading to homogeneous nucleation;
- the same (homogeneous) limit is obtained when the phase 2 does not wet the phase 3, i.e. $\theta_{eq} = \pi$, yielding from (15.11), (15.13)–(15.14): $f_V(-1, a) = 1$. An embryo in this case represents a sphere having a point contact with a foreign particle.

15.3 Flat Geometry

The case of flat geometry deserves special attention. This is a limiting case when the radius of an embryo is much smaller than the radius R_p of the foreign particle, or equivalently $a \to \infty$. Then, the embryo sees the particle as a flat wall while the particle radius becomes irrelevant. An embryo becomes a spherical segment (cup) resting on a plane as shown in Fig. 15.2. Taking the limit $R_p \to \infty$ in Eqs. (15.5)–(15.8), we find

$$\alpha = \theta_{eq}$$

$$V_2 = \frac{4}{3} \pi r^3 q(m) \tag{15.20}$$

$$A_{12} = 2\pi r^2 (1 - m) \tag{15.21}$$

$$A_{23} = \pi r^2 (1 - m^2) \tag{15.22}$$

where the function $q(m)$ is given by (15.12)

$$q(m) = \frac{1}{4} (2 - 3m + m^3) \tag{15.23}$$

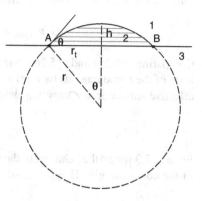

Fig. 15.2 Flat surface geometry: a sphere-cup-shaped embryo of radius r resting on a plane. The height of the cap is $h = r(1 - \cos\theta)$, where θ is the contact angle. Points A and B belong to the three phase contact line, which is a circle of radius $r_t = r \sin\theta$ located in the plane perpendicular to the plane of the figure

Fig. 15.3 Functions $q(\cos\theta)$
and $q^{1/3}(\cos\theta)$ given by
Eq. (15.23)

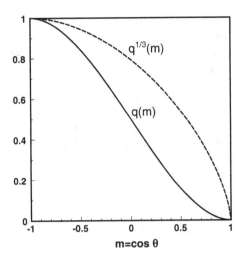

The height of the cup is

$$h = r(1 - \cos\theta_{eq})$$

and the $2 - 3$ interface becomes a circle bounded by the three-phase contact line of
the radius $r_t = r \sin\theta_{eq}$. From (15.2) and (15.20) we find the relation between the
number of molecules in the embryo and its radius:

$$r = \left(\frac{3}{4\pi}\right)^{1/3} \left(\rho^1\right)^{-1/3} q^{-1/3} n^{1/3} \qquad (15.24)$$

Using the general form (15.17) of the heterogenous free energy barrier together with
Eqs. (15.20)–(15.22) for the case of flat geometry we find:

$$\Delta G(n)_{het} = -n\Delta\mu + (\gamma_{12} q^{1/3}) s_1 n^{2/3} \qquad (15.25)$$

For *homogeneous* nucleation of a cluster with the same number of molecules at the
same temperature T and supersaturation S one would have:

$$\Delta G(n)_{hom} = -n\Delta\mu + \gamma_{12} s_1 n^{2/3} \qquad (15.26)$$

Comparing (15.25) and (15.26), one can see that the heterogeneous barrier has the
form of the homogeneous barrier in which the surface tension γ_{12} is replaced by the
effective surface free energy γ_{eff} defined by

$$\gamma_{eff} = \gamma_{12} q^{1/3}(\theta_{eq}) \qquad (15.27)$$

Figure 15.3 shows the behavior of the functions $q(m)$ and $q^{1/3}(m)$ for different values
of the contact angle. They increase monotonously from 0 at $\theta = 0$ to 1 at $\theta = \pi$.

Thus, $\gamma_{eff} \leq \gamma_{12}$ and therefore the heterogeneous barrier is always smaller or equal to ΔG_{hom} for the same values of T and S. Expressing ΔG in terms of the radius of the embryo, we have using (15.24):

$$\Delta G_{het}(r) = \Delta G_{hom}(r)\, q(m) \tag{15.28}$$

Note that Eq. (15.28) is solely based on macroscopic considerations. For nonwetting conditions ($m = -1$) we recover the homogeneous limit: $\Delta G_{het} = \Delta G_{hom}$.

15.4 Critical Embryo: The Fletcher Factor

Let us return to the case of arbitrary geometry. The *critical embryo* r_c satisfies

$$\left. \frac{\partial \Delta G}{\partial r} \right|_{r_c} = 0 \tag{15.29}$$

Finding the critical radius using the general form (15.17) of the Gibbs formation energy in the presence of a foreign particle requires a considerable amount of algebra. Fortunately, calculation can be substantially simplified if we take into account that irrespective of the presence or absence of the foreign particles, the critical cluster is in metastable equilibrium with the surrounding vapor (phase 1) and thus the chemical potentials of a molecule inside and outside the critical cluster are equal resulting in the Kelvin equation (3.61):

$$r_c = \frac{2\gamma_{12}}{\rho^l \Delta\mu} \tag{15.30}$$

This expression manifests an important feature of the heterogeneous problem: the radius of the critical embryo is determined solely by the temperature and the supersaturation of the phase 2 and does not contain any information about the foreign particles. At the same time, the *number of molecules* in the critical embryo depends on the wetting properties of the phases through Eqs. (15.2) and (15.7):

$$r_c = r_{c,hom}(T, S)$$
$$n_c = n_{c,hom}(T, S)\, f_V(m, a_c), \quad a_c = \frac{R_p}{r_c} \tag{15.31}$$

assuming the same bulk liquid density of phase 2 in the homogeneous and heterogeneous cases. Substituting (15.30) into (15.17) and using (15.5)–(15.14), we derive the nucleation barrier:

$$\Delta G^* = \Delta G_{hom}^*\, f_G(m, a_c) \tag{15.32}$$

Fig. 15.4 Fletcher factor $f_G(m, a)$ as a function of $a = R_p/r$. Labels are the corresponding values of $m = \cos\theta$

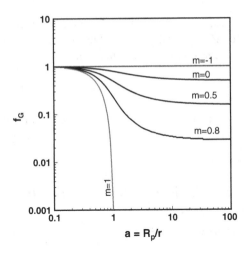

where

$$\Delta G^*_{\text{hom}} = \frac{1}{3}\gamma_{12}(4\pi\, r_c^2) = \frac{16\pi}{3}\frac{\gamma_{12}^3}{(\rho^1)^2\,(\Delta\mu)^2} \tag{15.33}$$

is the *homogeneous* nucleation barrier given by the CNT; the function

$$f_G(m, a) = \frac{1}{2} + \frac{1}{2}\left(\frac{1-ma}{w}\right)^3$$
$$+ \frac{1}{2}a^3\left[2 - 3\left(\frac{a-m}{w}\right) + \left(\frac{a-m}{w}\right)^3\right] + \frac{3}{2}m\,a^2\left(\frac{a-m}{w} - 1\right) \tag{15.34}$$

is called the *Fletcher factor* [6]. It varies between 0 and 1, depending on the contact angle and the relative size of foreign particles with respect to the embryo. Thus, heterogeneous nucleation *reduces* the nucleation barrier compared to the homogeneous case ΔG^*_{hom} due to the presence of foreign bodies; the Fletcher factor being the measure of this reduction. It is important to bear in mind that the Fletcher factor f_G refers exclusively to the *critical* embryo so that the expression (15.32) is not true for an *arbitrary* embryo. The behavior of $f_G(m, a)$ as a function of a for various values of the contact angle (parameter m) is shown in Fig. 15.4. The upper line $m = -1$ corresponds to the non-wetting conditions recovering the homogeneous limit: $f_G = 1$, $\Delta G^* = \Delta G^*_{\text{hom}}$ for all R_p. For the flat geometry the function $f_{G,\infty} = \lim_{a\to\infty} f_G(m, a)$ levels up yielding

$$f_{G,\infty} = q(m) \tag{15.35}$$

and $\Delta G^* = \Delta G^*_{\text{hom}}\, q(m)$ in accordance with Eq. (15.28). As it follows from Fig. 15.4, the flat geometry limit can be applied when $R_p/r > 10$.

15.5 Kinetic Prefactor

Recall that in *homogeneous* nucleation the kinetic prefactor takes the form (3.54)

$$J_0 = \mathscr{Z} \, (v \, A^*) \, \rho_1 \qquad (15.36)$$

where $v \, A^*$ is the rate of addition of molecules to the critical cluster of the radius r^*, $\mathscr{Z}$ is the Zeldovich factor (3.50):

$$\mathscr{Z} = \sqrt{\frac{\gamma_{12}}{k_B T}} \, \frac{1}{2\pi\rho^l (r^*)^2} \qquad (15.37)$$

and ρ_1 is the number of monomers (per unit volume) of the mother phase. The latter quantity coincides with the number of nucleation sites, since homogeneous nucleation can occur with equal probability at any part of the physical volume.

Kinetic prefactor in the *heterogeneous* case has the same form

$$K = \mathscr{Z} \, (v \, A^*) \, \rho_{1,s} \qquad (15.38)$$

with the number of nucleation sites $\rho_{1,s}$ being the number of molecules in contact with the substrate; clearly, $\rho_{1,s}$ is sufficiently reduced compared to ρ_1.

The value of the prefactor K depends on the particular mechanism of the cluster formation. Two main scenarios are discussed in the literature. The first one assumes that nucleation occurs by *direct deposition* of vapor monomers on the surface of the cluster [6]. Another possibility is *surface diffusion* [8–10]: vapor monomers collide with the CN surface and become adhered to it; the adsorbed molecules further migrate to the cluster by two-dimensional diffusion. Fundamental aspects of surface diffusion are discussed in the seminal monograph of Frenkel [11]. It was found theoretically and experimentally [12, 13] that the surface diffusion mechanism is more effective and leads to higher nucleation rates.

The dimensionality of K coincides with the dimensionality of the nucleation rate. The number of critical embryos in the process of heterogeneous nucleation depends on the amount of pre-existing foreign particles acting as CN. That is why the nucleation rate can be expressed as:

- number of embryos *per unit area* of the foreign particle per unit time; or
- number of embryos *per foreign particle* per unit time; or
- number of embryos formed per unit volume of the system per unit time

Let us discuss the "adsorption—surface diffusion" mechanism of an embryo formation. Adsorption can be visualized as a process in which molecules of phase 1 strike the surface of a foreign particle, remain on that surface for a certain adsorption time τ and then are re-evaporated. The motion of adsorbed molecules on the surface of a particle can not be a free one but reminds the 2D random walk which can be associated with the 2D diffusion process [8, 11].

Let N_{ads} be the surface concentration of adsorbed molecules, i.e. the number of molecules of phase 1 adsorbed on the unit surface of CN. It is difficult to determine a realistic value of this quantity for a general case. Fortunately, N_{ads} appears in the pre-exponential factor K and errors in evaluating it will not influence the nucleation rate as critically as errors in determination of ΔG^* (in particular, the uncertainty in the contact angle is much more significant than the uncertainty in N_{ads}). In view of these considerations we can use for N_{ads} an "educated estimate":

$$N_{ads} = v\tau \tag{15.39}$$

where v is the impingement rate of the monomers of the phase 1 per unit surface of CN; if phase 1 is the supersaturated vapor, v is given by the gas kinetics expression (3.38). The time of adsorption can be written in the Arrhenius form [14]

$$\tau = \tau_0 \exp(E_{ads}/k_B T) \tag{15.40}$$

where E_{ads} is the heat of adsorption (per molecule) and τ_0 is the characteristic time—the period of oscillations of a molecule on the surface of CN. The latter can be estimated as the inverse of the characteristic absorption frequency f_I of the substance. For most of the substances the absorption frequencies lie in the ultraviolet region [15]: $f_I = 3.3 \times 10^{15} \, \text{s}^{-1}$ implying that

$$\tau_0 \approx \frac{1}{f_I} \approx 3 \times 10^{-16} \, \text{s}$$

Obviously, E_{ads} depends on the nature of the adsorbing surface and the adsorbed molecules. One can find it from the data on diffusion coefficient D written in the form of an Arrhenius plot:

$$\ln D = \text{const} - \frac{E_{ads}}{k_B T}$$

where we associate E_{ads} with the activation energy for diffusion. Its value is given by the slope of $\ln D$ as a function of the inverse temperature. Typical *molar* values of E_{ads} in liquids are found to be $\approx 10 - 30 \, \text{kJ/mol}$.

If the heterogeneous nucleation rate is expressed as a number of embryos formed *per unit area* of a foreign particle per unit time [16], the prefactor takes the form:

$$K_S = \mathscr{Z} \, (v A_{12}^*) \, N_{ads} = \mathscr{Z} \, v^2 A_{12}^* \, \tau_0 \, e^{\beta E_{ads}} \, [\text{cm}^{-2} \, \text{s}^{-1}] \tag{15.41}$$

If nucleation rate is expressed as the number of embryos *per foreign particle* per unit time, Eq. (15.41) will be modified to

$$K_p = v^2 A_{12}^* \mathscr{L} \tau_0 e^{\beta E_{ads}} 4\pi R_p^2 [\text{s}^{-1}] \tag{15.42}$$

(assuming that a foreign particle is a sphere of the radius R_p). The nucleation rate

$$J_p = K_p \exp\left(-\frac{\Delta G^*}{k_B T}\right) [\text{s}^{-1}] \tag{15.43}$$

gives the rate at which a foreign particle is activated to growth.

Finally, if there are N_p foreign particles in the volume V of the system, one can define the nucleation rate as the *total amount of embryos* formed per unit volume of the system per unit time. In this case the prefactor reads:

$$K_V = v^2 A_{12}^* \mathscr{L} \tau_0 e^{\beta E_{ads}} 4\pi R_p^2 (N_p/V)[\text{cm}^{-3}\,\text{s}^{-1}] \tag{15.44}$$

15.6 Line Tension Effect

15.6.1 General Considerations

The presence of two or more bulk phases in contact with each other gives rise to the discontinuity of their thermodynamic properties and results in the corresponding interfacial tensions. Formation of an embryo on the surface of a foreign particle results in the occurrence of a line of three-phase contact. This line has an associated with it tension τ_t, which is the excess free energy of the system due to three-phase contact, per unit length of the contact line. A thermodynamic definition of the line tension τ_t can be given in a way analogous to the definition of the surface tension in Sect. 2.2 by introducing three dividing surfaces—gas-liquid, solid-liquid and solid-vapor—and using the methodology of Gibbs thermodynamics of nonhomogeneous systems. The excess Ω_t of the free energy (grand potential) associated with the three-phase contact line reads [17]:

$$\Omega_t = \tau_t L \tag{15.45}$$

where L is the length of the contact line. This relation defines τ_t and is the analogue of Eq. (2.25) for the surface tension. The line tension does not depend on the location of dividing surfaces like each of the 2D interfacial tensions γ_{12}, γ_{23}, γ_{13} [17]. However, *unlike* the 2D interfacial tensions, τ_t can be of either sign. A two-phase interfacial tension should be necessarily positive: if this would not be the case the increase of the interfacial area would become energetically favorable leading to the situation when two phases become mutually dispersed in each other on molecular scale, so that the interface between the phases disappears. Thus, the separation between any two phases requires a positive interfacial tension. The situation with the *three-phase line* is different. A negative line tension makes an increase of the length L of the triple line energetically favorable ($\Omega_t < 0$), however this increase inevitably changes the surface areas of the 2D interfaces—characterized by positive tensions—in a such a way that the total free energy of the system increases.[1]

The classical (Fletcher) theory, described in the previous sections, does not take into account the line tension effect. At the same time, several authors [19, 20] indicated that for highly curved surfaces (i.e. small critical embryos) it can have a substantial influence on the nucleation behavior. The presence of the line tension modifies the Gibbs free energy of an embryo formation (15.3)–(15.4):

$$\Delta G = -n\,\Delta\mu + \gamma_{12}A_{12} + (\gamma_{23} - \gamma_{13})A_{23} + \tau_t\,2\pi r_t \qquad (15.46)$$

where r_t is the radius of the contact line. Inclusion of the line tension changes the force balance at the contact line implying that the contact angle corresponding to the new situation, which we denote as θ_t and call the *intrinsic* (or *microscopic*) contact angle, will be different from its bulk value θ_{eq} given by the Dupre-Young equation.

For simplicity consider the flat geometry of Fig. 15.2. To derive the force balance at the contact line of radius r_t let us consider a small arc of this line seen from its center under a small angle α as shown in Fig. 15.5. The length of this arc is $l = r_t\,\alpha$. The line energy of the arc is $E_t = \tau_t\,l$. Consider a change of the contact line radius δr_t. The corresponding change of the arc length $\delta l = \alpha\,\delta r_t$ induces the change of the line energy

$$\delta E_t = \tau_t\,\delta l.$$

Then, the line force acting on the arc of length l in the radial direction (with the unit vector $\overrightarrow{e_r}$ pointing outwards from the center C of the contact line) is

$$\overrightarrow{F_t} = -\left(\lim_{\delta r_t \to 0}\frac{\delta E_t}{\delta r_t}\right)\overrightarrow{e_r} = -\tau_t\left(\lim_{\delta r_t \to 0}\frac{\delta l}{\delta r_t}\right)\overrightarrow{e_r} = -\tau_t\,\alpha\,\overrightarrow{e_r}$$

The line force per unit length of the contact line is

$$\overrightarrow{f_t} = \frac{\overrightarrow{F_t}}{l} = -\frac{\tau_t}{r_t}\,\overrightarrow{e_r} \qquad (15.47)$$

The balance of interfacial and line forces becomes:

$$\gamma_{13} - \gamma_{23} = \gamma_{12}\cos\theta_t + \frac{\tau_t}{r_t} \qquad (15.48)$$

---

[1] Note, that a possibility of a negative tension of the three-phase contact line was already mentioned by Gibbs [18].

Fig. 15.5 Line tension effect. A small arc of the contact line with the center in point C is characterized by the angle α. A change of the line radius δr_t causes the change in the length of the arc $\delta l = \alpha\, \delta r_t$. The line tension force is along the r-axis which points away from the center of curvature

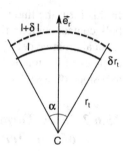

This expression defines the contact angle in the presence of line tension. An alternative derivation of this result stems from the observation that θ_t minimizes the Gibbs free energy (15.46) at a fixed embryo volume V_2. Comparing (15.48) with the Dupre-Young equation (15.15) we find

$$\cos\theta_t = \cos\theta_{eq} - \frac{\tau_t}{\gamma_{12}} \frac{1}{r\,\sin\theta_t} \tag{15.49}$$

This result is known as the *modified Dupre-Young equation*. A positive τ_t would mean that the intrinsic angle of a small embryo is larger than the bulk value; a negative τ_t leads to smaller contact angles compared to θ_{eq}. On a molecular level the line tension stems from the intermolecular interactions between the three phases in the vicinity of the contact line. This fact imposes the bulk correlation length ξ as a natural length scale of the line tension effect. Equation (15.49) suggests that the relevant energy scale (per unit area) is the gas-liquid surface tension γ_{12}. The order-of-magnitude estimate of τ_t is then

$$\tau_t \sim \gamma_{12}\,\xi$$

In liquids far from the critical temperature $\xi \sim 5 \div 10$ Å, typical gas-liquid surface tensions are $\gamma_{12} \sim 50 \div 80$ mN/m, yielding $\tau_t \sim 10^{-11} - 10^{-10}$ N. Though the actual measurements of τ_t meet serious difficulties, various theoretical studies [17, 21–23] indeed reveal that its value should be of the order of $10^{-12} - 10^{-10}$ N in agreement with our estimates. Hence, $\tau_t/\gamma_{12} \sim 0.1 - 1$ nm which implies that one can expect the line tension to have a measurable influence on nucleation behavior if embryos are nano-sized objects; for larger scales it becomes negligible compared to the interfacial tensions.

Derivation of the modified Dupre-Young equation assumed that τ_t does not depend on the radius of the contact line. At the same time it can not be independent on the contact angle, since the relative inclination of the phases determines the net effect of molecular interactions close to the contact line. Now we present (15.49) as

$$\cos\theta_t = \cos\theta_{eq} - \frac{\tau_t}{\gamma_{12}} \frac{1}{r\,\sin\theta_{eq}} + O\left(\frac{1}{r^2}\right) \tag{15.50}$$

In this form the line tension can be interpreted as the first order correction for the bulk contact angle in the inverse radius of the cap. Equation (15.50) can be viewed as an alternative definition of $\tau_t(\theta_{eq})$. If one can measure (or simulate) the intrinsic angle as a function of the cluster radius r and the bulk angle θ_{eq}, than τ_t can be found as a slope of $\cos\theta_t$ versus $1/(r\sin\theta_{eq})$.

15.6.2 Gibbs Formation Energy in the Presence of Line Tension

Consider implication of the line tension on the free energy of an embryo formation. Combining Eqs. (15.46) and (15.48), we write

$$\Delta G = -n\,\Delta\mu + \gamma_{12}\,A_{12}\left(1 - m_t\,\frac{A_{23}}{A_{12}}\right) - \frac{\tau_t\,A_{23}}{r_t} + \tau_t\,2\pi r_t$$

where $m_t = \cos\theta_t$. Using (15.22) for the surface area A_{23}, this expression reduces to

$$\Delta G = \left[-n\,\Delta\mu + \gamma_{12}\,A_{12}\left(1 - m_t\,\frac{A_{23}}{A_{12}}\right)\right] + \tau_t\,\pi\,r\,\sin\theta_t \qquad (15.51)$$

where we took into account that for flat geometry $r_t = r\sin\theta_t$. The expression in the square brackets represents the Fletcher model in which the bulk angle θ_{eq} is replaced by the intrinsic one θ_t:

$$\Delta G = \Delta G_{hom}(r)\,q(m_t) + \tau_t\,\pi\,r\,\sin\theta_t \qquad (15.52)$$

An important feature of this expression is that taking into account the line tension has a two-fold effect on the energy barrier:

- an additional term in ΔG appears which is proportional to the length of the contact line, and
- in the Fletcher factor q the bulk angle is replaced by the intrinsic contact angle which depends on τ_t and the radius of the embryo through the modified Dupre-Young equation

Denoting $m_{eq} = \cos\theta_{eq}$ and performing the first order perturbation analysis in $(1/r)$, we write

$$q(m_t) = q(m_{eq}) + \Delta q$$

$$\Delta q = \left.\frac{dq}{dm}\right|_{m_{eq}} \Delta m = -\frac{3}{4}\sin^2\theta_{eq}\,\Delta m \qquad (15.53)$$

where from (15.50)

$$\Delta m \equiv m_t - m_{eq} \approx -\frac{\tau_t}{\gamma_{12}} \frac{1}{r \, \sin \theta_{eq}} \tag{15.54}$$

In the same approximation the Gibbs energy reads:

$$\Delta G = \Delta G_{hom}(r) \left[q(m_{eq}) + \Delta q \right] + \tau_t \, \pi \, r \, \sin \theta_{eq} \tag{15.55}$$

The critical cluster corresponds to maximum of ΔG:

$$\Delta G'_{hom}(r) \left[q(m_{eq}) + \Delta q \right] + \Delta G_{hom}(r) \, \Delta q' + \tau_t \, \pi \, \sin \theta_{eq} = 0, \quad \text{where}' = \frac{d}{dr} \tag{15.56}$$

Following the same thermodynamic arguments, as those used in the derivation of the Fletcher theory, we can expect that the critical cluster size should not be affected either by the foreign particles or the line tension and corresponds to the maximum of the Gibbs energy for homogeneous case:

$$\Delta G'_{hom}(r)_{r=r_c} = 0$$

Indeed, substituting the homogeneous nucleation barrier

$$\Delta G_{hom}(r_c) = \frac{4\pi}{3} \gamma_{12} \, r_c^2$$

into Eq. (15.56) and taking into account (15.53) and (15.54), we find that Eq. (15.56) becomes an identity. This means that the critical cluster size in the presence of line tension is given by the classical Kelvin equation.[2] Setting $r = r_c$ in (15.55), we obtain the nucleation barrier in the presence of line tension:

$$\Delta G^* = \frac{4\pi}{3} \gamma_{12} \, r_c^2 \, q(m_{eq}) + 2 \, \pi r_c \, \tau_t \, \sin \theta_{eq} \tag{15.57}$$

A positive τ_t increases the nucleation barrier given by the Fletcher theory (first term), while a negative τ_t lowers it thus enhancing nucleation. Such an enhancement was observed experimentally in Refs. [24, 25].

15.6.3 Analytical Solution of Modified Dupre-Young Equation

The modified Dupre-Young equation (15.49) has an analytical solution $\theta_t(\theta_{eq}, r)$ for all values of the bulk contact angle θ_{eq}. However, the general form of the solution

---

[2] Note, that if in (15.52) the bulk Fletcher factor $q(m_{eq})$ is used instead of $q(m_t)$, the critical cluster size, maximizing ΔG, would violate the Kelvin equation.

is rather complex except for the special case of $\theta_{eq} = \pi/2$, for which Eq. (15.49) is simplified to:

$$\sin(2\theta_t) = -\frac{2p}{r}, \quad p \equiv \frac{\tau_t}{\gamma_{12}} \tag{15.58}$$

It is instructive to study this equation in order to verify the perturbation approach of the previous section. Obviously a solution of (15.58) exists if

$$r \geq \left| \frac{\tau_t}{2\gamma_{12}} \right| \tag{15.59}$$

This constraint gives the range of validity of the modified Dupre-Young equation. From the previous analysis (15.59) can be approximately expressed as

$$r > 0.5\,\text{nm} \tag{15.60}$$

When the radius of the embryo approaches the molecular size Eq. (15.58) fails.

At the same time *at large* r the classical expression should be recovered: $\lim_{r\to\infty} \theta_t = \theta_{eq}$, which in our case results in

$$\theta_t \to_{r\to\infty} \pi/2 \tag{15.61}$$

Solving (15.58) we find:

$$\theta_{t,1} = \frac{1}{2}\arcsin\left(-\frac{2p}{r}\right), \quad \text{and} \quad \theta_{t,2} = \frac{\pi}{2} - \frac{1}{2}\arcsin\left(-\frac{2p}{r}\right)$$

At $r\to\infty$: $\theta_{t,1}\to 0$, $\theta_{t,2}\to\pi/2$. Hence, the solution satisfying the asymptotic condition (15.61) is

$$\theta_t = \frac{\pi}{2} - \frac{1}{2}\arcsin\left(-\frac{2p}{r}\right) \tag{15.62}$$

$$m_t = \sin\left[\frac{1}{2}\arcsin\left(-\frac{2p}{r}\right)\right] \tag{15.63}$$

Expansion in $1/r$ yields

$$m_t = -\frac{p}{r} - \frac{p^3}{2r^3} - O\left(\frac{1}{r^5}\right) \tag{15.64}$$

Since the second order term is absent, the approximate solution

$$m_t(r) \approx -\frac{\tau_t}{\gamma_{12}}\frac{1}{r}$$

coincides with the exact one up to the terms of order $1/r^2$. One can expect that for angles close to $\pi/2$ the first order expansion in $1/r$ remains a good approximation.

15.6.4 Determination of Line Tension

To accomplish the model for the Gibbs formation energy (15.55) we need to have information about the line tension for a given system. The extreme smallness of τ_t makes its direct experimental measurement a very difficult task, that is why available experimental data remains scarce. To obtain reliable estimates of τ_t it is important that the droplets are of the same size as the deduced line tension values. Several advanced techniques were used recently aiming to satisfy this requirement. Pompe and Herminghaus [26] and Pompe [27] used Scanning Force Microscopy to study the shape of the sessile droplets on solid substrates near the contact line. From the droplet profile they deduced that τ_t lies in the range $10^{-12} - 10^{-10}$ N and can be either positive or negative depending on the system. In particular, it was found that line tension increases with lowering contact angle; at large θ_{eq} it is negative and changes sign at $\theta_{eq} \approx 6°$. Berg et al. [28] used Atomic Force Microscopy (AFM) to study nanometer-size sessile fullerene (C_{60}) droplets on the planar SiO_2 interface and observed the size-dependent variation of the contact angle which can be interpreted as the line tension. From the modified Dupre-Young equation they found the negative values

$$\tau_t = -(0.7 \pm 0.3) \times 10^{-10}\,\text{N} \tag{15.65}$$

and obtained a characteristic length scale of the effect $\tau_t/\gamma_{12} \approx 1.4\,\text{nm}$.

In most of the heterogeneous nucleation studies, which take into account the line tension effect, the value of τ_t is found from fitting to the experimental data on nucleation rates [24, 29, 30]. However, such fitting can not be considered reliable in view of a number of reasons. Homogeneous nucleation in the bulk has to be distinguished from heterogeneous nucleation on impurities, and small changes of parameters (e.g. substrate heterogeneities) can lead to considerable difference in measured nucleation rates. These and other artifacts can then be erroneously interpreted as line tension effects. In view of these reasons it is highly desirable to determine τ_t from an independent source: model/simulations/experiment.

The advantage of *computer simulations* is that the properties of interest are controllable parameters. The simplest model of a fluid is the lattice-gas with nearest neighbor interactions (Ising model) on the simple cubic lattice. The Ising Hamiltonian (discussed in Sect. 8.9) reads:

$$\mathcal{H} = -K \sum_{nn} s_i s_j \tag{15.66}$$

where the "spins" s_k are equal to ± 1, and K is the coupling parameter (interaction strength); summation is over nearest neighbors. The presence of the foreign substrate (a solid wall) is described by the corresponding boundary condition which is characterized by a surface field H_1 acting on the first layer of fluid molecules adjacent to the surface. Monte Carlo simulations of this system performed by Winter et al. [31] at temperatures far from T_c result in the appearance of a spherical cap-shaped (liquid) droplet surrounded by the vapor and resting on the adsorbing solid wall (favoring liquid). The surface free energy of the embryo formation ΔG_{MC}^{surf} is found in simulations using thermodynamic integration. Simulation results reveal that the difference between ΔG_{MC}^{surf} and the Fletcher model

$$\Delta G_{MC}^{surf} - \gamma_{12}\, 4\pi\, r^2\, q(\theta_{eq})$$

increases linearly with the droplet radius r. This difference can be attributed to the line tension contribution. In Ref. [31] this linear dependence is presented in the form

$$\Delta G_{MC}^{surf} - \gamma_{12}\, 4\pi\, r^2\, q(\theta_{eq}) = \tau_{MC}\, (2\pi\, r \sin\theta_{eq}) \tag{15.67}$$

which is used as a *definition* of the line tension τ_{MC}. We introduced the notation τ_{MC} to emphasize the difference between the latter and the previously defined quantity τ_t. We require that the simulated surface free energy ΔG_{MC}^{surf} be equal to its theoretical counterpart given by (15.55)

$$\Delta G_{th}^{surf} = \gamma_{12}\, 4\pi\, r^2\, q(\theta_{eq}) + \left(\gamma_{12}\, 4\pi\, r^2\, \Delta q + \tau_t\, \pi\, r\, \sin\theta_{eq}\right) \tag{15.68}$$

Comparing (15.67) and (15.68), we find

$$\tau_t = \frac{\tau_{MC}}{2} \tag{15.69}$$

Simulations, performed for the temperature

$$k_B T_{MC}/K = 3, \tag{15.70}$$

reveal that for all contact angles studied τ_{MC} is negative and its absolute value increases with θ_{eq} as shown in Fig. 15.6. The equilibrium contact angle is controlled by varying the surface field H_1. It is important to note, that for a fixed temperature different contact angles (obtained by tuning the surface field H_1) in Fig. 15.6 physically correspond to different solids.

Equation (15.66) being the simplest model of a magnet, can be also viewed as a model of a fluid. Mapping of the Ising model to a model of a fluid is not a unique procedure, therefore the results for τ_t for fluids can differ depending on the chosen procedure but most probably will be qualitatively the same. One of the strategies, used e.g. in the theory of polymers, is to equate the critical temperature of a substance to the

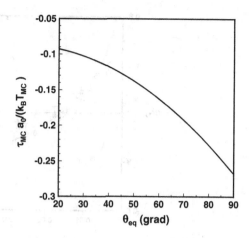

Fig. 15.6 Line tension τ_{MC} as a function of the equilibrium contact angle θ_{eq} for the Ising system at the temperature $k_B T_{MC}/J = 3$; a_0 is the lattice spacing. The *solid line* is a fit to the Monte Carlo results of Ref. [31]

critical temperature of 3D Ising model [32]:

$$k_B T_{c,\text{Ising}}/K \approx 4.51$$

Comparing this expression with (15.70), we find

$$T_{MC} = 0.665 \, T_{c,\text{Ising}}$$

Considering water as an example and setting $T_{c,\text{Ising}}$ equal to the critical temperature of water $T_{c,\text{water}} = 647\,K$, we find that MC simulations of Ref. [31] correspond to $T = 430.3\,K$. Considering a substance which at this temperature has the bulk contact angle $\theta_{eq} = 90°$, the simulation results of Fig. 15.6, give

$$\frac{\tau_{MC} \, \sigma}{k_B T_{MC}} = -0.26$$

where we replaced the lattice spacing a_0 in the Ising model by the water molecular diameter $\sigma = 2.64\,\text{Å}$ [33]. Then, taking into account (15.69)

$$\tau_t(T = 430.3\,K, \, \theta_{eq} = 90°) = -0.41 \times 10^{-11}\,N$$

One of the important issues that has to be considered is the temperature dependence of the line tension. Experimentally the latter can be deduced from the measurements of the microscopic contact angle using the modified Dupre-Young equation: τ_t is found from the slope of $\cos \theta_t$ as a function of $1/r$ at different temperatures. Such study was performed by Wang et al. [34] for n-octane and 1-octene in the temperature interval $301 < T < 316\,K$. Fig. 15.7 shows the plot of the line tension for n-octane (solid circles) and 1-octene (open circles) as a function of reduced temperature

$$t = (T_w - T)/T_w$$

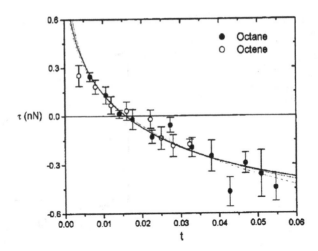

Fig. 15.7 Line tension as a function of reduced temperature $t = (T_w - T)/T_w$ for n-octane and 1-octene on coated silicon. (Reprinted with permission from Ref. [34], copyright (2001), American Physical Society.)

where T_w is the wetting temperature (corresponding to $\cos\theta_{eq} = 1$). The solid substrate in both cases is Si wafer.

For both liquids as temperature increases towards the wetting temperature T_w, the line tension changes from a negative to a positive value with an increasing slope $|d\tau_t/dT|$. This behavior qualitatively agrees with theoretical predictions [35, 36]. The wetting temperatures of n-octane and 1-octene on Si wafer were found to be

$$T_{w,\text{octane}} = 318.5 \,\text{K}, \quad T_{w,\text{octene}} = 324.3 \,\text{K}$$

Both of them lie well below the corresponding critical temperatures

$$T_{c,\text{octane}} = 568.7 \,\text{K} \quad T_{w,\text{octene}} = 566.7 \,\text{K}$$

15.6.5 Example: Line Tension Effect in Heterogeneous Water Nucleation

For illustration purposes let us analyze the implication of the line tension for water nucleating on a large seed particle with the bulk contact angle $\theta_{eq} = 90°$ at $T = 285\,\text{K}$ and supersaturation $S = 2.93$. At these conditions the critical cluster radius according to CNT is $r_c = 1.03\,\text{nm}$. If the radius of a seed particle $R_p \gg 1\,\text{nm}$, it can be considered as a flat wall for the critical embryo and the Fletcher factor is given by the function $q(m)$. Figure 15.8 shows $\Delta G(r)$ for 3 different models:

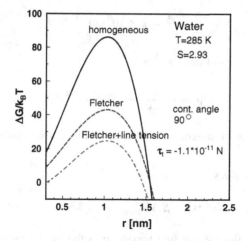

Fig. 15.8 Gibbs free energy of a cluster formation for water nucleating on a large seed particle at $T = 285$ K and supersaturation $S = 2.93$. *Solid line*: classical homogeneous nucleation theory (CNT). *Dashed line*: the classical heterogeneous (Fletcher) theory with the bulk the contact angle $\theta_{eq} = 90°$. *Dashed-dotted line*: the Fletcher theory corrected with the line tension effect according to Eq. (15.55); the value of line tension is $\tau_t = -1.1 \times 10^{-11}$ N

- classical homogeneous nucleation theory (CNT),
- classical heterogeneous (Fletcher) theory, and
- Fletcher theory corrected with the line tension effect according to Eq. (15.55).

Using previous considerations we choose a typical value of the line tension

$$\tau_t = -1.1 \times 10^{-11} \text{ N}$$

The homogeneous nucleation barrier is $\approx 86 \, k_B T$, the kinetic prefactor

$$J_0 = 4 \times 10^{25} \text{ cm}^{-3} \text{ s}^{-1}$$

so that homogeneous nucleation is suppressed:

$$J_{hom} \sim 10^{-12} \text{ cm}^{-3} \text{ s}^{-1}$$

The Fletcher correction reduces the barrier to $\Delta G^*_{Fletcher} \approx 43 \, k_B T$. The line tension leads to further reduction $\Delta G^* \approx 25 \, k_B T$ resulting in considerable enhancement of nucleation. Setting $R_p = 10$ nm, $\tau_0 = 2.55 \times 10^{-13}$ s [7], $E_{ads} = 10.640$ kcal/mol [37] we find for the heterogenous nucleation prefactor (15.42)

$$K_p \approx 5.4 \times 10^{13} \text{ s}^{-1}$$

Using a typical concentration of aerosol particles in experiments [38]

$$\frac{N_p}{V} = 2 \times 10^4 \, \text{cm}^{-3}$$

the prefactor in the units of $\text{cm}^{-3}\text{s}^{-1}$ is

$$K_V = K_p \frac{N_p}{V} \approx 10^{18} \, \text{cm}^{-3} \, \text{s}^{-1}$$

which is 7.5 orders of magnitude lower than the corresponding homogeneous quantity J_0.

The Fletcher theory gives

$$J_{\text{Fletcher}} = K_V \, \exp(-\beta \Delta G^*_{\text{Fletcher}}) \approx 3.6 \times 10^{-1} \, \text{cm}^{-3} \, \text{s}^{-1}$$

while the incorporation of the line tension into the model yields a considerable increase of the nucleation rate

$$J_{\text{Fletcher+line tension}} = K_V \, \exp(-\beta \Delta G^*_{\text{Fletcher+line tension}}) \approx 1.5 \times 10^7 \, \text{cm}^{-3} \, \text{s}^{-1}$$

15.7 Nucleation Probability

Activation of a foreign particle occurs when the first critical embryo is formed on its surface. This is a random event and as such can be studied using the methodology of the theory of random processes. This approach to heterogeneous nucleation can be particularly useful when analyzing the experimental data.

Let us choose some characteristic time t, during which heterogeneous nucleation is observed. Typical value of t in experiments is ~ 1 ms. Let $P_k(t)$ be the probability that *exactly k* activation events occurred during time t. The average number of such events per unit time is given by the nucleation rate J_p. A probability of activation during an infinitesimally small interval Δt is $J_p \, \Delta t$ (assuming that two simultaneous activation events during Δt are highly unlikely). Then the probability that *no events* happened during the same interval is $1 - J_p \, \Delta t$. Consider the quantity $P_k(t + \Delta t)$ which is the probability that exactly k activation events occurred during time $t + \Delta t$. Straightforward probabilistic considerations yield:

$$P_k(t + \Delta t) = P_k(t)\,(1 - J_p \, \Delta t) + P_{k-1}(t)\, J_p \, \Delta t \tag{15.71}$$

The first term on the right-hand side refers to the situation when all k events happened during time t and no events occurred during time Δt. The second term gives the probability that exactly $k - 1$ events took place during time t and one event happened during time Δt. Dividing both sides by Δt and taking the limit at $\Delta t \to 0$ we obtain

$$\frac{dP_k(t)}{dt} = -P_k(t)\,J_p + P_{k-1}(t)\,J_p, \quad k = 0, 1, 2, \ldots \tag{15.72}$$

Consider the first equation of this set, corresponding to $k = 0$; $P_0(t)$ describes the probability that no events happened during time t. Obviously, we must set $P_{-1}(t) = 0$ which yields

$$\frac{dP_0(t)}{dt} = -J_p\,P_0(t) \tag{15.73}$$

Integration of Eq. (15.73) gives

$$P_0(t) = P_0(0)\,e^{-J_p t}$$

where $P_0(0)$ is the probability that no events happened in zero time. Obviously, $P_0(0) = 1$ resulting in

$$P_0(t) = e^{-J_p t}$$

Then, the quantity

$$P_{het}(t) = 1 - e^{-J_p t} \tag{15.74}$$

is the probability that at least one foreign particle was activated to growth during time t. For large number of events $P_{het}(t)$, termed the *nucleation probability* [37, 39, 40], describes the fraction of foreign particles activated to growth during time t. The latter quantity is measured in heterogeneous nucleation experiments. Setting P_{het} to 0.5 we refer to the situation when half of the foreign particles are activated to growth. This can be viewed as the onset conditions for heterogeneous nucleation. Since the nucleation rate is a very steep function of the supersaturation (activity), the nucleation probability is expected to be close to the step-function centered around the onset activity.

For illustration we use the example of water nucleation considered in Sect. 15.6.5. Setting the characteristic experimental time to $t = 1\,\text{ms}$ [37] we find:

$$P_{het,Fletcher} = 0, \quad P_{het,Fletcher+line\,tension} = 0.52$$

This result implies that $S = 2.93$ is the onset condition if the line tension effect is taken into account; at the same S the Fletcher theory predicts no nucleation. From experimental data (see e.g. [37]) it follows that the onset conditions are not much sensitive to the choice of t.

References

1. M. Ganero-Castano, J. Fernandez de la Mora, J. Chem. Phys. **117**, 3345 (2002)
2. J.L. Katz, J. Fisk, M. Chakarov, J. Chem. Phys. **101**, 2309 (1994)
3. A.B. Nadykto, F. Yu, Atmosd. Chem. Phys. **4**, 385 (2004)
4. H. Rabeony, P. Mirabel, J. Phys. Chem. **91**, 1815 (1987)
5. R.J. Charlson, T. Wigley, Sci. Am. **270**, 48 (1994)
6. N.N. Fletcher, J. Chem. Phys. **29**, 572 (1958)
7. M. Lazaridis, M. Kulmala, A. Laaksonen, J. Aerosol Sci. **22**, 823 (1991)
8. M. Lazaridis, J. Coll. Interface Sci. **155**, 386 (1993)
9. G.M. Pound, M.T. Simnad, L. Yang, J. Chem. Phys. **22**, 1215 (1954)
10. H.R. Pruppacher, J.C. Pflaum, J. Coll. Int. Sci. **52**, 543 (1975)
11. J. Frenkel, *Kinetic Theory of Liquids* (Clarendon, Oxford, 1946)
12. J.S. Sheu, J.R. Maa, J.L. Katz, J. Stat. Phys. **52**, 1143 (1988)
13. H.R. Pruppacher, J.D. Klett, *Microphysics of Clouds and Precipitation* (Reidel, Dordrecht, 1978)
14. P. Hamill et al., J. Aerosol Sci. **13**, 561 (1982)
15. J. Israelashvili, *Intermolecular and Surface Forces* (Cambridge University Press, Cambridge, 1992)
16. M. Lazaridis, I. Ford, J. Chem. Phys. **99**, 5426 (1993)
17. J.S. Rowlinson, B. Widom, *Molecular Theory of Capillarity* (Clarendon Press, Oxford, 1982)
18. J.W. Gibbs, *The Scientific Papers* (Ox Bow, Woodbridge, NJ, 1993)
19. R.D. Gretz, J. Chem. Phys. **45**, 3160 (1966)
20. L.F. Evans, J.E. Lane, J. Atmos. Sci. **30**, 326 (1973)
21. J. Indekeu, Physica A **183**, 439 (1992)
22. R. Lipowsky, J. Phys. II (France) **2**, 1825 (1992)
23. L. Schimmele, M. Napiorkowski, S. Dietrich, J. Chem. Phys. **127**, 164715 (2007)
24. A. Sheludko, V. Chakarov, B. Toshev, J. Coll. Int. Sci. **82**, 83 (1981)
25. B. Lefevre, A. Saugey, J.L. Barrat, J. Chem. Phys. **120**, 4927 (2004)
26. T. Pompe, S. Herminghaus, Phys. Rev. Lett. **85**, 1930 (2000)
27. T. Pompe, Phys. Rev. Lett. **89**, 076102 (2002)
28. J.K. Berg, C.M. Weber, H. Riegler, Phys. Rev. Lett. **105**, 076103 (2010)
29. A.I. Hienola, P.M. Winkler, P.E. Wagner et al., J. Chem. Phys. **126**, 094705 (2007)
30. P. Winkler, Ph.D. Thesis, University of Vienna, 2004
31. D. Winter, P. Virnau, K. Binder, Phys. Rev. Lett. **103**, 225703 (2009)
32. R.J. Baxter, *Exactly Solved Models in Statistical Mechanics* (Academics Press, London, 1982)
33. R.C. Reid, J.M. Prausnitz, B.E. Poling, *The Properties of Gases and Liquids* (McGraw-Hill, New York, 1987)
34. J.Y. Wang, S. Betelu, B.M. Law, Phys. Rev. **E63**, 031601 (2001)
35. J. Indekeu, Int. J. Mod. Phys. B **8**, 309 (1994)
36. I. Szleifer, B. Widom, Mol. Phys. **75**, 925 (1992)
37. P.E. Wagner, D. Kaller, A. Vrtala et al., Phys. Rev. **E67**, 021605 (2003)
38. M. Kulmala, A. Lauri, H. Vehkamaki et al., J. Phys. Chem. B **105**, 11800 (2001)
39. M. Lazaridis, M. Kulmala, B.Z. Gorbuniv, J. Aerosol Sci. **23**, 457 (1992)
40. H. Vehkamäki, *Classical Nucleation Theory in Multicomponent Systems* (Springer, Berlin, 2006)

Chapter 16
Experimental Methods

Throughout this book we compared the predictions of theoretical models with available experimental data. This chapter is aimed at providing a reader with a flavor of experimental methods used in nucleation research. As in the previous chapters, we focus on vapor to liquid nucleation as most of the experimental studies refer to this type of transition.

Prior to 1960–1970s most experiments dealt with the *critical supersaturation* measurements, or more generally, the conditions accompanying the *onset of nucleation* at various temperatures (for review see [1]). This research was pioneered by Wilson in the end of the nineteenth century [2] who studied the behavior of water vapor in expansion chamber and observed the onset of the condensation process and the associated with it light scattering. The main conclusion drawn from Wilson's experiments is that if the vapor is sufficiently supersaturated, thermal density fluctuations trigger droplet formation in the chamber in the absence of impurities—the process we now refer to as homogeneous nucleation. Starting with 1970s a number of newly developed techniques appeared which make it possible to measure not only the onset conditions but the *nucleation rates* themselves at various temperatures and supersaturations. This big step in experimental research opened the way for quantitative tests of nucleation theories (within the accessible range of temperatures and pressures). At present quantitative nucleation rate measurements, using various experimental techniques, span the range of nucleation rates from 10^{-3} cm^{-3} s^{-1} to 10^{18} cm^{-3} s^{-1}. Also combination of these measurements with nucleation theorems, studied in Chap. 4, provides direct information on the properties of critical cluster.

Among the variety of methods (for a review see e.g. [3]) we describe the four most widely used techniques:

- thermal diffusion cloud chamber
- expansion cloud chamber
- shock tube
- supersonic nozzle

V. I. Kalikmanov, *Nucleation Theory*, Lecture Notes in Physics 860,
DOI: 10.1007/978-90-481-3643-8_16, © Springer Science+Business Media Dordrecht 2013

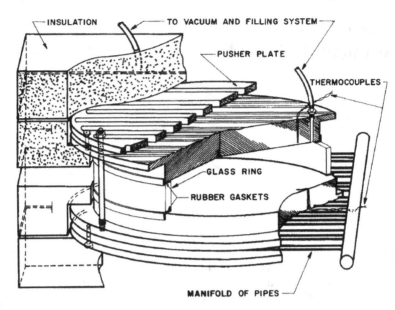

Fig. 16.1 Cutaway view of diffusion cloud chamber. (Reprinted with permission from Ref. [4], copyright (1975), American Institute of Physics.)

16.1 Thermal Diffusion Cloud Chamber

The thermal diffusion cloud chamber consists of two metallic cylindrical plates separated by the optically transparent cylindrical ring. The region between the plates forms the working volume of the chamber. The substance under study is placed as a shallow liquid pool on the lower plate of the chamber and the working volume is filled by the carrier gas (aiming at removal of the latent heat emerging in the process of condensation). The lower plate is heated while the upper plate is cooled. Due to the temperature difference ΔT between the plates, vapor[1] evaporates from the top surface of the liquid pool, diffuses through a noncondensable carrier gas (usually helium, argon or nitrogen), and condenses on the lower surface of the top plate. Construction of the diffusion cloud chamber is illustrated in Fig. 16.1.

Thermal diffusion gives rise to the profiles of temperature, density, pressure and supersaturation inside the chamber. These profiles can be calculated from the one-dimensional energy and mass transport equations using an appropriate equation of state for the vapor/carrier gas mixture as shown in Fig. 16.2. At certain values of ΔT the supersaturation in the chamber becomes sufficiently large to cause nucleation of droplets which are subsequently detected by light scattering using a laser and a photo-multiplier.

---

[1] The term "vapor" in this chapter is used for the condensible component; while the term "gas" refers to the carrier gas.

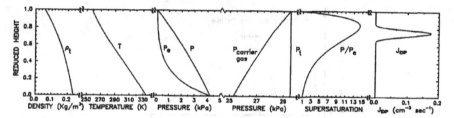

Fig. 16.2 Profiles of density, temperature, supersaturation and the nucleation rate inside the chamber. (Reprinted with permission from Ref. [5], copyright (1989), American Institute of Physics.)

Diffusion cloud chamber can operate in the temperature range from near the triple point (of the substance under study) up to the critical temperature and in the pressure range from below the ambient to elevated pressures. A typical range of accessible nucleation rates is $10^{-3} - 10^3$ cm^{-3} s^{-1}. The growing droplets are removed by gravitational sedimentation or by convective flow which ensures the steady-state self-cleaning operational conditions. Due to this feature and to the relatively low nucleation rates, e.g. relatively small number of droplets to be counted, the quantitative nucleation rate measurements are straightforward [4–8]. Measuring nucleation rate as a function of supersaturation at a constant temperature, one can determine the size of the critical cluster using the nucleation theorem (Chap. 4). The experimentally determined critical cluster can then be compared to the nucleation models.

Note, however, that nonlinear temperature and pressure profiles inside the chamber can lead to substantial nonuniformities of temperatures and supersaturations in the working volume making it difficult to assign particular values to supersaturations and temperatures corresponding to the observed nucleation rates.

16.2 Expansion Cloud Chamber

Compared to the diffusion chamber, functioning of the *expansion* cloud chamber relies upon a different mechanism: rapid adiabatic expansion of the vapor/gas mixture which produces supersaturation and subsequent nucleation. The device can be generally described as a cylinder piston-like structure containing the vapor/gas mixture in the region above the cylinder and bounded by the piston walls [9] (or valves connecting additional volumes to the chamber, as in the nucleation pulse chamber of Ref. [10]). Initially the mixture has the temperature of the piston wall and the vapor may or may not be saturated. After rapid withdrawal of the piston adiabatic cooling occurs: the pressure and temperature of the mixture decrease. As a result, the supersaturation of the vapor

$$S = \frac{p^v}{p_{sat}(T)}$$

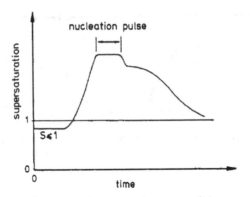

Fig. 16.3 Time dependence of the supersaturation in the expansion cloud chamber during a single nucleation experiment. Experiment starts when vapor is supersaturated. As a result of adiabatic expansion vapor becomes supersaturated and nucleation occurs during the time of nucleation pulse. After that a slight recompression terminates nucleation process; condensational growth of droplets results in further reduction of the supersaturation due to vapor depletion. (Reprinted with permission from Ref. [10], copyright (1994), American Chemical Society.)

increases because the decrease of its partial pressure p^v during the isentropic expansion is slower than the exponential decrease of the saturation pressure p_{sat} with temperature, which is given by the Clapeyron equation (2.14). As opposed to the diffusion chamber, nucleation of the supersaturated vapor in the working volume of the expansion chamber takes place at uniform conditions.

Among various modifications of the expansion camber—single-piston chamber [11, 12], piston-expansion tube [13]—we will describe in a somewhat more detail the nucleation pulse chamber (NPC) [10, 14, 15]. In order to ensure the constant conditions during the nucleation period a small recompression pulse is issued in NPC after the completion of the adiabatic expansion, which terminates the nucleation process after a short time, of the order of 1 ms, called the *nucleation pulse*. Condensational growth of droplets results in further reduction of the supersaturation due to vapor depletion. Schematically evolution of the supersaturation in the NPC during the nucleation experiment is depicted in Fig. 16.3. The cooling rate in the NPC is of the order 10^4 K/s.

The values of supersaturation and temperature corresponding to the measured nucleation rate are calculated from the following considerations. If p_0 is the initial total pressure of the vapor/gas mixture, T_0 is the initial temperature and y is the vapor molar fraction, then

$$p_0^v = y \, p_0$$

is the partial vapor pressure at the initial conditions. After adiabatic expansion the total pressure drops by Δp_{expt} becoming equal to

$$p = p_0 - \Delta p_{expt}$$

The nucleation temperature follows the Poisson law:

$$\frac{T}{T_0} = \left(\frac{p}{p_0}\right)^{(\kappa-1)/\kappa} \tag{16.1}$$

where $\kappa = c_p/c_v$ is the ratio of specific heats for the vapor/gas mixture. Then, from the saturation vapor pressure at temperature T, $p_{sat}(T)$, given by the Clapeyron equation (2.14), the supersaturation is found to be

$$S_{expt} = \frac{y(p_0 - \Delta p_{expt})}{p_{sat}(T)} \tag{16.2}$$

During the short ($\sim$1 ms) nucleation pulse only a negligible fraction of the vapor is consumed, i.e. depletion effects are negligible which leaves the supersaturation practically constant. After recompression supersaturation drops, nucleation is suppressed so that only particle growth at constant number density occurs (while no new droplets are formed). Thus, the nucleation pulse method realized in the expansion tube makes it possible to decouple nucleation and growth processes.

The last step is to determine the number density ρ_d of droplets formed during the nucleation pulse. In the NPC the nucleated droplets grow to the sizes $\sim$1 μm when they are detected by the constant angle Mie scattering (CAMS) technique leading to determination of ρ_d. CAMS, which uses a laser operating in the visual, is based on the Mie theory of scattering of electromagnetic waves by dielectric spherical particles [16].

The basic idea behind the technique is straightforward: (i) analyzing the time evolution of the intensity of light *scattered* by the droplets and comparing it with the Mie theory, one finds the size r_d of the droplet at time t; (ii) analyzing the evolution of the intensity of the *transmitted* light one determines the number density of droplets using the value of extinction coefficient corresponding to the droplet size r_d.

16.2.1 Mie Theory

To clarify CAMS, we briefly formulate the main results of the Mie theory (for details the reader is referred to Refs. [16, 17]) relevant for the analysis of nucleation experiments. Consider a single dielectric spherical particle of radius r_d emerged in the vacuum and having the refractive index m. The particle is illuminated by the incident light with a wavelength λ. Let us introduce the dimensionless droplet radius[2]

$$\alpha = \left(\frac{2\pi}{\lambda}\right) r_d = k\,r_d$$

---

[2] If instead of vacuum the particle is emerged in a homogeneous medium with the refractive index m_{medium}, the wavelength should be replaced by $\lambda_{vacuum}/m_{medium}$.

If I_0 is the intensity of the incident light (watt/m^2), the sphere will intercept $Q_{ext} \pi r_d^2 I_0$ watt from the incident beam, independently of the state of polarization of the latter. The dimensionless quantity $Q_{ext}(m, \alpha)$ is called the *extinction efficiency*. In the Mie theory it is given by

$$Q_{ext}(m, \alpha) = \frac{2}{\alpha^2} \sum_{n=1}^{\infty} (2n + 1) \, \Re(a_n + b_n) \qquad (16.3)$$

Here the complex Mie coefficients a_n and b_n are obtained from matching the boundary conditions at the surface of the spherical droplet. They are expressed in terms of spherical Bessel functions evaluated at α and $y = m\alpha$:

$$a_n = \frac{\psi_n'(y) \, \psi_n(\alpha) - m \, \psi_n(y) \, \psi_n'(\alpha)}{\psi_n'(y) \, \zeta_n(\alpha) - m \, \psi_n(y) \, \zeta_n'(\alpha)}$$

$$b_n = \frac{m \, \psi_n'(y) \, \psi_n(\alpha) - \psi_n(y) \, \psi_n'(\alpha)}{m \, \psi_n'(y) \, \zeta_n(\alpha) - \psi_n(y) \, \zeta_n'(\alpha)}$$

where

$$\psi_n(z) = (\pi z/2)^{1/2} \, J_{n+1/2}(z)$$
$$\zeta_n(z) = (\pi z/2)^{1/2} \, H_{n+1/2}^{(2)}(z)$$

and $J_{n+1/2}(z)$ is the half-integer-order Bessel function of the first kind, $H_{n+1/2}^{(2)}(z)$ is the half-integer-order Hankel function of the second kind [18].

The intensity of the incident beam decreases with the a distance L (called the optical path) as it proceeds through the cloud of droplets. The transmitted light intensity is given by the Lambert-Beer law [19]

$$I_{trans} = I_0 e^{-\beta_{ext} L} \qquad (16.4)$$

where β_{ext} is the extinction coefficient computed from

$$\beta_{ext} = \rho_d \pi r_d^2 Q_{ext}(m, \alpha) \qquad (16.5)$$

(here we assumed that all ρ_d dielectric spheres in the unit volume are identical). The behavior of the extinction efficiency Q_{ext} as a function of the size parameter α is illustrated in Fig. 16.4 for the two substances with the values of refractive index $m = 1.33$ (water) and $m = 1.55$ (silicone oil).

Consider now the light *scattered* by a single sphere. The direction of scattering is given by the polar angle θ and the azimuth angle ϕ. The intensity $I_{scat,1}$ of the scattered light in a point located at a large distance r from the center of the particle has a form

$$I_{scat,1} = \frac{I_0}{k^2 r^2} F(\theta, \phi; m, \alpha) \qquad (16.6)$$

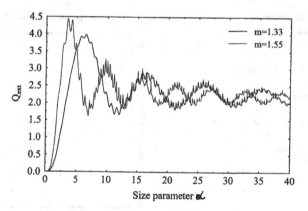

Fig. 16.4 Mie extinction efficiency versus size parameter α for water ($m = 1.33$) and silicone oil ($m = 1.55$). For small particles $Q_{ext} \sim \alpha^4$ (Rayleigh limit); for big particles $Q_{ext} \to 2$ (limit of geometrical optics $\alpha \to \infty$). The largest value of Q_{ext} is achieved when the particle size is close to the wavelength

where F is the dimensionless function of the direction (not of r). For the linearly polarized incident light

$$F = i_1 \sin^2 \phi + i_2 \cos^2 \phi \tag{16.7}$$

Here i_1 and i_2 refer, respectively, to the intensity of light vibrating perpendicularly and parallel to the plain through the directions of propagation of the incident and scattered beams. The quantities i_1 and i_2 are expressed in terms of the *amplitude functions* $S_1(m, \alpha; \theta)$ and $S_2(m, \alpha; \theta)$:

$$i_1 = |S_1(m, \alpha; \theta)|^2, \quad i_2 = |S_2(m, \alpha; \theta)|^2$$

The amplitude functions are given by

$$S_1(m, \alpha; \theta) = \sum_{n=1}^{\infty} \frac{2n+1}{n(n+1)} [a_n \pi_n(\cos\theta) + b_n \tau_n(\cos\theta)] \tag{16.8}$$

$$S_2(m, \alpha; \theta) = \sum_{n=1}^{\infty} \frac{2n+1}{n(n+1)} [b_n \pi_n(\cos\theta) + a_n \tau_n(\cos\theta)] \tag{16.9}$$

where

$$\pi_n(\cos\theta) = \frac{1}{\sin\theta} P_n^1(\cos\theta) \tag{16.10}$$

$$\tau_n(\cos\theta) = \frac{d}{d\theta} P_n^1(\cos\theta) \tag{16.11}$$

and $P_n^1(\cos\theta)$ is the associated Legendre polynomial [18].

Fig. 16.5 Time dependence of the normalized scattered light intensity (*left y-axis*) and the total pressure (*right y-axis*) for CAMS (scattering angle is 15°). It is clearly seen that scattering is detected *after* the nucleation pulse. (Reprinted with permission from Ref. [10], copyright (1994), American Chemical Society.)

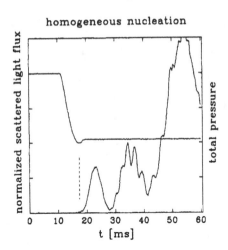

Averaging F over the azimuth angle using the identity

$$\frac{1}{2\pi} \int_0^{2\pi} \sin^2 \phi \, d\phi = \frac{1}{2\pi} \int_0^{2\pi} \cos^2 \phi \, d\phi = \frac{1}{2}$$

we have

$$\langle F \rangle_\phi = \frac{i_1 + i_2}{2} = \frac{|S_1|^2 + |S_2|^2}{2} \tag{16.12}$$

If multiple scattering can be avoided, the total intensity of light scattered in the direction θ by *all* spheres in the volume V is

$$I_{\text{scat}} = I_{\text{scat},1} \, \rho_d \, V = \frac{I_0 \, \rho_d \, V}{2 \, k^2 r^2} \left(|S_1|^2 + |S_2|^2 \right) \tag{16.13}$$

At fixed m and θ the quantity in the round brackets as a function of α has a distinct pattern of maxima and minima [16].

During the single nucleation experiment one measures the time dependence of the scattering intensity $I_{\text{scat}}^{\text{expt}}$ (see Fig. 16.5) which has a form of a sequence of maxima and minima. Assuming that on the time scale of experiment droplets, nucleated during the pulse, grow without coagulation and Ostwald ripening, one can state that their number density ρ_d remains constant. That is why in order to determine ρ_d it is sufficient to compare the first peak of the function $I_{\text{scat}}^{\text{expt}}$ with the first maximum of the scattering intensity from the Mie theory. This comparison yields the droplet radius r_d at the first peak, which after substitution into (16.4) and (16.5) yields

$$\rho_d = \frac{\frac{1}{L} \ln(I_0/I)}{\pi \, r_d^2 \, Q_{\text{ext}}(m, \alpha)} \tag{16.14}$$

Note, that this procedure does not provide information about the droplet growth $r_d(t)$—the latter can be obtained from the analysis of *series of peaks* in the scattering intensity.

16.2.2 Nucleation Rate

The nucleation rate is calculated as

$$J = \frac{\rho_d}{\Delta t} \qquad (16.15)$$

where Δt is the duration of the nucleation pulse. In various versions of expansion chambers accessible range of nucleation rates is approximately $10^2 - 10^9$ cm^{-3} s^{-1} which nicely complements the range achieved in diffusion cloud chambers.

For studies of homogeneous nucleation it is important to provide the particle-free operational regime of the chamber excluding heterogeneous effects. It has to be noted that expansion chamber is not a self-cleaning device (as the previously considered static diffusion chamber) and care has to be taken to avoid significant contamination prior to nucleation experiment.

16.3 Shock Tube

Shock tube realizes the same idea of a short nucleation pulse, providing the separation in time of the nucleation and growth processes, which we discussed in Sect. 16.2. In the shock tube this is achieved by means of the shock waves. The tube consists of two sections: the driver-, or High-Pressure Section (HPS), and the driven-, or the Low Pressure Section (LPS). The two sections are separated by the diaphragm. A small amount of condensable vapor is added to the driver section. During the nucleation experiment the diaphragm is rapidly ruptured and the high-pressure vapor/gas mixture from the driver section sets up a nearly one-dimensional, unsteady flow and the shock wave traveling from the diaphragm into the driven section. At the same time the expansion wave travels back—from the diaphragm into the driver section. Cooling of the rapidly expanding gas in the driver section imposes nucleation.

The construction of the shock tube for nucleation studies was proposed by Peters and Paikert [21, 22] and further developed by van Dongen and co-workers [20, 23–26]. The scheme of the experimental set-up [20] is shown in Fig. 16.6. The HPS has a length of 1.25 m, the length of the LPS is 6.42 m. The local widening in the LPS plays an important role in creating the desired profile of pressure and, accordingly, the supersaturation: after the rupture of the polyester diaphragm between HPS and LPS the initial expansion wave traveling from LPS to HPS is followed by a set of reflections of the shock wave at the widening (see Fig. 16.7). These reflections travel

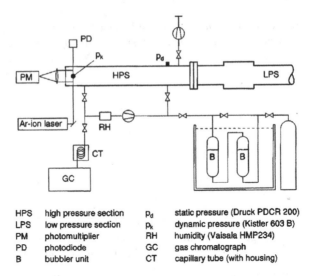

Fig. 16.6 Pulse expansion wave tube set-up. (Reprinted with permission from Ref. [20], copyright (1999), American Institute of Physics.)

back into the HPS and create the pulse-shaped expansion at the end wall of the HPS. After the short pulse and a small recompression the pressure remains constant for a longer period of time during which no nucleation occurs but the already nucleated droplets are growing to macroscopic sizes to be detected by means of the scattering technique. The temperature profile follows the adiabatic Poisson law (16.1).

Similar to the nucleation-pulse chamber, discussed in Sect. 16.2, the number density of droplets, ρ_d, is obtained by means of a combination of the constant-angle Mie scattering and the measured intensity of transmitted light—the procedure described in Sect. 16.2.1. In the experiments of Refs. [20, 23–25] the droplet cloud in the HPS was illuminated by the Ar-ion laser with a wavelength $\lambda = 514.2$ nm. Since the observation section in the shock tube is located near the endwall of the HPS, an obvious choice of the scattering polar angle is $\theta = 90°$. Figure 16.8 shows the optical set-up of the device. The laser beam passes the tube through two conical windows. The transmitted light is focused by lens L_2 onto photodiode D_2. The scattered intensity is recorded by the photomultiplier PM.

Because of the nature of the nucleation pulse method, the value of ρ_d should be approximately constant in time. The steady-state nucleation rate is given by Eq. (16.15)

$$J = \frac{\rho_d}{\Delta t}$$

where Δt is the duration of the pulse.

As pointed out in the previous section, besides the steady-state nucleation rate one can obtain from the same experimental data the growth law of the droplets. At each moment of time during the nucleation experiment for which the measured scattered

Fig. 16.7 Profiles of pressure and temperature and the wave propagation in the pulse expansion shock tube. (Copied from Ref. [27])

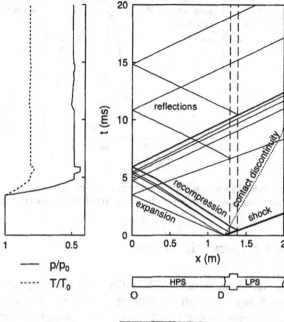

Fig. 16.8 Optical set-up used for measurements of droplet size and number density of droplets in the endwall of the shock tube. (Copied from Ref. [27])

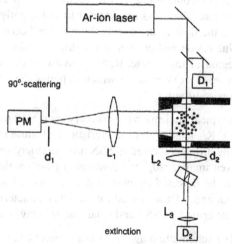

signal is at maximum, one can find the value of the droplet radius by comparison with the corresponding maximum of the theoretical scattering intensity given by the Mie theory of Sect. 16.2.1 as illustrated in Fig. 16.9. This gives the droplet growth curve $r_d(t)$.

The absence of moving parts in the shock-tube (as opposed to the expansion chamber) opens a possibility to study nucleation at sufficiently high nucleation pressures—up to 40 bar—and reach nucleation rates in the range of $10^8 - 10^{11}$ cm^{-3} s^{-1} [20, 28].

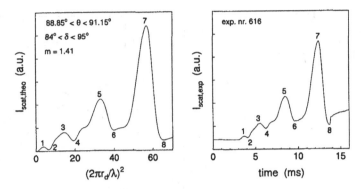

Fig. 16.9 Theoretical and experimental scattering patterns for n-nonane droplets. From mutual correspondence of extrema the time-resolved droplet radius $r_d(t)$ is found. (Copied from Ref. [27])

16.4 Supersonic Nozzle

The supersonic nozzle (SSN) relies upon adiabatic expansion of the vapor/gas mixture flowing through a nozzle of some sort. The most widely used type of these devices contain the *Laval* (converging/diverging) nozzle [29–32]. The vapor/gas mixture is undersaturated prior to and slightly after entering the nozzle region. During the flow in the nozzle the mixture becomes saturated and then supersaturated. Nucleation and growth of the droplets takes place when the flow passes the throat region of the nozzle. Rapid increase of the supersaturation results in spontaneous onset of condensation which depletes the vapor and subsequently terminates the supersaturation.

A typical nucleation pulse is very short $\sim 10\,\mu s$, and cooling rates are very high: $\sim 5 \times 10^5$ K/s leading to characteristic nucleation rates as high as $10^{16} - 10^{18}$ cm^{-3} s^{-1}. The droplets formed in SSN are extremely small $\sim 1 - 20$ nm; *critical* embryos are even smaller ~ 0.1 nm, containing 10–30 molecules. Clearly, droplets of this size can not be detected by optical devices operating in the visual—shorter wavelength is required. Methods used for particle characterization in SSN are small-angle neutron scattering (SANS) and small-angle x-ray scattering (SAXS).

The schematic diagram of the experimental set-up with the supersonic nozzle and SAXS unit is shown in Fig. 16.10. The experiment consists of the pressure trace measurements during the expansion and the SAXS measurements. A movable pressure probe measures the pressure profile of the gas $p(x)$ along the axis of the nozzle. The condensible vapor mole fraction y is determined from the mass flow measurements. Using the stagnation conditions p_0, T_0 of the vapor/gas mixture, one determines the pressure profile of the vapor along the nozzle

$$p^v(x) = y\,p_0 \left[\frac{p(x)}{p_0}\right]\left[1 - \frac{g(x)}{g_\infty}\right]$$

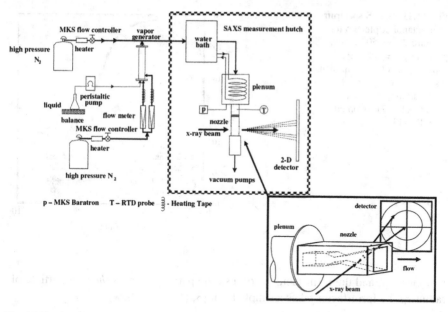

Fig. 16.10 Schematic diagram of the experimental set-up with supersonic nozzle and SAXS unit. (Copied from Ref. [33])

where $g(x)$ is the condensate mass fraction at point x,

$$g_\infty = \frac{\dot{m}_v}{\dot{m}_v + \dot{m}_{gas}}$$

where $\dot{m}_v$, $\dot{m}_{gas}$ are the mass flow rates of the vapor and gas, respectively. Then, the supersaturation profile is

$$S(x) = \frac{p^v(x)}{p_{sat}(T(x))}$$

where $T(x)$ is the temperature at point x found from the Poisson equation.

Using SAXS technique, one studies elastic scattering of X-rays by a cloud of droplets. Scattering leads to interference effects and results in a pattern, which can be analyzed to provide information about the size of droplets and their number density. Let us define a scattering vector (length) according to

$$q = \frac{4\pi}{\lambda} \sin(\theta/2)$$

where θ is the scattering angle, λ is wavelength of the incident beam; for SAXS $\lambda \approx 1 \text{Å}$. The scattering intensity I_{scat} is proportional to the number density of

Fig. 16.11 SAXS spectrum of n-butanol at plenum temperature $T_0 = 50°$ C and pressure $p_0 = 30.2$ kPa. The *solid line* is a fit to Gaussian distribution of spherical sizes with parameters given by (16.16)–(16.17). (Copied from Ref. [33])

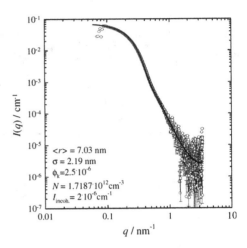

$\langle r \rangle = 7.03$ nm
$\sigma = 2.19$ nm
$\phi_k = 2.5 \cdot 10^{-6}$
$N = 1.7187 \cdot 10^{12}$ cm^{-3}
$I_{\text{incoh.}} = 2 \cdot 10^{-6}$ cm^{-1}

particles, ρ_d, and the form-factor P of a single particle. For a *spherical* particle of radius r_d the form-factor takes a simple form [31]

$$P(q, r_d) = \left[\frac{4\pi \left(\sin(qr_d) - qr_d \cos(qr_d) \right)}{q^3} \right]^2 \rho_{SLD}^2$$

where ρ_{SLD} is the *contrast factor*, being the difference in the scattering length density between the liquid droplet and surrounding bulk gas. Assuming Gaussian distribution of droplet sizes with the mean $\langle r_d \rangle$ and the width σ, I_{scat} can be written as

$$I_{\text{scat}}(q) = \rho_d \frac{1}{\sigma \sqrt{2\pi}} \int \exp \left[- \frac{(\langle r_d \rangle - r_d)^2}{2\sigma^2} \right] P(q, r_d) \, dr_d$$

Fitting the measured scattering intensity to this expression, one finds the desired quantities ρ_d and $\langle r_d \rangle$.

Figure 16.11 from Ref. [33] illustrates this procedure for SSN experiments with n-butanol at plenum temperature $T_0 = 50°$ C and pressure $p_0 = 30.2$ kPa. The solid line gives the fit to the Gaussian distribution with the following set of parameters

$$\langle r_d \rangle = 7 \, \text{nm}; \quad \sigma = 2.19 \, \text{nm} \tag{16.16}$$

and the number density

$$N \equiv \rho_d = 1.7 \times 10^{12} \, \text{cm}^{-3} \tag{16.17}$$

Taking into account the pulse duration $\Delta t \approx 10 \, \mu s$, this leads to the nucleation rate

$$J \approx 1.7 \times 10^{17} \, \text{cm}^{-3} \, \text{s}^{-1}$$

Thus, quantitative nucleation rate measurements, using various techniques discussed in this chapter, cover the range of more than 20 orders of magnitude. These measurements, in combination with nucleation theorems of Chap. 4, also provide direct information about the properties of the critical clusters which can be compared to predictions of theoretical models.

References

1. G.M. Pound, J. Phys. Chem. Ref. Data **1**, 119 (1972)
2. C.R.T. Wilson, Phil. Trans. R. Soc. London A **189**, 265 (1897)
3. R.H. Heist, H. He, J. Phys. Chem. Ref. Data **23**, 781 (1994)
4. J.L. Katz, C. Scoppa, N. Kumar, P. Mirabel, J. Chem. Phys. **62**, 448 (1975)
5. C. Hung, M. Krasnopoler, J.L. Katz, J. Chem. Phys. **90**, 1856 (1989)
6. J.L. Katz, M. Ostermeier, J. Chem. Phys. **47**, 478 (1967)
7. J.L. Katz, J. Chem. Phys. **52**, 4733 (1970)
8. R. Heist, H. Riess, J. Chem. Phys. **59**, 665 (1973)
9. P.E. Wagner, R. Strey, J. Chem. Phys. **80**, 5266 (1984)
10. R. Strey, P.E. Wagner, Y. Viisanen, J. Phys. Chem. **98**, 7748 (1994)
11. G.W. Adams, J.L. Schmitt, R.A. Zalabsky, J. Chem. Phys. **81**, 5074 (1984)
12. J.L. Schmitt, G.J. Doster, J. Chem. Phys. **116**, 1976 (2002)
13. T. Rodemann, F. Peters, J. Chem. Phys. **105**, 5168 (1996)
14. J. Wölk, R. Strey, J. Phys. Chem. B **105**, 11683 (2001)
15. K. Iland, J. Wölk, R. Strey, D. Kashchiev, J. Chem. Phys. **127**, 154506 (2007)
16. H.C. van de Hulst, *Light Scattering by Small Particles* (Dover, New York, 1981)
17. M. Kerker, *The Scattering of Light and Other Electromagnetic Radiation* (Academic Press, New York, 1969)
18. K.F. Riley, M.P. Hobson, S.J. Bence, *Mathematical Methods for Physics and Engineering* (Cambridge University Press, Cambridge, 2007)
19. J.D.J. Ingle, S.R. Crouch, *Spectrochemical Analysis* (Prentice Hall, New Jersey, 1988)
20. C.C.M. Luijten, P. Peeters, M.E.H. van Dongen, J. Chem. Phys. **111**, 8535 (1999)
21. F. Peters, B. Paikert, J. Chem. Phys. **91**, 5672 (1989)
22. F. Peters, B. Paikert, Exp. Phys. **7**, 521 (1989)
23. K.N.H. Looijmans, P.C. Kriesels, M.E.H. van Dongen, Exp. Fluids **15**, 61 (1993)
24. K.N.H. Looijmans, C.C.M. Luijten, G.C.J. Hofmans, M.E.H. van Dongen, J. Chem. Phys. **102**, 4531 (1995)
25. K.N.H. Looijmans, C.C.M. Luijten, M.E.H. van Dongen, J. Chem. Phys. **103**, 1714 (1995)
26. D.G. Labetski, Ph.D. Thesis, Eindhoven University, 2007
27. C.C.M. Luijten, Ph.D. Thesis, Eindhoven University, 1999
28. C. Luijten, M.E.H. van Dongen, J. Chem. Phys. **111**, 8524 (1999)
29. A. Khan, C.H. Heath, U.M. Dieregsweiler, B.E. Wyslouzil, R. Strey, J. Chem. Phys. **119**, 3138 (2003)
30. C.H. Heath, K.A. Streletzky, B.E. Wyslouzil, J. Wölk, R. Strey, J. Chem. Phys. **118**, 5465 (2003)
31. Y.J. Kim, B.E. Wyslouzil, G. Wilemski, J. Wölk, R. Strey, J. Phys. Chem. A **108**, 4365 (2004)
32. S. Tanimura, Y. Zvinevich, B. Wyslouzil et al., J. Chem. Phys. **122**, 194304 (2005)
33. D. Ghosh, Ph.D. Thesis, University of Cologne, 2007

Appendix A
Thermodynamic Properties

Water

Molecular mass: $M = 39.948$ g/mol

Critical state parameters: $p_c = 221.2$ bar, $T_c = 647.3$ K, $\rho_c = 17.54 \times 10^{-3}$ mol/cm^3 [1]

Equilibrium liquid mass density [1]:

$$\rho^l_{mass} = 0.08 \tanh y + 0.7415\, x^{0.33} + 0.32 \text{ g/cm}^3$$
$$x = 1 - \frac{T}{T_c}, \quad y = (T - 225)/46.2$$

Saturation vapor pressure [1]:

$$p_{sat} = \exp\left[77.3491 - 7235.42465/T - 8.2 \ln T + 0.0057113\, T\right] \text{ Pa}$$

Surface tension [1]:

$$\gamma_\infty = 93.6635 + 9.133 \times 10^{-3}\, T - 0.275 \times 10^{-3}\, T^2 \text{ mN/m}$$

Lennard-Jones interaction parameters [2]:

$$\sigma_{LJ} = 2.641 \text{Å}, \quad \varepsilon_{LJ}/k_B = 809.1 \text{ K}$$

Second virial coefficient [3]:

$$B_2(T) = 17.1 - 102.9/(1-x)^2 - 33.6 \times 10^{-3}\,(1-x)\,\exp\left[5.255/(1-x)\right], \text{ cm}^3/\text{mol}$$

V. I. Kalikmanov, *Nucleation Theory*, Lecture Notes in Physics 860, 293
DOI: 10.1007/978-90-481-3643-8, © Springer Science+Business Media Dordrecht 2013

Nitrogen

Molecular mass: $M = 28.0135$ g/mol
Critical state parameters: $p_c = 33.958$ bar, $T_c = 126.192$ K, $\rho_c = 0.3133$ g/cm^3 [4]

Equilibrium liquid mass density [4]:

$$\ln \frac{\rho_{\text{sat}}^{\text{l}}}{\rho_c} = 1.48654237\, x^{0.3294} - 0.280476066\, x^{4/6}$$
$$+ 0.0894143085\, x^{16/6} - 0.119879866\, x^{35/6},$$
$$x = 1 - \frac{T}{T_c}$$

Saturation vapor pressure [4]:

$$\ln \frac{p_{\text{sat}}}{p_c} = \frac{T_c}{T} \left(-6.12445284\, x + 1.2632722\, x^{3/2} - 0.765910082\, x^{5/2} - 1.77570564\, x^5 \right)$$

Surface tension [4]:
$$\gamma_\infty = 29.324108\, x^{1.259} \text{ mN/m}$$

Lennard-Jones interaction parameters [2]:

$$\sigma_{\text{LJ}} = 3.798 \text{ Å}, \quad \varepsilon_{\text{LJ}}/k_{\text{B}} = 71.4 \text{ K}$$

Pitzer's acentric factor: $\omega_P = 0.037$ [2].

Mercury

Molecular mass: $M = 200.61$ g/mol
Critical state parameters: $p_c = 1510$ bar, $T_c = 1765$ K, $\rho_c = 23.41 \times 10^{-3}$ mol/cm^3
[5].

Saturation vapor pressure [5]

$$\log_{10} p_{\text{sat}}[\text{Torr}] = -\frac{a}{T} + b + c \lg_{10} T$$

with
$$a = 3332.7, \quad b = 10.5457, \quad c = -0.848$$

Equilibrium liquid mass density [5]

$$\rho_{\text{mass}}^{\text{l}}[\text{g/cm}^3] = 13.595\,[1 - 10^{-6}(181.456\, T_{\text{Cels}} + 0.009205\, T_{\text{Cels}}^2$$
$$+ 0.000006608\, T_{\text{Cels}}^3 + 0.000000067320\, T_{\text{Cels}}^4)]$$

where $T_{Cels} = T - 273.15$ is the Celsius temperature.

Surface tension [6]

$$\gamma_\infty[\text{mN/m}] = 479.4 - 0.22\, T_{Cels}$$

Second virial coefficient [7]:

$$B_2\, N_A = \frac{T_B}{T} \left\{ c_1 \left[\exp\left(\frac{\varepsilon_1}{T_B}\right) - 1 \right] + c_2 \left[\exp\left(-\frac{\varepsilon_2}{T_B}\right) - 1 \right] \right\}$$
$$- \left\{ c_1 \left[\exp\left(\frac{\varepsilon_1}{T}\right) - 1 \right] + c_2 \left[\exp\left(-\frac{\varepsilon_2}{T}\right) - 1 \right] \right\} \quad \text{cm}^3/\text{mol} \qquad \text{(A.1)}$$

where $T_B = 4286$ K is the Boyle temperature of mercury, $c_1 = 69.87$ cm^3/mol, $c_2 = 22.425$ cm^3/mol, $\varepsilon_1 = 655.8$ K, $\varepsilon_2 = 7563$ K.

The value of coordination number N_1 can be obtained from the measurements of the static structure factor. For fluid mercury it was studied over the whole liquid-vapor density range by Tamura and Hosokawa [8] and Hong et al. [9] using X-ray diffraction measurements. Their data show that the first peak of the pair correlation function $g(r)$ in the liquid phase is located at ≈ 3 Å and is relatively insensitive to the mass density in the range 10-13 g/cm^3. The packing fraction in the liquid mercury at this range of densities is $\eta \approx 0.581$. Using (7.68) one finds $N_1 \approx 6.7$.

Argon

Molecular mass: $M = 39.948$ g/mol
Critical state parameters: $p_c = 48.6$ bar, $T_c = 150.633$ K, $\rho_c = 13.29 \times 10^{-3}$ mol/cm^3 [4]

Equilibrium liquid mass density [4]:

$$\rho^l_{\text{mass}} = M \left(13.290 + 24.49248\, x^{0.35} + 8.155083\, x \right) \times 10^3 \text{ g/cm}^3, \quad x = 1 - \frac{T}{T_c}$$

Saturation vapor pressure [4]:

$$\ln \frac{p_{\text{sat}}}{p_c} = \frac{T_c}{T} \left(-5.904188529\, x + 1.125495907\, x^{1.5} - 0.7632579126\, x^3 - 1.697334376\, x^6 \right)$$

Surface tension [4]:

$$\gamma_\infty = 37.78\, x^{1.277} \text{ mN/m}$$

Lennard-Jones interaction parameters [10]:

$$\sigma_{\text{LJ}} = 3.405 \text{ Å}, \quad \varepsilon_{\text{LJ}}/k_B = 119.8 \text{ K}$$

Pitzer's acentric factor: $\omega_P = -0.002$ [2].

The second virial coefficient is given by the Tsanopoulos correlation for nonpolar substances Eq. (F.2).

N-nonane

Molecular mass: $M = 128.259$ g/mol
Critical state parameters: $p_c = 22.90$ bar, $T_c = 594.6$ K, $\rho_c = 1.824 \times 10^{-3}$ mol/cm^3 [2]

Equilibrium liquid mass density [2]:

$$\rho^l_{mass} = 0.733503 - 7.87562 \times 10^{-4}\, T_{Cels} - 9.68937 \times 10^{-8}\, T^2_{Cels} - 1.29616 \times 10^{-9}\, T^3_{Cels} \ \ g/cm^3$$

where $T_{Cels} = T - 273.15$.

Saturation vapor pressure [3]:

$$p_{sat} = \exp\left(-17.56832 \ln T + 1.52556\, 10^{-2}\, T - 9467.4/T + 135.974\right) \ dyne/cm^2$$

Surface tension [3]:

$$\gamma_\infty = 24.72 - 0.09347\, T_{Cels} \ \ mN/m$$

The second virial coefficient [3]:

$$B_2\, N_A = 369.2 - 705.3/T_r + 17.9/T_r^2 - 427.0/T_r^3 - 8.9/T_r^8 \ \ cm^3/mol$$

where $T_r = T/T_c$.

Appendix B
Size of a Chain-Like Molecule

As one of the input parameters MKNT and CGNT use the size of the molecule. For a chain-like molecule, like nonane, it can be characterized by the *radius of gyration* R_g—the quantity used in polymer physics representing the mean square length between all pairs of segments in the chain [11]:

$$R_g^2 = \frac{1}{2N_{segm}^2} \sum_{i,j=1}^{N_{segm}} \langle (\mathbf{R}_i - \mathbf{R}_j)^2 \rangle$$

where N_{segm} is the number of segments. Equivalently R_g can be rewritten as

$$R_g^2 = \frac{1}{N_{segm}} \sum_{i=1}^{N_{segm}} \langle (\mathbf{R}_i - \mathbf{R}_0)^2 \rangle$$

where $\mathbf{R}_0$ is the position of the center of mass of the chain. The latter expression shows that the chain-like molecule can be appropriately represented as a sphere with the radius R_g. The radius of gyration can be found using the Statistical Associating Fluid Theory (SAFT) [12]. Within the SAFT a molecule of a pure n-alkane can be modelled as a homonuclear chain with N_{segm} segments of equal diameter σ_s and the same dispersive energy ε, bonded tangentially to form the chain. The soft-SAFT correlations for pure alkanes read [13]:

$$N_{segm} = 0.0255\, M + 0.628 \tag{B.1}$$

$$N_{segm}\, \sigma_{segm}^3 = 1.73\, M + 22.8 \tag{B.2}$$

where M is the molecular weight (in g/mol). Thus, the number of segments and the size of a single segment depend only on the molecular weight. Note, that within this approach N_{segm} is not necessarily an integer number. Having determined N_{segm}, the radius of gyration can be calculated using the Gaussian chain model in the theory of

V. I. Kalikmanov, *Nucleation Theory*, Lecture Notes in Physics 860,
DOI: 10.1007/978-90-481-3643-8, © Springer Science+Business Media Dordrecht 2013

polymers [11]:

$$R_g = \sigma_{segm} \sqrt{\frac{N_{segm}}{6}} \tag{B.3}$$

Then, the effective diameter of the molecule can be estimated as

$$\sigma = 2\,R_g \tag{B.4}$$

For n-nonane Eqs. (B.1)–(B.2) give:

$$R_g = 3.202\,\text{Å}, \quad \sigma = 6.404\,\text{Å} \tag{B.5}$$

Appendix C
Spinodal Supersaturation for van der Waals Fluid

In reduced units $\rho^* = \rho^{\mathrm{v}}/\rho_c$, $T^* = T/T_c$, $p^* = p^{\mathrm{v}}/p_c$ the van der Waals equation of state reads [14]

$$p^* = -3\rho^{*2} + \frac{8\rho^* T^*}{3 - \rho^*} \tag{C.1}$$

The spinodal equation $\partial p^*/\partial \rho^* = 0$ is:

$$T^* = \frac{\rho^*}{4}(3 - \rho^*)^2 \tag{C.2}$$

Solving Eq. (C.2) for the spinodal vapor density $\rho_{\mathrm{sp}}^{*\mathrm{v}}$ we obtain using the standard methods [15]:

$$\rho_{\mathrm{sp}}^{*\mathrm{v}}(T^*) = 2 - 2\cos\left(\frac{1}{3}\beta\right), \qquad \beta = \arccos(1 - 2T^*) \tag{C.3}$$

Substitution of (C.3) into the van der Waals equation (C.1) yields the vapor pressure at the spinodal:

$$p_{\mathrm{sp}}^{*\mathrm{v}} = \frac{8\left[4T^* - 3\cos\left(\frac{1}{3}\beta\right) + 3\cos\left(\frac{2}{3}\beta\right)\right]\sin^2\left(\frac{1}{6}\beta\right)}{1 + 2\cos\left(\frac{1}{3}\beta\right)} \tag{C.4}$$

from which the supersaturation at spinodal is

$$\cdot S_{\mathrm{sp}}(T^*) = \frac{p_{\mathrm{sp}}^{*\mathrm{v}}(T^*)}{p_{\mathrm{sat}}^*(T^*)} = \left[\frac{1}{p_{\mathrm{sat}}^*(T^*)}\right]$$
$$\left\{\frac{8\left[4T^* - 3\cos\left(\frac{1}{3}\beta\right) + 3\cos\left(\frac{2}{3}\beta\right)\right]\sin^2\left(\frac{1}{6}\beta\right)}{1 + 2\cos\left(\frac{1}{3}\beta\right)}\right\} \tag{C.5}$$

V. I. Kalikmanov, *Nucleation Theory*, Lecture Notes in Physics 860,
DOI: 10.1007/978-90-481-3643-8, © Springer Science+Business Media Dordrecht 2013

Appendix D
Partial Molecular Volumes

D.1 General Form

In this section we present a general framework for calculation of partial molecular volumes of components in a mixture. These quantities for a liquid phase are involved in the Kelvin equations. We consider here a general case of a two-phase m-component mixture v_i^α, $i = 1, \ldots, m$; $\alpha = \mathrm{v}$, l. The partial molecular volume of component i in the phase α is defined as

$$v_i^\alpha = \left.\frac{\partial V^\alpha}{\partial N_i^\alpha}\right|_{p^\alpha, T, N_{j, j\neq i}^\alpha}, \quad i = 1, 2, \ldots, \quad \alpha = \mathrm{v}, \mathrm{l} \tag{D.1}$$

where the quantities with the superscript α refer to the phase α. Since

$$V^\alpha = \frac{N^\alpha}{\rho^\alpha}$$

the change of the total volume due to the change of N_i^α is

$$dV^\alpha = \frac{1}{\rho^\alpha}\, dN_i^\alpha - \frac{N^\alpha}{(\rho^\alpha)^2}\, d\rho^\alpha$$

where we took into account that $dN^\alpha = dN_i^\alpha$. Then

$$v_i^\alpha = \frac{1}{\rho^\alpha}\left[1 - N^\alpha \left.\frac{\partial \ln \rho^\alpha}{\partial N_i^\alpha}\right|_{p^\alpha, T, N_{j, j\neq i}^\alpha}\right] \tag{D.2}$$

The density ρ^α is an intensive property which can be expressed as a function of the set of intensive quantities—molar fractions of components in the phase α

V. I. Kalikmanov, *Nucleation Theory*, Lecture Notes in Physics 860,
DOI: 10.1007/978-90-481-3643-8, © Springer Science+Business Media Dordrecht 2013

$$y_j^\alpha = \frac{N_j^\alpha}{\sum_k N_k^\alpha} \tag{D.3}$$

In view of normalization $\sum_{k=1}^m y_k^\alpha = 1$, ρ^α is a function of $m - 1$ variables y_k^α and one is free to choose a particular component to be excluded from the list of independent variables. Discussing the partial molecular volume of component i, it is convenient to exclude this component from the list, i.e. to set

$$\rho^\alpha = \rho^\alpha(y_1^\alpha, \ldots, y_{i-1}^\alpha, y_{i+1}^\alpha, \ldots, y_m)$$

Then the right-hand side of (D.2) can be expressed using the chain rule:

$$\frac{\partial \ln \rho^\alpha}{\partial N_i^\alpha}\bigg|_{p^\alpha, T, N_{j, j \neq i}^\alpha} = \sum_{j \neq i} \frac{\partial \ln \rho^\alpha}{\partial y_j^\alpha} \frac{\partial y_j^\alpha}{\partial N_i^\alpha} \tag{D.4}$$

From (D.3) we find

$$\frac{\partial y_j^\alpha}{\partial N_i^\alpha}\bigg|_{N_{j, j \neq i}^v} = -\frac{y_j^\alpha}{N^\alpha} \tag{D.5}$$

Substituting (D.5) into ((D.2) and D.4) we obtain the general result

$$v_i^\alpha = \frac{1}{\rho^\alpha} \eta_i^\alpha, \quad \eta_i^\alpha = 1 + \sum_{j \neq i} y_j^\alpha \frac{\partial \ln \rho^\alpha}{\partial y_j^\alpha}\bigg|_{p^\alpha, T}, \quad \alpha = \text{v}, 1 \tag{D.6}$$

In particular, for the *binary* $a - b$ mixture ($y_a + y_b = 1$) we have:

$$\eta_a^\alpha = 1 - y_b^\alpha \frac{\partial \ln \rho^\alpha}{\partial y_a^\alpha}\bigg|_{p^\alpha, T} \tag{D.7}$$

$$\eta_b^\alpha = 1 + y_a^\alpha \frac{\partial \ln \rho^\alpha}{\partial y_a^\alpha}\bigg|_{p^\alpha, T} \tag{D.8}$$

Mention a useful identity resulting from (D.7)–(D.8):

$$y_a \eta_a^\alpha + y_b \eta_b^\alpha = 1 \tag{D.9}$$

D.2 Binary van der Waals Fluids

Here we present calculation of the partial molecular volumes of components in binary mixtures (vapor or liquid) described by the van der Waals equation of state:

$$p = \frac{\rho k_B T}{1 - b_m \rho} - a_m \rho^2 \tag{D.10}$$

where the van der Waals parameters a_m, b_m for the mixture read [2]:

$$a_m = (y_a \sqrt{a_a} + y_b \sqrt{a_b})^2 \tag{D.11}$$
$$b_m = y_a b_a + y_b b_b \tag{D.12}$$

where y_i is the molar fraction of component i in vapor or liquid; a_i and b_i are the van der Waals parameters for the pure fluid i. According to the definition of the partial molecular volume consider a small perturbation in a number of molecules of component a at a fixed pressure, temperature and the number of molecules of component b. This perturbation results in the change of ρ, a_m and b_m:

$$\widetilde{a_m} = a_m + \Delta a_m \tag{D.13}$$
$$\widetilde{b_m} = b_m + \Delta b_m \tag{D.14}$$
$$\widetilde{\rho} = \rho + \Delta \rho \tag{D.15}$$

Substituting (D.13)–(D.15) into the van der Waals Eq. (D.10) and linearizing in Δa_m, Δb_m and $\Delta \rho$ we find

$$p - \frac{\rho k_B T}{1 - b_m \rho} + a_m \rho^2 = \frac{\Delta \rho k_B T}{1 - b_m \rho} + \frac{\rho k_B T}{(1 - b_m \rho)^2} (\Delta b_m \, \rho + \Delta \rho \, b_m) - \rho^2 \Delta a_m - 2 a_m \rho \Delta \rho$$

The left-hand side vanishes in view of (D.10) resulting in

$$\Delta \rho = A_0 \left[\Delta a_m - \frac{\Delta b_m k_B T}{(1 - b_m \rho)^2} \right], \quad A_0 \equiv \frac{\rho^2 (1 - b_m \rho)^2}{k_B T - 2 a_m \rho (1 - b_m \rho)^2} \tag{D.16}$$

Van der Waals parameters a_m and b_m change due to the variation in molar fractions satisfying $\Delta y_a = -\Delta y_b$:

$$\Delta a_m = 2 \Delta y_a \sqrt{a_m} (\sqrt{a_a} - \sqrt{a_b}) \tag{D.17}$$
$$\Delta b_m = \Delta y_a (b_a - b_b) \tag{D.18}$$

Substituting (D.18) into (D.16) we obtain:

$$\Delta \rho = \Delta y_a \left\{ A_0 \left[2 \sqrt{a_m} (\sqrt{a_a} - \sqrt{a_b}) - \frac{(b_a - b_b) k_B T}{(1 - b_m \rho)^2} \right] \right\} \tag{D.19}$$

Thus, for the binary van der Waals system

Table D.1 Reduced partial molecular volumes in the liquid phase, η_i^l, $i = a, b$, for the mixture n-nonane (a)/methane (b) at $T = 240$ K and various total pressures

p^v(bar)	η_a^l	η_b^l
1	1.005	0.289
10	1.057	0.305
25	1.142	0.289
40	1.228	0.365

$$\left(\frac{\partial \ln \rho}{\partial y_a}\right)_{p,T}^{\text{vdW}} = \left[\frac{(1 - b_m \rho)^2}{1 - \frac{2a_m \rho}{k_B T}(1 - b_m \rho)^2}\right]\left[\frac{2\rho \sqrt{a_m}(\sqrt{a_a} - \sqrt{a_b})}{k_B T} - \frac{\rho(b_a - b_b)}{(1 - b_m \rho)^2}\right]$$

(D.20)

Substituting this result into (D.7)–(D.8) we obtain the expression for η_i and hence for the partial molecular volumes

$$v_a = \frac{1}{\rho}\left\{1 - y_b \left(\frac{\partial \ln \rho}{\partial y_a}\right)_{p,T}^{\text{vdW}}\right\}$$

(D.21)

$$v_b = \frac{1}{\rho}\left\{1 + y_a \left(\frac{\partial \ln \rho}{\partial y_a}\right)_{p,T}^{\text{vdW}}\right\}$$

(D.22)

The reduced partial molecular volumes of the components in the liquid phase, η_i^l, $i = a, b$ play an important role for binary nucleation, especially at high pressures. Table D.1 shows the values of these parameters for the mixture n-nonane (a)/methane (b) for the nucleation temperature $T = 240$ K and various total pressures.

Appendix E
Mixtures of Hard Spheres

This appendix summarizes the relations describing the thermodynamic properties of a binary mixture of hard spheres [16–18]. The hard-sphere diameters of the components are d_1 and d_2; their respective number densities are ρ_1 and ρ_2.

We start with defining the first three "moments" of the hard-sphere diameters:

$$R_i = d_i/2$$
$$A_i = \pi d_i^2$$
$$V_i = \pi d_i^3/6$$

Note that A_i can be regarded as a molecular surface area, whereas V_i denotes the molecular volume of component i. Next, the parameters $\xi^{(k)}$ are defined as

$$\xi^{(0)} = \rho_1 + \rho_2$$
$$\xi^{(1)} = \rho_1 R_1 + \rho_2 R_2$$
$$\xi^{(2)} = \rho_1 A_1 + \rho_2 A_2$$
$$\xi^{(3)} = \rho_1 V_1 + \rho_2 V_2.$$

Note that $\xi^{(3)}$ is the total volume fraction occupied by hard spheres. For notational convenience, we also introduce

$$\eta = 1 - \xi^{(3)}.$$

On the basis of $\xi^{(k)}$, new parameters $c^{(k)}$ are calculated according to

$$c^{(0)} = -\ln \eta$$
$$c^{(1)} = \frac{\xi^{(2)}}{\eta}$$

V. I. Kalikmanov, *Nucleation Theory*, Lecture Notes in Physics 860,
DOI: 10.1007/978-90-481-3643-8, © Springer Science+Business Media Dordrecht 2013

$$c^{(2)} = \frac{\xi^{(1)}}{\eta} + \frac{\left(\xi^{(2)}\right)^2}{8\pi\eta^2}$$

$$c^{(3)} = \frac{\xi^{(0)}}{\eta} + \frac{\xi^{(1)}\xi^{(2)}}{\eta^2} + \frac{\left(\xi^{(2)}\right)^3}{12\pi\eta^3}.$$

The pressure p_d of the hard-sphere mixture follows from

$$p_3 = c^{(3)}k_BT \tag{E.1}$$

$$p_2 = p_3 - \frac{\left(\xi^{(2)}\right)^3 \xi^{(3)}}{12\pi\eta^3}k_BT \tag{E.2}$$

$$p_d = \frac{2p_3 + p_2}{3}. \tag{E.3}$$

Let us introduce for brevity of notations two additional quantities:

$$Y^{(1)} = 3\left[\frac{1 - \frac{3}{2}\xi^{(3)}}{\eta^2} + \frac{\ln\eta}{\xi^{(3)}}\right]$$

$$Y^{(2)} = \frac{\xi^{(2)}}{6\xi^{(3)}}\left[\frac{\eta + \left(1 - 2\xi^{(3)}\right)^2}{\eta^3} + \frac{2\ln\eta}{\xi^{(3)}}\right].$$

The chemical potentials $\mu_{d,i}$ can be derived from the virial equation using standard thermodynamic relationships:

$$\mu_i^{(3)} = c^{(0)} + c^{(1)}R_i + c^{(2)}A_i + c^{(3)}V_i \tag{E.4}$$

$$\mu_i^{(2)} = \mu_i^{(3)} + \frac{\left(R_i\xi^{(2)}\right)^2}{3\xi^{(3)}}\left(Y^{(1)} - 2R_iY^{(2)}\right) \tag{E.5}$$

$$\mu_i^{ex} = k_BT\frac{2\mu_i^{(3)} + \mu_i^{(2)}}{3} \tag{E.6}$$

$$\mu_{d,i} = k_BT\left[\ln\left(\rho_i\Lambda_i^3\right)\right] + \mu_i^{ex} \tag{E.7}$$

The last expression presents the chemical potential of a species as a sum of an ideal and excess contributions.

Appendix F
Second Virial Coefficient for Pure Substances and Mixtures

Calculation of the second virial coefficient

$$B_2(T) = \frac{1}{2} \int \left[1 - e^{-\beta u(r)} \right] d\mathbf{r} \tag{F.1}$$

from first principles requires the knowledge of the intermolecular potential $u(\mathbf{r})$ which is in most cases is not available. With this limited ability second virial coefficient is calculated from appropriate corresponding states correlations. For *nonpolar* substances such correlation has the form due to Tsanopoulos [2, 19]:

$$\frac{B_2 \, p_c}{k_B T_c} = f_0 + \omega_P \, f_1 \tag{F.2}$$

where

$$f_0 = 0.1445 - 0.330/T_r - 0.1385/T_r^2 - 0.0121/T_r^3 - 0.000607/T_r^8 \tag{F.3}$$
$$f_1 = 0.0637 + 0.331/T_r^2 - 0.423/T_r^3 - 0.008/T_r^8 \tag{F.4}$$

$T_r = T/T_c$ is the reduced temperature and ω_P is Pitzer's acentric factor.

For normal fluids van Ness and Abbott [20] suggested simpler expressions for f_0 and f_1

$$f_0 = 0.083 - 0.422/T_r^{1.6} \tag{F.5}$$
$$f_1 = 0.139 - 0.172/T_r^{4.2} \tag{F.6}$$

Expressions (F.5)–(F.6) agree with (F.3)–(F.4) to within 0.01 for $T_r > 0.6$ and $\omega_P < 0.4$, while for lower T_r the difference grows rapidly.

For the mixtures the second virial coefficient is written using the mixing rule

$$B_2 = \sum_i \sum_j y_i y_j B_{2,ij} \tag{F.7}$$

where $B_{2,ii}$ are the second virial coefficient of the pure components. For the cross term $B_{2,ij}$ the combining rules should be devised to obtain $T_{c,ij}$, $p_{c,ij}$ and $\omega_{P,ij}$ which then are substituted into the pure component expression (F.2), with the coefficients f_0 and f_1 satisfying (F.3)–(F.4) or (F.5)–(F.6). For typical applications the following combining rules are used [2]

$$T_{c,ij} = (T_{c,i} T_{c,j})^{1/2} \tag{F.8}$$

$$V_{c,ij} = \left(\frac{V_{c,i}^{1/3} + V_{c,j}^{1/3}}{2} \right)^3 \tag{F.9}$$

$$Z_{c,ij} = \frac{Z_{c,i} + Z_{c,j}}{2} \tag{F.10}$$

$$\omega_{P,ij} = \frac{\omega_{P,i} + \omega_{P,j}}{2} \tag{F.11}$$

$$p_{c,ij} = \frac{Z_{c,ij} k_B T_{c,ij}}{V_{c,ij}} \tag{F.12}$$

where $V_{c,i} = 1/\rho_{c,i}$.

Appendix G
Saddle Point Calculations

In Chap. 13 we search for the saddle-point of the free energy of cluster formation in the space of total numbers of molecules of each species in the cluster. For calculation of $g(n_a, n_b)$ we choose an arbitrary *bulk* composition n_i^1 and find the excess numbers n_i^{exc} according to Eqs. (11.84)–(11.85). The *total* numbers of molecules are: $n_i = n_i^1 + n_i^{exc}$. Then, $g(n_a, n_b)$ is found from Eq. (13.91). Is easy to see that although we span the entire space of (nonnegative) bulk numbers (n_a^1, n_b^1), the space of total numbers (n_a, n_b) contains "holes", i.e. the points, to which no value of $g(n_a, n_b)$ is assigned. This feature complicates the search of the saddle point of $g(n_a, n_b)$.

To overcome this difficulty we apply a smoothing procedure aimed at elimination of the holes in (n_a, n_b)-space by an appropriate interpolation procedure between the known values. The simplest procedure for the 2D space is the bilinear interpolation which presents the function g at an arbitrary point (n_a, n_b) as

$$g(n_a, n_b) = a\, n_a + b\, n_b + c\, n_a n_b + d \qquad (G.1)$$

where coefficients a, b, c, d are defined by the known values of g around the point (n_a, n_b). However, due to randomness of the location of the "holes", the straightforward application of bilinear interpolation is quite complicated. This difficulty can be avoided if we notice that (G.1) is the solution of the 2D Laplace equation

$$\Delta g(n_a, n_b) = 0, \quad \Delta \equiv \frac{\partial^2}{\partial n_a^2} + \frac{\partial^2}{\partial n_b^2} \qquad (G.2)$$

Thus, filling the holes in (n_a, n_b) space by bilinear interpolation is equivalent to solving the Laplace equation (G.2), which turns out to be a quick and efficient procedure. Discretizing (G.2) on the 2D grid with the grid-size $\delta n_a = \delta n_b = 1$, we have

V. I. Kalikmanov, *Nucleation Theory*, Lecture Notes in Physics 860,
DOI: 10.1007/978-90-481-3643-8, © Springer Science+Business Media Dordrecht 2013

$$\frac{\partial^2 g}{\partial n_a^2} = g(n_a - 1, n_b) - 2g(n_a, n_b) + g(n_a + 1, n_b)$$

$$\frac{\partial^2 g}{\partial n_b^2} = g(n_a, n_b - 1) - 2g(n_a, n_b) + g(n_a, n_b + 1)$$

The discrete version of Eq. (G.2) becomes

$$g(n_a, n_b) = \frac{g(n_a - 1, n_b) + g(n_a, n_b - 1) + g(n_a + 1, n_b) + g(n_a, n_b + 1)}{4}$$

(G.3)

An iterative procedure of finding $g(n_a, n_b)$ satisfying Eq. (G.3) is known as "Laplacian smoothing" [21]. Among various possibilities of performing iterations the Gauss-seidel relaxation scheme [22] seems most computationally efficient:

$$g^{i+1}(n_a, n_b) = \frac{g^{i+1}(n_a - 1, n_b) + g^{i+1}(n_a, n_b - 1) + g^i(n_a + 1, n_b) + g^i(n_a, n_b + 1)}{4}$$

(G.4)

where g^i is the value of g at i-th iteration step. The procedure is repeated until $g^{i+1}(n_a, n_b) \approx g^i(n_a, n_b)$.

The saddle point of the smoothed Gibbs function satisfies

$$\left.\frac{\partial g}{\partial n_a}\right|_{n_b} = \left.\frac{\partial g}{\partial n_b}\right|_{n_a} = 0$$

Note, that computationally it is preferable to search for the saddle point by solving the equivalent variational problem:

$$\left(\left.\frac{\partial g}{\partial n_a}\right|_{n_b}\right)^2 + \left(\left.\frac{\partial g}{\partial n_b}\right|_{n_a}\right)^2 \to \min$$

(G.5)

References

1. J. Wölk, R. Strey, J. Phys. Chem. B **105**, 11683 (2001)
2. R.C. Reid, J.M. Prausnitz, B.E. Poling, *The Properties of Gases and Liquids*, 4th edn. (McGraw-Hill, New York, 1987)
3. A. Dillmann, G.E.A. Meier, J. Chem. Phys. **94**, 3872 (1991)
4. K. Iland, Ph.D. Thesis, University of Cologne, 2004
5. International critical tables of numerical data, vol. 2, p. 457 (McGraw-Hill, New York, 1927)
6. L.E. Murr, *Interfacial Phenomena in Metals and Alloys* (Addison-Wisley, London, 1975)
7. A. Kaplun, A. Meshalkin, High Temp. High Press. **31**, 253 (1999)
8. K. Tamura, S. Hosogawa, J. Phys. Condens. Matter **6**, A241 (1994)
9. H. Hong et al., J. Non-Cryst. Solids **312—314**, 284 (2002)
10. A. Michels et al., Physica **15**, 627 (1949)
11. M. Sano, S.F. Edwards, *The Theory of Polymer Dynamics* (Oxford University Press, Oxford, 1988)

12. W.G. Chapman, K.E. Gubbins, G. Jackson, M. Radosz, Fluid Phase Equilib. **52**, 31 (1989)
13. J. Pamies, Ph.D. Thesis, Universitat Rovira i Vrgili, Tarragona, 2003
14. V.I. Kalikmanov, *Statistical Physics of Fluids. Basic Concepts and Applications* (Springer, Berlin, 2001)
15. G.A. Korn, T.M. Korn, *Mathematical Handbook* (McGraw-Hill, New York, 1968)
16. G.A. Mansoori, N.F. Carnahan, K.E. Starling, T.W. Leland, J. Chem. Phys. **54**, 1523 (1971)
17. Y. Rosenfeld, J. Chem. Phys. **89**, 4272 (1988)
18. C.C.M. Luijten, Ph.D. Thesis, Eindhoven University, 1999
19. C. Tsanopoulos, AICHE J. **20**, 263 (1974)
20. H.C. van Ness, M.M. Abbott, *Classical Thermodynamics of Non-electrolyte Solutions* (McGraw, New York, 1982)
21. F. O'Sullivan, J. Amer. Stat. Assoc. **85**, 213 (1990)
22. J. Stoer, R. Bulirsch, *Introduction to Numerical Analysis* (Springer, New York, 1993)

Index

i, ν-cluster, 71

A

Absorption
  frequency, 262
Activation
  barrier, 141
  energy for diffusion, 262
Activity
  coefficients, 188, 193
  gas-phase, 192
  liquid-phase, 194
Adiabatic expansion, 279, 280, 288
Adiabatic system, 120
Aerosol particles
  concentration, 273

B

Barrier
  nucleation, 55, 56, 67, 145, 150, 151, 162, 164, 260
Binary interaction parameter, 200, 209, 210
Binding energy, 84
Binodal, 146

C

Capillarity approximation, 21, 29, 30, 32, 55, 72, 77, 154, 182, 215, 230, 231, 254
Cluster
  distribution function, 173
  growth law, 34, 35, 37
Cluster definition
  live-time criterion, 128
Cluster distribution function

equilibrium, 33, 80, 110, 192, 245
  nonequilibrium, 24
Coarse-grained
  configuration integral, 227, 232
  nucleation theory (CGNT), 233
Coarse-graining, 215, 228, 235
Coexistence
  pressure, 190, 242
Compensation pressure
  effect, 200, 201, 235
Compressibility factor, 18, 30, 91, 104, 191
  critical, 91
  vapor, 89
Condensation nuclei
  in heterogeneous nucleation, 253
Configuration integral, 84, 224
Constant Angle Mie Scattering (CAMS), 281, 286
Constrained equilibrium, 25, 26, 173, 174, 216–219, 232, 246
Contact
  angle, 141, 254, 256–258, 260, 262, 264–267, 269–273
  line, 254, 257, 258, 263–266, 269
Continuity equation, 134, 177
Cooling rate, 280, 288
Coordination number, 92, 95, 97, 129, 155, 162, 166, 230
Core-shell structure, 236
Courtney correction, 32
Critical cluster, 23, 27, 28, 30, 32, 36, 37, 43–45, 47–51, 55, 73, 74, 77, 97–99, 103, 137, 138, 145, 150, 151, 152, 154, 155, 158, 174, 181, 183, 185, 186, 196, 198, 202, 211, 215, 220, 234–236, 243, 259, 260, 267, 272
Cut-off radius, 164

V. I. Kalikmanov, *Nucleation Theory*, Lecture Notes in Physics 860,
DOI: 10.1007/978-90-481-3643-8, © Springer Science+Business Media Dordrecht 2013